FUNDAMENTALS OF POWDER METALLURGY

By

Leander F. Pease, III
President
Powder-Tech Associates, Inc.
North Andover, MA

And

William G. West
President
National Sintered Alloys, Inc.
Clinton, CT

105 College Road East
Princeton, New Jersey 08540-6692 USA

ISBN: 1-87895-486-5

The Authors

Leander F. Pease, III

Born and raised in the Providence, Rhode Island area, Pease enjoyed a country boyhood. His interest in the sciences and mathematics was born while attending Providence Classical High School. He earned himself a place at the Massachusetts Institute of Technology (MIT) at 17. In his senior year in the Physics Department at MIT, Pease enrolled in a metallurgy elective class and realized how important metallurgy was to manufacturing. He received a M.S. degree in metallurgy in 1962 and Sc.D the following year.

Upon graduating he went to work for Wakefield Bearing Corp. and continued his education, studying part time at MIT's Sloan School and then transferring to Northeastern University where he earned an MBA degree. He left Wakefield in 1968 to join Kennecott Copper Corp's Ledgemont Lab as program manager in P/M. While there he met William Hooper and the idea of starting a P/M parts business was born. The two budding entrepreneurs teamed up with Raymond Collette in 1970 to form Sinterbond Corp. located in Gloucester. MA.

Pease and Collette severed ties with Sinterbond in 1973 and formed Powder-Tech Associates Inc. immediately thereafter.

Pease has been very active in the P/M industry via technical publications and standards work. He was a director of APMI international and even passed the Powder Metallurgy Technologist (PMT) Certification test in 1996!

William G. West

Bill West grew up in New Jersey but spent most of his adult life in Connecticut. He is a graduate of Stevens Institute of Technology with a degree in Mechanical Engineering.

After graduation, he was employed by Norwalk Powdered Metals as an engineer. He was chief engineer when he left in 1964 to start his own company, National Sintered Alloys Inc.. He has been active in the Powder Metal Industry for over 40 years. He has served the Powder Metal Parts Association as a Director and President. He has worked on many committees including the Industry Development Board. He was active in promoting material standards for the industry. He has also served on the Board of Directors of the APMI International. He received the Distinguished Service Award in 1991.

Dedications

To my wife Barbara and my children: Jennifer, David, and Douglas.

Leander F. Pease, III

To My wife Jane who did most of my data entry and my children Steve, Greg, Jim and Sharon for their encouragement and support.

William G. West

Table of Contents

Foreword

The Rise of the American P/M Industry

[A thumb nail retrospective according to Kempton Roll]

When a new technology graduates from the comfortable environment of the laboratory and moves gingerly to the manufacturing floor prepared, it hopes, to face the real world of free enterprise [aka, competition], strange, unpredictable things can happen. At the end of World War II the American powder metallurgy industry was just such a new technology and, ready or not, it was destined to graduate into the mature, well established, rough and tumble metalworking world where competition was more than an expression of free enterprise. It was cut-throat.

Those were interesting times for everyone who lived through them. Now more than half a century later, it is easier to understand and appreciate the impact of mundane past events on the present. And which ones the passage of time proved to be genuinely significant benchmarks; worthy of recollection, if not preservation. Sadly, too soon those whose careers played a role in the rise of America's powder metallurgy industry are fading into oblivion. So, any effort to recapture some of the history of what actually took place back then and the kinds of people who helped initiate those first difficult steps is to be applauded. This, it seems, is what Lanny Pease and Bill West and MPIF may be attempting to do for the American powder metallurgy industry.

It takes a special breed of people to initiate and nurture a new technology's beginnings. They dance to a tune different than that played by the Harvard Business School. No 'cash cows' to milk at this early stage. Besides the classical motivation to make a fortune, these special people needed to prove to themselves - and anyone else who cared - that they knew what they're doing. They suffered from the ingrained urge to meet and conquer challenges and were convinced that the goals they set could be met. Nor could they be deterred from the pursuit of their quest, even if deep in their hearts, they were not quite sure they really did know what they were doing, much less where their ambitions were taking them. For them, it was ... "Damn the torpedoes of ambivalence! Full speed ahead!"

The French have a word for these kinds of people: Entrepreneurs. The powder metallurgy industry has been from the outset, and surely still is, the quintessential entrepreneurial birthing ground.

Without realizing it, everyone involved with this technology in the last century was to some degree an entrepreneur. The nine-fingered press operators and graphite-smudged, latter-day alchemists disguised as powder metallurgists were all driven by entrepreneurial instincts inspired by a new way of making small parts out of metal powder. It was called powder metallurgy. They were all caught up in the same flush of enthusiasm attempting to do what had never been done before with a new and untested, yet intriguing way to fashion metals into finished products possessing properties unachievable - almost unimaginable - by any other known methods.

They weren't even sure in those days what to call the products of this process. Powdered metal parts? Powder metal parts? Metal powder parts? Powder metallurgy parts? Sinterings? Sintered metal parts? All names were being used. Customers and the trade press were thoroughly confused until the Metal Powder Industries Federation created and promoted the pseudo-acronym, "P/M". Vestiges of that early uncertainty is evident today in the names of some of the P/M parts companies still in business.

The supreme entrepreneurs in this virgin world of powder metallurgy were the company founders who were willing to stake their futures, their financial resources - sometimes even their families - to find out all they could about this process, the details of which were closely guarded by a handful of large companies. Applying what little knowledge they could acquire - more often based on 'trial and error' logic than on the application of scientific principles - they began their new careers into the unknown. What they were doing was still considered an art. And a secret one at that. After the war both the science and the technology of powder metallurgy began to pick up momentum at the same time - the science in Europe; the technology in the United States. But communication between the two was hampered by fears of process disclosure; a paranoidian attitude particularly rampant on this side of the Atlantic.

Powder metallurgy parts manufacturing up until the end of the war was a closed industry dominated by five large companies [Moraine Products Div., GMC; Amplex Div., Chrysler; Boundbrook Oilless Bearing Co.; U.S. Graphite Co.; and Keystone Carbon Co.], all serving one market - automotive, with one product - self-lubricating bearings. It was a big enough business to spawn and support a metal powder producing industry consisting of obscure off-shoots of several major non-ferrous metal companies, e.g., American Metal Climax, New Jersey Zinc Co., Metals Refining Co., National Lead Co. Two others, Metals Disintegrating Co. and Greenback Industries, Inc., had been producing pyrotechnic powders for the military so their peace-time transition to copper and tin powders was a natural. No iron powder producers existed in the United States at this time. Nor was there any interest because there was only one possible market: iron powder rotating bands for artillery shells. It worked for the defeated Germans because they were desperate for copper. We weren't. That market never materialized.

In the mid 40s and early 50s, the design engineering community began talking about this intriguing new technology - much of it disparagingly: "How can you possibly make a solid, strong bronze bushing out of tiny particles of metal without any melting? Impossible." In fact, it was the powder producers who took the lead in dispelling misconceptions about their customers products by addressing such questions. Their motivation was obvious and rational: Broaden the market base. Raw material suppliers traditionally do whatever is needed to make it easier for their customers to make and sell their products. The more self-lubricating bearings manufactured; the more copper and tin powders would be needed.

These companies were able to do this because they had one very important advantage: Their own trade association, the Metal Powder Association. It was established in 1944 by the metal powder producers who welcomed as members any company having any connection whatsoever with powder metallurgy; a significant distinction in trade association operation at the time. Besides powder producers, membership was opened to makers of P/M parts, users of P/M parts, equipment makers, tool and die makers, metalworking magazines, in-plant P/M parts makers, powder metallurgy consultants, even iron powder core makers. Every conceivable business entity that had the remotest connection with powder metallurgy was invited to sit around the same table set by the Association.

In the mid 50s this arrangement had become a little awkward. There were times when customers felt uncomfortable sitting next to their suppliers. Membership began declining. In 1956 a new, full-time Executive Director was hired and shortly after the Metal Powder Industries Federation was created out of the remnants of the Metal Powder Association. It was this concept, also unheard of before then, that enabled everyone connected in some fashion with the powder metallurgy industry to work together as a total, all-encompassing industry focused on one technology and to do so without jeopardizing their right to work independently as raw material suppliers or parts manufacturers or equipment makers. Using the Federation and its professional staff as the vehicle and driving force they were able to expand their businesses by 'selling' the acceptance of powder metallurgy as a viable metalworking alternative. To accomplish this, the Federation added a full-time staff person to help the Executive Director handle public relations and began utilizing every legitimate means available and affordable such as, exhibiting at the Design Engineering Shows, issuing a P/M design guidebook, promulgating manufacturing and quality control standards, organizing industry development programs, sponsoring national and international technical conferences, publishing books about powder metallurgy, etc., etc. Many such programs were funded largely by the powder producers and equipment makers but controlled content-wise by their customers, the P/M parts makers.

It was in that period that the Federation succeeded in convincing most, but not all of the industry's leaders that keeping their processing techniques hidden was holding back the entire industry; that their 'closed door' policy was only keeping more information out than in and stifling growth. When the industry as a whole agreed to open the 'doors' - or, at least, let the Federation open them - it soon became apparent that revealing the true facts about the complexities of the powder metallurgy process was a blessing. It was helping a skeptical engineering world accept the indisputable fact that useful products possessing unique properties could not only be manufactured out of strange combinations of metals and non-metals but could be mass produced by the thousands, each an exact replica of the other, with high precision, no scrap loss and at amazingly low cost. And all starting with particles of powder! The design engineering public began to understand, at last, that making a powder metallurgy part was, indeed, a complex process, one not to be undertaken frivolously or by the faint of heart. Customers began to empathize rather than criticize. It became much easier to convince customers that making quality P/M parts was not that simple and; therefore, worth every penny they were being charged. Sometimes it worked.

In retrospect, one of the keys to the success of the powder metallurgy industry in the United States was the creation of the Metal Powder Industries Federation. The federation/association concept enabled every business having any connection with the production or consumption of metal powders [now particulate materials] to work in concert with their colleagues, their competitors and their customers - foreign or domestic - without fear of ending up behind anti-trust bars. Of course, this being a free enterprise society, they still competed vigorously with one another yet without diminishing their ability to collaborate in efforts to penetrate markets dominated by other well-entrenched technologies - die casting, foundry casting, investment casting, machining, forging, slip casting, injection molding, etc. Their trade association gave them in addition to strength in numbers and dollars, an objective credibility in their efforts to eliminate public ignorance and engineering rejection of a metalworking process that for decades was considered more art than science.

The hindsight of history shows that in the last half of the previous century the powder metallurgy industry succeeded in meeting and conquering the challenges of other technologies and eliminating the dark shroud of process secrecy. America's powder metallurgy industry today is a product of the dedication, foresight and courage possessed by those hardy souls who, having survived World War II, saw in powder metallurgy the possibility for making a good living in a highly competitive but rapidly growing post-war manufacturing world. Little could they have known then how far this technology would advance and how important it would become to the whole world.*

Here in this publication, those many entrepreneurs whose courage and talents led powder metallurgy into the future can refresh their recollections of those who went before and who led the way on a path that has yet to see its end. These were real people, real pioneers, real visionaries just as are those working in the myriad, perhaps still secret, fields of powder metallurgy today. They, too, are pioneering the creation of new knowledge and setting their sights on new goals, lighting the way for new generations of entrepreneurs seeking to stretch out for powder metallurgy horizons still unimagined. This ancient art/modern science which uses tiny particles of metal and all kinds of other stuff to make life easier and better for all humankind has also made the struggles and sacrifices of those early entrepreneurs worthwhile.

...Kempton Roll

**The names of all those associated with the rise of America's powder metallurgy industry are far too numerous to detail here; at least not without fear of omission. A great many have been recognized via the Federation's Powder Metallurgy Pioneer award and, perhaps to a lesser degree, by the Distinguished Service to Powder Metallurgy award. All are listed in Who's Who in Powder Metallurgy.*

Preface

Most people don't read prefaces anyway, so it is OK to skip this one.

How did this little text come about? In 1996, Peter Johnson of MPIF mentioned to us that Dr. Gordon Dowson did not intend to update his excellent little (167 page) monograph, "Powder Metallurgy, the Process and Its Products". Written in 1990, it had served as a good introduction to the P/M process, a kind of overview. How much trouble would it be to put together something like it? A lot! Gordon was a terrific writer, and since 1990, there had been several comprehensive and serious works published, e.g. the 1146 pages of Volume 7 of the "ASM Handbook" in 1998, and at least five books by Prof. Randall German of the Pennsylvania State University and MPIF and APMI, through their publications and short courses, were adding several hundred technical articles each year. The assignment was to look over this wealth of information and distill out a few interesting items that would give a map to the new entrant into the P/M world. However, while doing this, we had great fun inserting a few thoughts that we hope will be of interest to our colleagues who are anything but new to P/M! You may come to think that more than a few trees would have been saved had there not been so many asides and wry remarks. True, but it would not have been as much fun to write!

It will soon be apparent to the reader that WW and LP have different styles of writing and that we have made absolutely no attempt to homogenize them. To justify this unusual feature, we would quote the Preface from John S. Major and Clifton Fadiman who authored The *New Lifetime Reading Plan,* 4th edition, 1997, Harper Collins Publishers, New York, N.Y: "Nevertheless we have decided to sign each chapter with the initials of its author; part of the value of this book should derive from the freedom of its authors to be opinionated, and we did not want to strive for an artificial homogeneity either of style or judgment". Well, WW and LFP are also quite opinionated!

J. Major goes on to explain that it is difficult to include science books in his reading plan. He notes: "Scientific writing is frequently highly technical, making demands of knowledge that most readers cannot meet. Scientists have often, in addition, not been very good writers, nor has the culture of science necessarily encouraged an agreeable literary style; so many scientific works give little pleasure to the non-specialist reader." Mr. Major certainly lays out the challenge for us. When do you suppose he will publish his 5th edition?

Preface

CHAPTER 1

Why is Powder Metallurgy Important?

When people outside our industry ask us what we do, they are usually surprised to learn that anyone would even try to make something strong and useful, starting with metal powders. They can see that you could pour liquid metal into a mold and get a strong casting, or you could machine parts from a bar of solid steel. And everyone has seen a movie of a blacksmith with fire and tongs, heating and beating and forming iron. But, in everyday life you don't see metal powders. And why would pressed metal powders be any stronger than sand castles? How can the metal powder be made to stick together? Well, we have here a small book that explains how to stick the powder together!

If you will temporarily suspend your disbelief about the process, we would like to show you some of the things that are made from the powders and their economic importance. (Dollar values help separate big deals from little deals.) Now, this book is being printed for the first time in 2002. If you are reading it in say, 2012, how will you make use of the dollar values? Well, this year the U.S. Bureau of Labor Statistics Consumer Price Index stands at 175.1. You could use the ratio of your future CPI to our present one, and multiply up the dollar figures we cite to make an estimate for your time.

Now, if you really have to find out why metal powders stick together to form the gears in your washing machine, leap ahead and read Chapter 2, Visual Basics. But please, mark this page, and don't forget to come back to see why powder metallurgy is important. We try to convey the importance of P/M by giving you lots of examples of where it is used. In doing this, it soon becomes very clear that many of the uses involve pretty sophisticated technology. Sometimes the company using the P/M material really does not want you to know the reasons why P/M was used or the details of its manufacture: they regard it as a trade secret! Nonetheless, you will still see a great many places where P/M is used, and we will do our best to make the technology understandable.

You will often read that P/M parts are used for their unique properties;

and, in some cases that is actually true! But, by and large, P/M parts are used because they are the least expensive way to get the minimum acceptable function in your electric drill or car. Consumers want to get what they pay for but they don't want to pay for more than they need. This forces design engineers to find the least expensive manufacturing processes to get the job done.

The automotive engineers are past masters at getting the most out of each process. We are rightly proud that in 2000 the average U.S. car had more than 37.5 pounds of P/M parts. Some SUV's and light trucks had 60 pounds! Ford Motor Company targeted 50 lbs./car for 2000. Figure 1.1 (Courtesy of Cincinnati Inc.) shows where some of the parts are used and Figure 1.2 shows some excellent automotive parts, as well as parts used in other industries, courtesy of Metal Powder Industries Federation, year 2000, Grand Prize Winners in the MPIF International Design Competition. You can visit MPIF on the internet at "www.MPIF.org." Russell Chernenkoff of Ford Motor Company (at the MPIF Short Course, Raleigh Durham, NC, November 1998) indicated how usage of some P/M was expected to grow between 1998 and 2005: the connecting rods would go from being used in 30% of the engines to 55%, bearing caps from 8% up to 35%, transmission gears from 15 to 30% and valve seat inserts from 60% up to 80%.

The Commerce Department keeps track of our industry with its North American Industry Classification System (NAICS), formerly called SIC. The P/M parts manufacturing code is 332117 and the making of iron and steel parts is 3321170211. (It is a humbling feeling to think that our terrific industry can be summarized as a single 10 digit number, one among 10 billion!) In 1997 the Department of Commerce reported that the value of all iron and steel powder sold in the U.S. was $520 million. (For metal powders of all kinds, the dollar value was $1.73 billion) By 2000, there were 445,000 tons or 890 million lbs of iron powder produced in North America. Plain iron powder sold for $.41/lb. Of that iron powder, 413,000 tons or 826 million lbs were used to make iron and steel parts. On average, the iron and steel parts sold for $2.80/lb., so there was $2.3 billion in steel part sales in North America alone. On a world wide basis, there were 5 billion dollars worth of steel parts sold. The 2000 iron powder sales were running about .4% above 1999. In North America, we have the capacity to make 704,000 tons/yr. and will have 912,000 tons of capacity for Fe (iron) by 2002.

Is auto usage a big deal? Yes! In 1998 about 525 million pounds of iron powder were used to make auto parts. For parts then selling at an average of $2.80/lb., this amounted to $1.47 billion. (The $2.80/lb. figure is just an average, many smaller parts or highly machined parts sold for many more $/lb. and vice versa.) Auto usage is 70% of the total

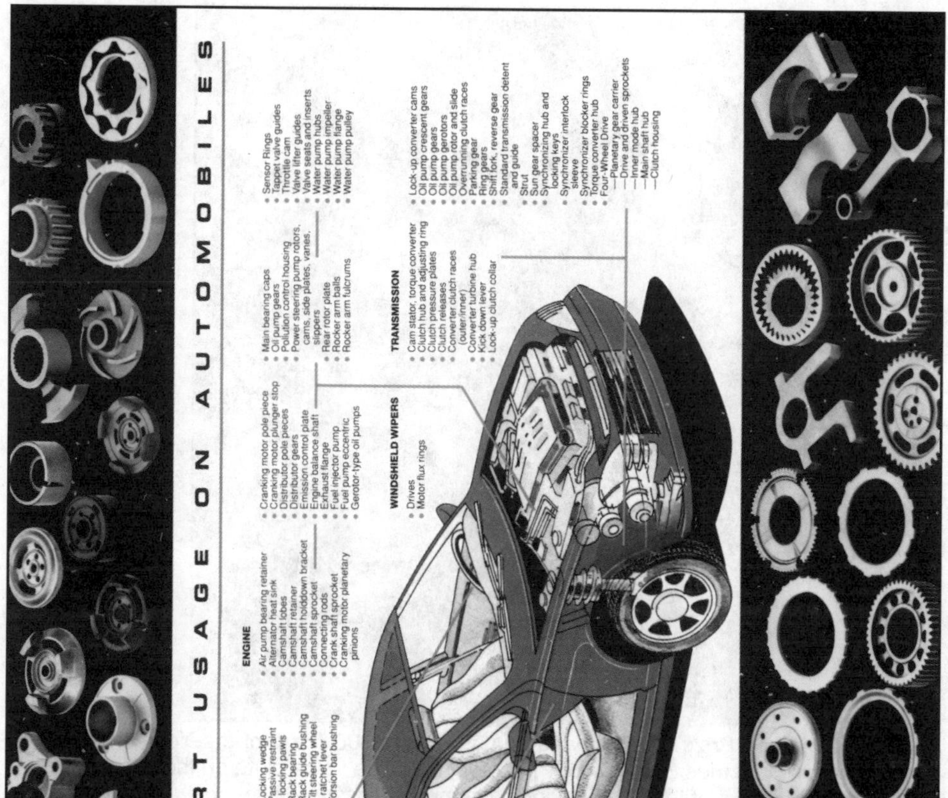

Figure 1.1. Powder Metal Part Usage in Cars. Courtesy of Cincinnati Inc.

Figure 1.2. Grand Prize winning parts from the year 2000 "Part -of-the-Year" Contest, Metal Powder Industries Federation, Princeton, N.J. The valve plate on the left is a pressure tight component of the air conditioning compressor for a diesel truck. The sprocket at the top drives a diesel engine fuel injection pump. The part with nine legs is made with special pressing technology to assure secure bonding of the legs to the base. It is a soft magnetic part, an axial flux stator used in a brushless DC pump motor. The part lower right is metal injection molded and is part of an instrument sensor control valve for chemical and oil flow.

usage of iron P/M parts. Of the iron and steel that went into automotive in 1998, about 40,000 tons or 80 million pounds were forged to near full density. Automobiles also consumed most of the 8,500 tons of stainless steel powder produced in 2000. It went into things like exhaust system flanges and bosses, as well as the mount on the windshield of your car, that holds the rear view mirror in place.

One of the earliest uses of metal powder was the making of porous bearings. In 1921, Metals Disintegrating (now AcuPowder International) worked with Harry Williams at General Motors. He was pressing mixtures of copper and tin powder. After heating to melt the tin so it would dissolve in the copper, the result was a porous bronze bearing. The advantage over a solid cast bronze was that the pores could be filled with

Figure 1.3. Sintered bronze bearings, Courtesy of Symmco, Inc.

oil. Then, when the bearing ran against a shaft and heated up, the oil oozed out and lubricated the shaft. When the rotation stopped, and the bearing cooled, the oil was sucked back into the bearing. Porosity and oil retention are genuine technical advantages for P/M! However, P/M bearings are also less expensive than ball bearings or Teflon lined bearings. A sintered bronze bearing 1/2 in. O.D., 5/16 in. I.D. and 1/2 in. long sells for $.11 each in quantities of 5,000, with a tooling charge of $1,500. Typical sintered bronze bearings are shown in Figure 1.3. You can see that a lot of bearings are being made, when the annual sales of sintered iron and bronze bearings is about $100 million in North America and about $300 million world wide. In 2000, there were 25,279 tons of copper and copper based alloy powders produced as powder or flakes. The powders were sold for $150 million, but they did not all go into bronze bearings. Paul Matthews, formerly V.P. at U.S. Bronze Powders, says that 45-50% go into self lubricating bearings. Quite a bit of copper powder was mixed with iron powder to form alloys, or went into friction materials (35%) and the rest went into other things like the flakes used to impart bronze color to ink and paints.

Would anyone ever pay more for a P/M product than for the cast or wrought part of the same chemistry? Tool steels are an interesting example. T15 tool steel is strong and tough, and is used to make the punches and dies that press metal powder into shapes. T15 is made from iron with 5% chromium, 15% vanadium, 1% molybdenum, and 3.4% carbon If the foundry melts the T15 at 2900°F, the alloying elements are all uniformly dissolved in the liquid steel. When the melt is poured into a large casting, the elements don't stay uniformly distributed. During the slow cooling process, the chromium and the vanadium have time to react with carbon to form carbide inclusions (lumps) as large as 40 micrometers. These lumpy carbides lower the strength and make the wrought tool steel hard to machine and grind.

In powder metallurgy, the liquid tool steel is gas atomized (disintegrated) into small particles which cool rapidly. The carbides don't have time to grow and are very small, 1 to 3 micrometers, compared to those in a regular casting. The tool steel particles are then joined together by heat and pressure, (Chapter 12), to form large billets up to 19 in. diameter by 8 feet long, weighing 3 tons. The billets still have the same small carbides that the powder had, and they are much stronger and easier to grind than the cast product. P/M T15 has a bend fracture strength of 678,000 psi (pounds per square inch), more than twice the conventional wrought or ingot-based tool steel. The toughness is 35% higher and grindability is three times better. Tool makers pay $15.57/lb. for the P/M T15, 3-5X the

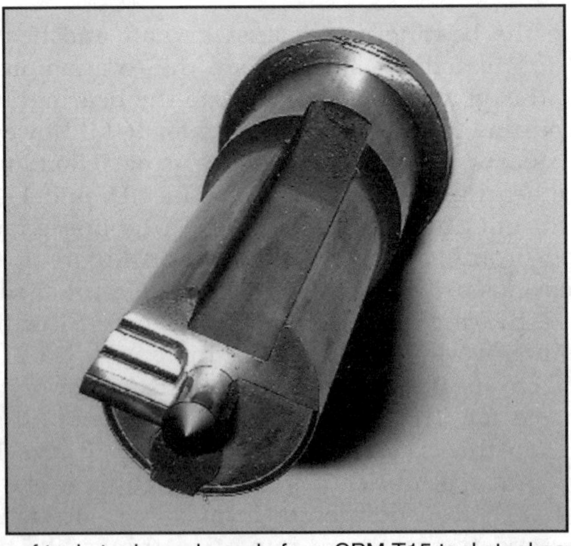

Figure 1.4. Photo of tool steel punch made from CPM T15 tool steel powder. Courtesy of National Sintered Alloys.

price of the cast product. Figure 1.4 shows tooling made from P/M tool steels, used to press metal powders. The only reason that sensible and frugal toolmakers and P/M houses pay this premium is because the punches last a lot longer, and because the steel is usually only about 5-10% of the finished tool's cost. In 1999, world production of P/M tool steels was about 10,000 tons with 6,500 tons made in the U.S. The P/M tool steels represented about 10% of the total amount produced by casting and forging. If you want to cold press and sinter tool steel powders directly into parts, the water atomized and annealed powders sell for $3-8/lb. Figure 1.5 shows some parts that were cold pressed from tool steel powder, and then sintered to near full density.

Filters and Porous Media

If you pour uniformly sized rounded particles of bronze into a graphite or stainless steel mold, and vibrate the mold to settle the particles, the mass of powder will be about 60% metal and have 40% air space or porosity surrounding the particles. Figure 1.6 is a two dimensional view showing how the round particles pack together. The mold that contains the powder is shown in Figure 1.7. Pass the mold and powder through a

Figure 1.5. Tool steel parts made from P/M tool steel powder, courtesy of Krupp P/M Technology a unit of Pacific Sintered Metals.

furnace and the particles will bond at their points of contact. The bonds are good enough that you can shake the assembly out of the mold and have a bronze filter of reasonable strength, say 2,500 to 7,000 psi. When liquids flow through a metal filter, particles larger than the spaces between the particles will be trapped, and the cleaned fluid will pass through. As an example, uniform spherical particles settled by tapping could accommodate other smaller spheres which are no larger than 1/7 their size, in the interior spaces among them. So, filters will certainly remove anything larger than 1/7 the diameter of the particles used in its

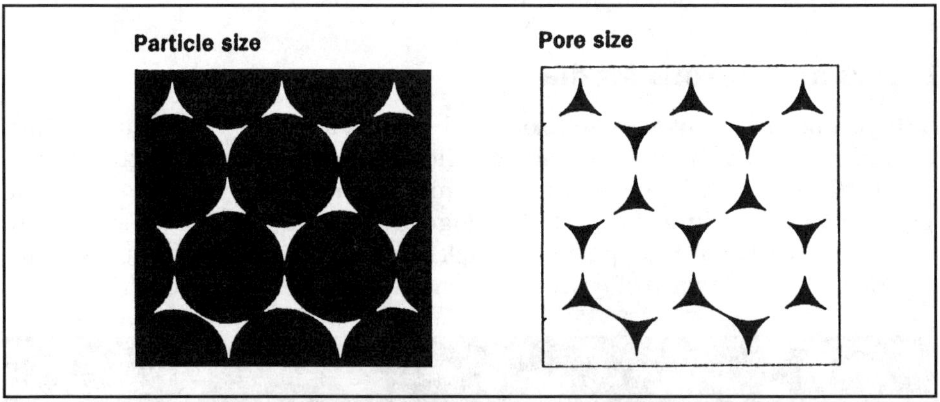

Figure 1.6. Principle of filtration, showing the pore spaces through which coarser debris could not pass. Pre space = 15% of particle diameter.

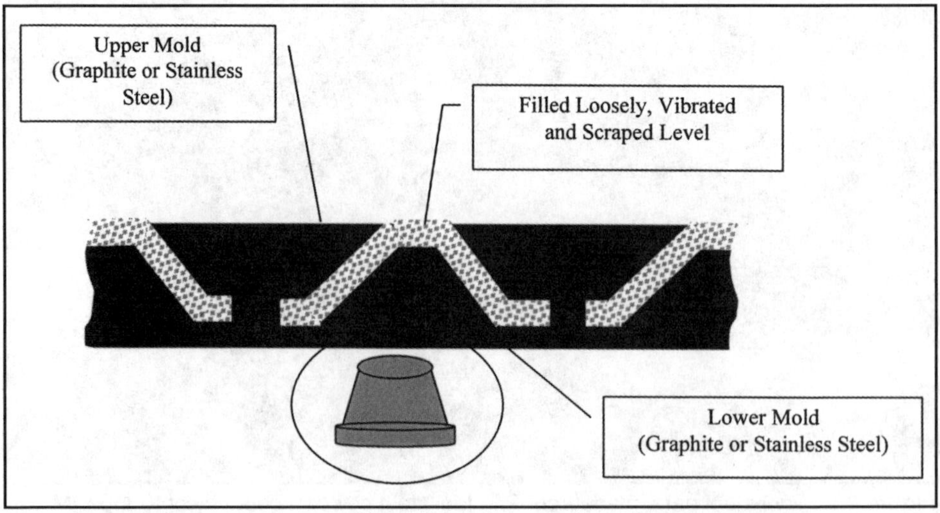

Figure 1.7. Gravity Sintering of Conical Filters with Multi cavity Molds (plates).

manufacture. Figure 1.6 shows the sizes of the pores that are left in between the round particles. In practice, filters remove even smaller particles, because filter particles are not exactly round. For example, a bronze filter made from particles .006 in. to .010 in. will remove particles larger than .0015 in.. Figure 1.8 shows stainless steel filters that trap particles of debris in a liquid polymer, as well as an assortment of bronze filters, courtesy of AcuPowder International.

Filters are not always made with the smallest possible round particles. Small particles result in excellent removal of fine debris from the fluid, but they also restrict the flow of fluid. And thick filters result in

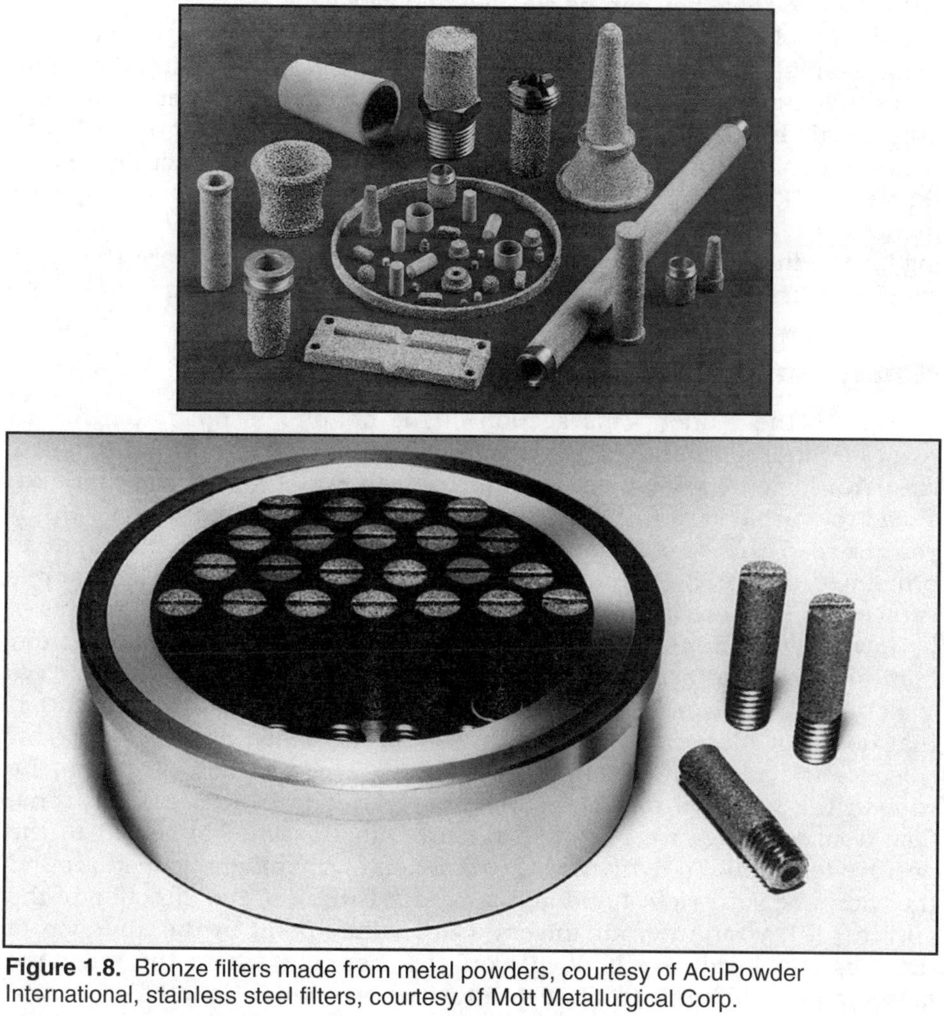

Figure 1.8. Bronze filters made from metal powders, courtesy of AcuPowder International, stainless steel filters, courtesy of Mott Metallurgical Corp.

restriction of fluid flow. Filters are designed by starting with the size of the debris particles to be removed and the amount of fluid flow restriction that is tolerable. The debris size determines the size of the filter powder, and the flow restriction is then minimized by going to thinner filters and possibly filters of greater area or larger diameter.

Stainless steel powders are often pressed into filter shapes, rather than just forming them by pouring them into a mold. Again, passing the pressed shape through a furnace yields a reasonably strong product. The stainless filters are corrosion resistant. Both the stainless steel and bronze filters have the advantage of being strong and temperature resistant versus plastic or paper filters. The P/M filters have an economic advantage in that they can be cleaned and reused.

Filters and porous media are used in ways you might never suspect! Some examples are removing particles from the gases used in processing semiconductors, protecting hydraulic valves, removing water from compressed air lines, vents and breathers in injection molding dies, liquid wicks, and sound attenuation. Although we never have exact figures, in the year 2000, about 750,000 lbs. of bronze filters were produced by four manufacturers in the U.S. The powders were sold for $3-$4/lb., depending on the fineness of the particle size. The finest powders are the most expensive. The sales value of the bronze filters is about $12 million.

Magnetism and Soft Magnetic Materials

(Note to the reader: This section of the chapter is not much fun to read! It wanders around in the fringes of the technical world of magnetism. You have been warned, and no one will ever know if you just skip ahead to "carbides" on p.20! Carbides are "easier"., but not softer.) But if you do read this section and want a further explanation of magnetic principles, the 1973 reference by B. D. Cullity is a little gem. General references for P/M are listed at the end of this chapter,

How do you create a magnetic field? Either you own a permanent horseshoe magnet or you connect a wire to a battery and let some electric current run through the wire as in Figure 1.9. (Note: by a convention that does not please me very much, something called electrical current, I, leaves the + (positive) terminal on the battery and flows through the wire to the negative terminal. Something I really believe in, electrons, flow from the negative terminal, up through the wire, attracted to the positive terminal. To figure out the direction of the magnetic field around the wire, use your right hand and aim your thumb in the direction of the current, I. If you curl your fingers a little, they point in the direction of the magnetic field. Try it. Well, you can visualize where the magnetic fields is, but you can't see it or feel it with your hand. Nonetheless, there

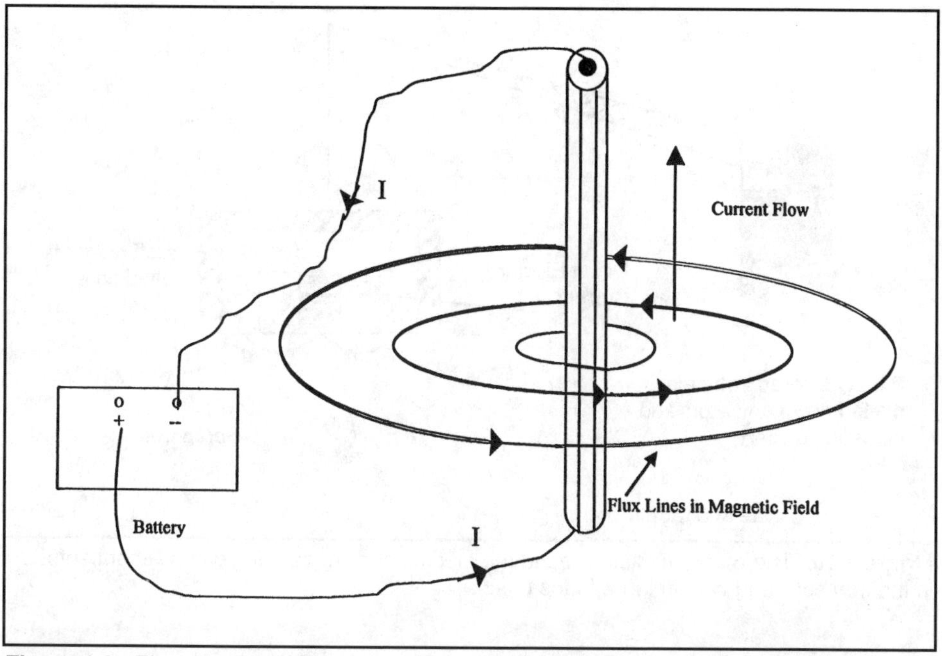

Figure 1.9. Electric current flowing in a wire causes a magnetic field around the wire.

is a magnetic field surrounding the wire, or, there is a magnetic field running between the tips of the horseshoe magnet. How do you know its really there, or put scientifically, how do you detect it? Every kid knows the answer, you bring the horseshoe magnet near a small piece of iron, and it grabs it. But, you soon learn, the magnet does not grab your aluminum frying pan or the glass in the kitchen window. What's going on? It looks like only a few materials really sense and react to the magnetic field. Why does the little piece of iron sense it, and why does the iron move toward the magnet? And would the iron really feel a magnetic field near the piece of wire with the current running in it? First of all, yes, the iron really would notice the field around the wire.

In scrap yards, there are cranes that move piles of scrap iron around. Instead of a bucket on the end of the crane's cable, they have a huge electro-magnet. Such a magnet is simply made out of a lot of electric wires carrying steady, direct current, Figure 1.10. The wires are wrapped around a big round chunk of iron called a core. The magnetic field from the wires gets concentrated in the core, and comes out through the end of the core. The scrap iron metal gets attracted to the core, and sticks to it. When the metal has been moved to another pile or a rail car, turn off the electric current and the iron falls right off. It was only temporarily

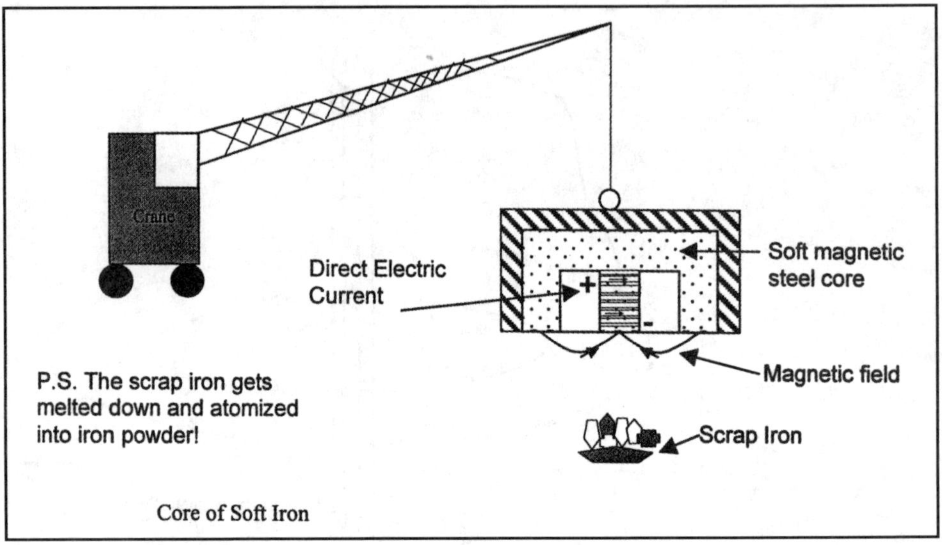

Figure 1.10. The magnetic field around the electric current carrying wires is concentrated in the iron core and can lift large loads of scrap.

magnetized. If they built cranes with a bunch of horseshoe magnets on the end, they wouldn't need electrical current, because those magnets are permanent. Of course you only build one such crane because there is no way to get the scrap iron pieces off the horseshoe magnets except by pulling them off, and you are a quick study!

Well, what about the aluminum frying pan? It turns out that only a few elements, notably iron, cobalt and nickel, have atoms that react strongly to a nearby magnetic field. If you believed the idea that electric current flowing in a wire gave rise to a magnetic field, then you might be able to be convinced that spinning and revolving electrons in atoms might have small magnetic fields around them, Figure 1.11. (Note that we have followed the convention I do not like: some negative charges move in one direction and we must think of current moving the opposite way so that we can apply the right hand screw rule above, and determine the direction of the magnetic field.) You would think that in most atoms, these little magnetic fields would cancel each other out, with one electron's field being opposed to the next one's. But in the ferromagnetic materials, (iron, cobalt and nickel), they don't all cancel out, and there is a net, small magnetic field, left over. Small groups of iron atoms align themselves with their magnetic fields in the same direction. These magnetically aligned groups of atoms are called domains, and are smaller than the metal grains, i.e., domains are about .00007 in. (1.7 micrometers) across. In most iron, the magnetic fields from the domains are

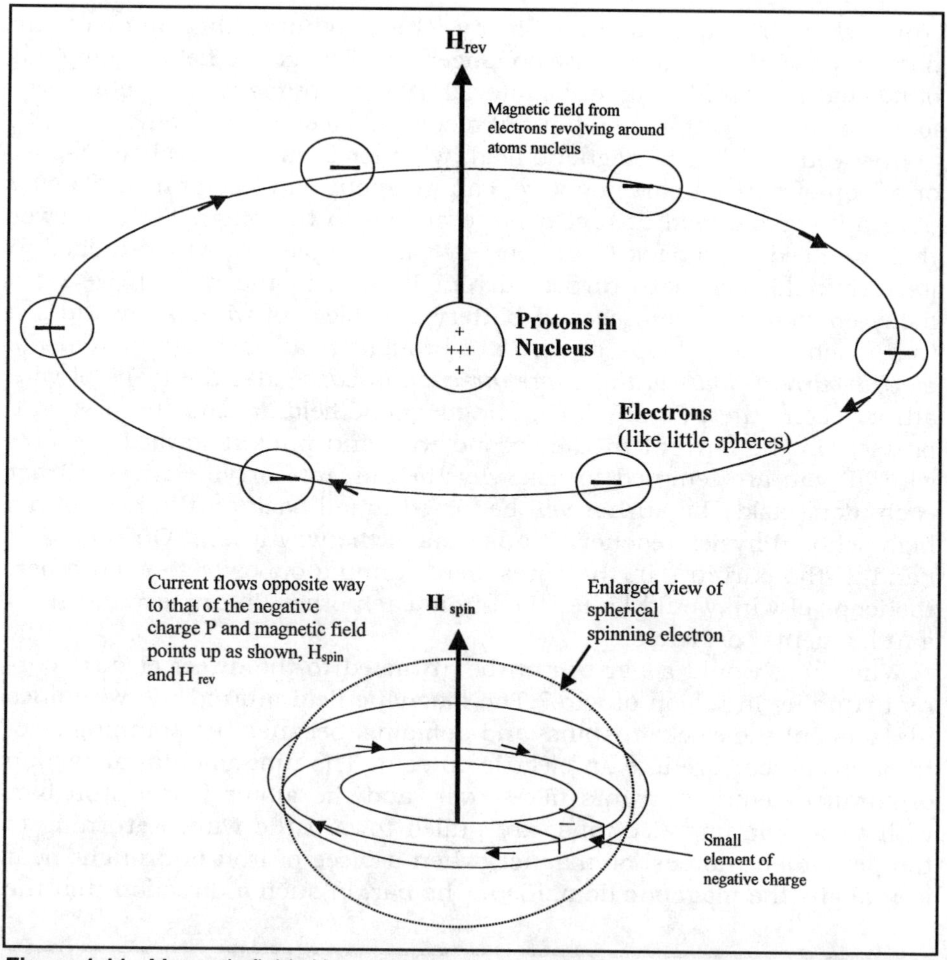

Figure 1.11. Magnetic fields H around an atom, contributions from electrons revolving and spinning on their own axes.

pointing in all different directions, and the magnetic field from one domain cancels out the one in the next domain. The bar of iron just sits there, not acting like a magnet. However, if you bring a magnetic field near the bar of iron, the magnetic field will act on the revolving and spinning electrons and cause the domains to line up in one direction, they will no longer be random. The bar of iron is then magnetized internally, and you will have a magnetic field coming out of the ends. If the iron bar is hardened steel, when you take away the imposed magnetic field, the bar stays magnetized. It is a "hard magnet", and the domains do not easily return to their random position . If the bar were made of pure soft iron, when the field is taken away, the magnetic alignment of the domains dissipate and the bar no longer has a magnetic field coming out of it. That is a "soft" magnetic material. (So, do we make the metal cores for the crane's electromagnet out of nice hard steel or soft iron?)

Now you see that a magnetic field, whether from a horseshoe magnet or a loop of current carrying wire, can magnetize a piece of iron. So why does a magnetic field attract a piece of iron to the magnet? To answer that, we need to go back to the battery with the piece of wire attached to its terminals, and with direct current flowing in the wire. If we were lucky enough to have a second battery and piece of wire, we would see that when the two loops of wire are brought near each other, *and the electric current flows in the same direction in each,* that the loops of wire attract each other, Figure 1.12. The magnetic field around the first wire presses on the current in the second wire and pulls it toward the first wire. (If you are tempted to ask why the current carrying wires attract each, don't ask! The author will be forced to fall back on the logic of his high school Physics teacher, "Cause that's the way it is!"). On the other hand, if the currents in the wires were running opposite to each other, the loops of wire would be repelled from each other. In any event, this is a fundamental principle.

Why then, would a bar of iron be attracted to the direct electric current running in a loop of wire? The magnetic field around the wire lines up (acts on) the electron spins (and domains) because the spinning electrons in the bar are like an electric current. The fundamental attraction of parallel electric currents takes over, and the atoms in the iron bar, with their spinning electrons, are pulled toward the wire. Returning to the permanent horseshoe magnet, when a piece of iron is brought near it, it aligns the magnetic domains in the bar, in such a direction that the

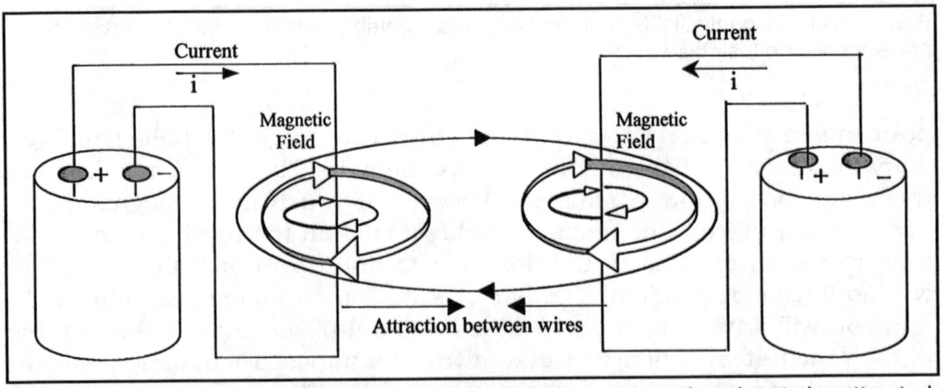

Figure 1.12. Electric current flowing in two parallel wires causes the wires to be attracted to each other: a fundamental observation on the effect of magnetic fields.

magnetic field around the iron piece is in the same direction as that of the horseshoe. How does it do that? The horseshoe magnet is just like the magnetized bar of iron in the previous paragraph in that one can imagine aligned electron spins and an effective electric current in it. The magnetic field aligns the electron spins in the iron bar near the horseshoe and the spins are effectively a current. Currents going the same way attract each other.

There is one other aspect of electro-magnetism that is definitely worth describing. Up to now, we have talked only about direct current flowing in a wire, such as what you would get from a battery. But your house voltage is 60 cycles per second (60 Hertz) alternating current. If you run it through a loop of wire, the magnetic field around the loop is always changing. It increases in one direction, then decreases and reverses its direction. This has an effect which is central to most electrical devices. The changing current and its magnetic field can, in turn, cause (induce) an electrical current to flow in the opposite direction in a nearby (non-magnetic) copper wire. Finally, the magnetic field from the first current physically pushes on the copper wire carrying the induced current. That pushing action is what is harnessed in an electric motor, to make it spin.

Why on earth did we have to say so much about the fundamentals of magnetism? Because its the only way to make any sense out of the different ways that magnetic materials are used. In Volume 7 of the "ASM Handbook", Ken Moyer of Magna-Tech P/M Labs points out that there are 20-30 magnets used in a typical car, and that there may soon be 100. He supplies the following list of uses:

- Door locks
- Windows
- Mirror
- Starter
- Radiator fan
- Idle speed
- Cruise control
- Air pump
- Brake assist
- Steering
- Transmission control
- Antilock braking
- Suspension
- Fuel pump
- Seat base
- Seat recliners
- Seat supports
- Heater blower
- Water pump
- Oil pump
- Air conditioning
- Electric braking
- Differential control
- Windshield wipers
- Spoilers
- Headlight wipers
- Headlight direction
- Entertainment
- Power sunroof
- Traction control

On the electric door locks, there is a solenoid actuator that moves the lock buttons up and down, Figure 1.13. In a solenoid, there is a little bar of magnetically soft iron inside a coil of electric wire. Switch on the

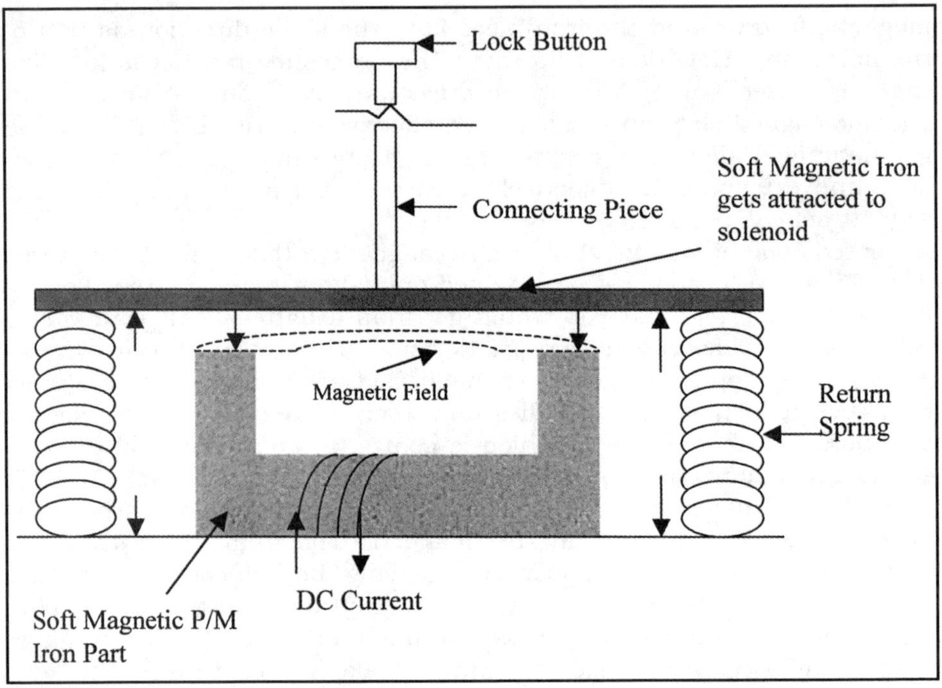

Figure 1.13. A soft magnetic iron material in a soleneoid. Turning the current in the solenoid "on", attracts a piece of iron. The iron moves something else, such as a lock.

current in the wire, and magnetically attract the bar of iron into the coil, thus pulling down the lock button. Or the coil could be wrapped around a C-shaped powder metallurgy part (core). When the current is turned on, there is a strong magnetic field connecting the two ends of the "C", and that field pulls an iron bar that then pulls on the door locks. The power seats and windows are driven by little permanent magnet motors. The permanent magnet attached to the shaft supplies the magnetic field. Send some DC electric current from the car's battery through the coil of wires surrounding the shaft, and the magnetic field pushes on the wires that drives the motor shaft that moves the seat. The more powerful the permanent magnet, the smaller and stronger will be the little drive motors.

Figure 1.14 shows typical soft magnetic parts. Figure 1.15 shows a sensor ring for speed control in car. Sensor rings are part of the control system for your anti-lock braking or cruise control. If you think of a ring attached to the axle of the car, you will see it rotating each time the tire goes around. The rings are soft magnets with little teeth or projections on the outside. The control system will have a little coil of copper wire placed near the O.D. (outer diameter) of the ring. There will be a small

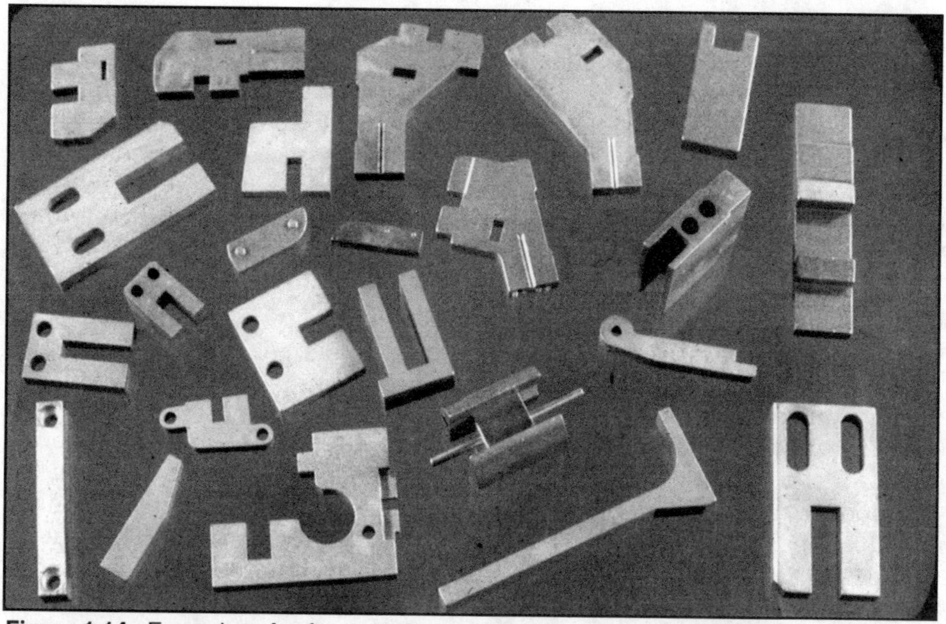

Figure 1.14. Examples of soft magnetic parts that are parts of magnetic circuits that cause other parts of the system to move. Courtesy of L. Baum, Remington Arms.

Figure 1.15. Soft magnetic iron sensor ring to measure automobile speed through rotation of the wheels or axles.

electric current flowing in the coil, creating a magnetic field. When a tooth on the sensor ring comes by the coil, the coil's magnetic field tries to magnetize the ring's tooth. The coil "feels" the energy flowing out into the sensor ring tooth. By "feel", we mean that the current flowing in the coil changes a little, and the control system sees these changes as happening periodically but at a constant rate. The constant rate means that the axle and wheel are running at a constant speed. If the wheel stops moving, the ABS system sees no pulses from the coil, and so it eases up on the brakes. The wheel starts to rotate and the skid is controlled.

Hard Magnetic Materials

Hard magnets are the old childhood horseshoe magnets or the ones on the outside of your refrigerator that hold the next veterinary appointment for the cat. They are also in the form of a strip on the inside edge of the door, to hold the door shut and sealed. (They replace the old fashioned refrigerator door latches so that someone having the misfortune to get locked in the refrigerator has merely to give the door a good shove and escape. Also, does anyone outside of Rhode Island or Emeril Lagasse still refer to these food storage devices as "ice boxes"? (We don't have to explain who Emeril is, do we?) The old horseshoe magnet was not P/M, it was hardened solid steel. But all the really strong ones since then have been based on metal or oxide powders, or alloys of neodymium iron, and boron, or alloys of samarium and cobalt.

According to Dowson, (see ref.) the naturally occurring double oxide of iron, ferrous ferrite, has been known for its magnetism for over 3,000 years. It is commonly called lodestone and small pieces of it are permanently magnetized. A piece of the lodestone, balanced on a sharp point, becomes a compass with its length aligned with the Earth's magnetic field. One end will point North and the other will point South. The word "lode" is from the old English, meaning "way". B. D. Cullity offers the suggestion that lightning strikes near the ferrous ferrite mineral had an associated magnetic field that permanently magnetized it, creating the lodestone. Around 1600, an Englishman named Gilbert discovered that you could rub lodestone on a bar of relatively hard iron, and cause the iron to be permanently magnetized. (We will leave it to your imagination to ponder whether Gilbert then bent the bar of iron into a horseshoe magnet or whether some later fellow rubbed lodestone on an actual horse's shoe!)

Nowadays, you can buy a nice alloy magnet from your local industrial distributor: in quantities of one each, $3.50 will lift 2.2 lbs., or for $66. you could lift 50 lbs. However, the real P/M action is in magnets made from other more exotic materials. In 2000, there were about $6 billion/yr worth of hard magnets sold at the component level, on a world-wide basis. This breaks down as ferrites, ($3.0 billion), neodymium-iron-boron (NdFeB) ($2.4 billion), and samarium-cobalt (SmCo $300 million), with a smaller amount of P/M going into aluminum-nickel-cobalt (Alnico $60 million). The ferrites really are oxide ceramics, rather than metals, but they are interesting anyway, and besides, they are the ones that end up on your refrigerator. The iron in the ferrite can come from pickling waste, that is, the residual acid from immersing steel in acid to clean up the surface. To this one adds strontium carbonate, wet mills it, calcines it (roasts it in air), and gets a powder that can be pressed. The powder is pressed either wet or dry into the magnet shape. Sintering is done in air, in a kiln like a rail car. The resulting parts are then magnetized by putting them in a high magnetic field. The flexible refrigerator magnets are made by mixing some of the unmagnetized powder with plastic as a bonding agent and forming it into sheets. Later, pieces of the sheets are magnetized in a strong field. These are referred to as bonded magnets.

Permanent magnets have the characteristic that when the magnetizing field is removed, the material still stays magnetized. You can, however, get rid of that residual magnetism with a strong enough opposite field. The higher the magnetic field needed to get rid of the residual magnetism in the permanent magnet, the stronger the magnet was. However, in ordinary life, permanent magnets don't see intense reversing fields, and they stay permanently magnetized. The rare earth** magnets, NdFeB and SmCo, are as strong as they are because once their domains are lined up magnetically, it is very hard to push them out of alignment. (**No one should put this book down without committing to memory the names of the rare earth elements, in order of their atomic numbers, 58-71,: Cerium, Praesodymium, Neodymium, Promethium, Samarium, Europium, Gadolinium, Terbium, Dysprosium, Holmium, Erbium, Thulium, Ytterbium, Lutetium, plus Yttrium, Scandium, Thorium!) In any event, how strong are they? We like to think of magnets in terms of lifting or holding power, that is, something real, tangible, and accessible to the unaided eye:

Material	**$ to lift 50 lbs.**	**Magnet Size, in.**	**$ to lift 200 lbs.**	**Magnet Size, in.**
Ferrite	$ 19	3 x 1 x 1½	$ 38	5 x 2 x 2
NdFeB, 2 pole	$ 40	1 x 1 x 1	$148	1½ x 1½ x 3
Alnico, horseshoe	$ 65	2 x 3 x 1	$210	4 x 5 x 1¼
SmCo	$158	2 x 2 x ½	$NA	NA

The selling prices for finished magnets and the raw material powder costs are approximately as follows:

Material →	NdFeB	SmCo	Alnico	Ferrite
Selling price magnet, $/lb.	$ 50.	$200.	$15.	$1.
Powder selling price raw material $/lb.	$8-9	$ 50.	$ 5	$ .08

Carbides

(If you survived magnets, great! If you skipped ahead, it will be our little secret.) Powder metallurgy and carbides remind us of pulling yourself up by your bootstraps! Cemented carbides are hard and wear resistant mixtures of tungsten carbide and cobalt, first developed in 1922 by OSRAM as a substitute for diamond dies in the drawing of fine tungsten wire for light bulb filaments. In 1925, F. Krupp in Germany bought the patent rights and began promoting carbide for wear resistance. There are about 55,000 tons of tungsten produced each year, on a world wide basis (B. Lux, Vol. 7 ASM, p.188). Of that amount, about 25,000 tons are converted into tungsten carbide powder. In the U.S., In 1998, 6,124 tons of tungsten carbide powder were produced . (D. White, *1998 State of the Industry* as reported in Vol 35 number 5, IJPM) Tungsten carbide powder sells for $10.72/lb. in moderate quantities. Carbide parts are made by pressing the mixture of tungsten carbide, about 10% cobalt and some organic binder. The real bonding together of the powders occurs by heating them in a furnace to about 3000°F. The cobalt melts, but the tungsten carbide does not. Starting with very fine tungsten carbide particles (200 nm) (1 nm = 1 nanometer = 4 atoms laid end to end) and 10% Cobalt, the pressed powders shrink 15% during the furnace treatment. The final part has a hardness of Vickers 1950 kg/mm^2. Heat treated steels at Rockwell C 63 are about 800 kg/mm^2 Vickers, so the carbides can be twice as hard as steel. These metal-bonded carbides are hard enough to make excellent dies in which to press other P/M parts, and they are wear resistant enough to make the cutting tools that can do finish machining on sintered steel parts. (Hence the analogy of pulling oneself up by one's bootstraps!)

Carbides are so hard that they can only be finished by diamond grinding or electrical discharge machining (EDM). The tool makers who make the P/M tools buy tungsten carbide blanks from specialist carbide companies. These blanks are made by cold isostatic pressing (you could see Chapter 12 if this is a new term to you, but it probably makes more sense to be patient 'til you get there). Because tungsten carbide is so hard to machine after sintering, it is usually rough machined in the green or as-pressed state. The blanks are then sintered with the 15% shrinkage noted above. If the sintered blanks are later to be finished by wire electrical discharge machining, they are first stress relieved by a special heat treatment. (Without that stress relief or extra sintering, when the wire cuts into the tungsten carbide, the blank can crack and spoil the toolmaker's day!)

A much larger use for tungsten carbide is the production of cutting tools and other wear components. The cutting tools are things like the tungsten carbide inserts that go into holders on lathes, for turning down metal parts (including P/M parts!), Figure 1.16a. The cutting tools come with special nitride and oxide coatings on them, which are even more wear resistant than the tungsten carbide itself. Figure 1.16b shows some carbide tips attached to a large drill bit used to cut through rock, in drilling an oil well.

Figure 1.16a. Selection of tungsten carbide, WC, cutting tool inserts, courtesy of Kennametal, Latrobe, P.A.

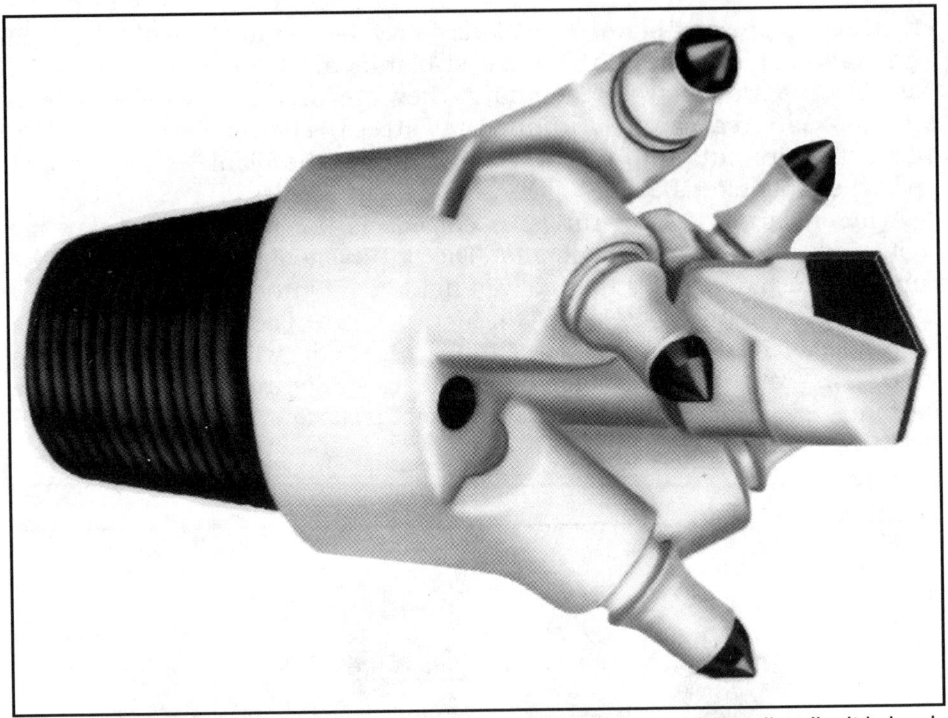

Figure 1.16b. Tungsten carbide tips affixed to a drill bit used in drilling oil wells. It is hard enough to cut through the solid rock layers. Courtesy of Kennametal, Latrobe, PA.

Thermal Spray Coatings

The idea of plating one metal on another is familiar to us all. For instance, hard chromium plating on steel gives it better wear resistance and helps prevent corrosion. Well, metal powders can also be applied to the surface of new parts to impart wear and corrosion resistance. That way you don't have to make the whole part out of something expensive, just the surface. Also, if a large or expensive metal part has suffered enough wear that it no longer functions, the worn area can be built up with metal powder and re-machined or ground back to the original size. The general term for this build-up process is hard facing, and one of the methods is thermal spraying of metal powders.

Figure 1.17 shows a combustion spray gun or flame spray gun. A mixture of 10-20 psi oxygen and acetylene gas are fed into the gun. The flame burns intensely at 5600°F. When gases are heated by about 500°F, their pressure doubles, when you heat them to 5600°F, by a factor of 11, the pressure goes up by a factor of 11. It is a modest understatement to say that the burning gases exit the gun very quickly! Metal powder particles, .002 in. to .004 in. in diameter are fed into the back of the gun, get mixed in with the burning gases, and melt to some degree. The droplets are hurled forward by the burning gases at 70 ft./sec. and aimed at the part that needs to be built up. The molten droplets hit the "target" ...splat!...and stick to it. The powder is fed at 20 lbs./hr. and soon builds up a layer up to .10 in. thick. Not all of the metal particles melt completely, but they stick on the target anyway. The built up layer will have some porosity, perhaps 10-15%. The particles travel line-of-sight, so you can only coat areas at which you can aim the gun, typically the outside surfaces. There is an amazing variety of metal powders, carbides, and oxides that can be used to repair or protect parts. For instance, you can coat with tungsten carbide, nickel based alloys (Haynes), cobalt based alloys (Stellites), iron based alloys and oxides. Tungsten carbide is the best for abrasion resistance. Cobalt base is best for corrosion resistance and hot hardness while nickel base is better for wear resistance by abrasion and adhesion. Steel powders are cheap, and are often used when you need to build up a large area such as on earth-moving equipment or crushing and grinding mills. Carbide powders sell for $20-40/lb., cobalt base for $15-30/lb., nickel base for $5-10/lb. and steel powders like 300 series stainless steel for $5-10/lb.

The combustion spray guns in Figures 1.17 and 1.18 are the simplest thermal spray devices for applying metal powders, but there are several

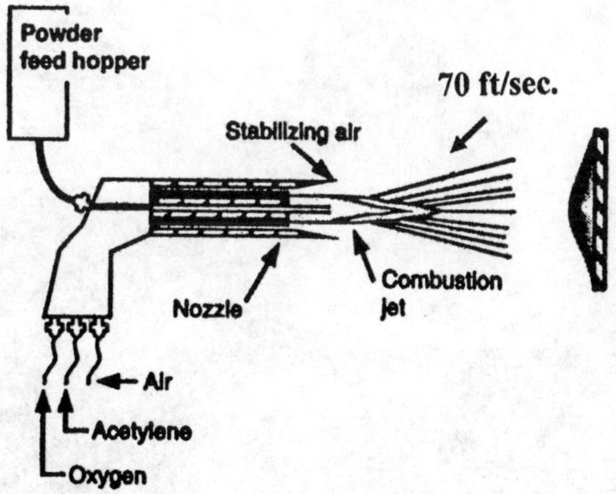

Figure 1.17. Combustion spray gun for spraying molten metal powders to build up a layer of material on a part. Used to add wear or corrosion resistant surface layers. Courtesy of ASM Handbook, Volume 7.

other types; more, in fact, than need to be covered here. The examples we want to mention are the High Velocity Oxy-Flame (HVOF) device in Figure 1.19 and the plasma gun in Figure 1.20. The high velocity oxy-flame gun burns propylene with oxygen at 140 psi. Again, the metal powder gets injected into this hot burning gas mixture (3500°F) and the molten or semi-molten droplets exit the gun at supersonic velocity (2000 - 2400 feet/sec.1500 mph), The target is 6 to 12 in. away. A .002 in. thick layer can be deposited in one minute, and in 20 minutes one could apply a coating .004 in. thick on the outside of a 4 in. diameter cylinder, 20 inches long. The deposit would have less than 1% porosity, because the particles hit so hard that they form very dense layers. Because hard chrome plating has environmental issues and because it lowers the fatigue strength of the high strength steel under it, it seems likely that HVOF spraying of tungsten carbide on steel will gradually supplant the hard chrome plating, wherever the shape of the underlying part allows it. Figure 1.19 shows a large ball valve being coated with chromium carbide and nickel-chrome alloy.

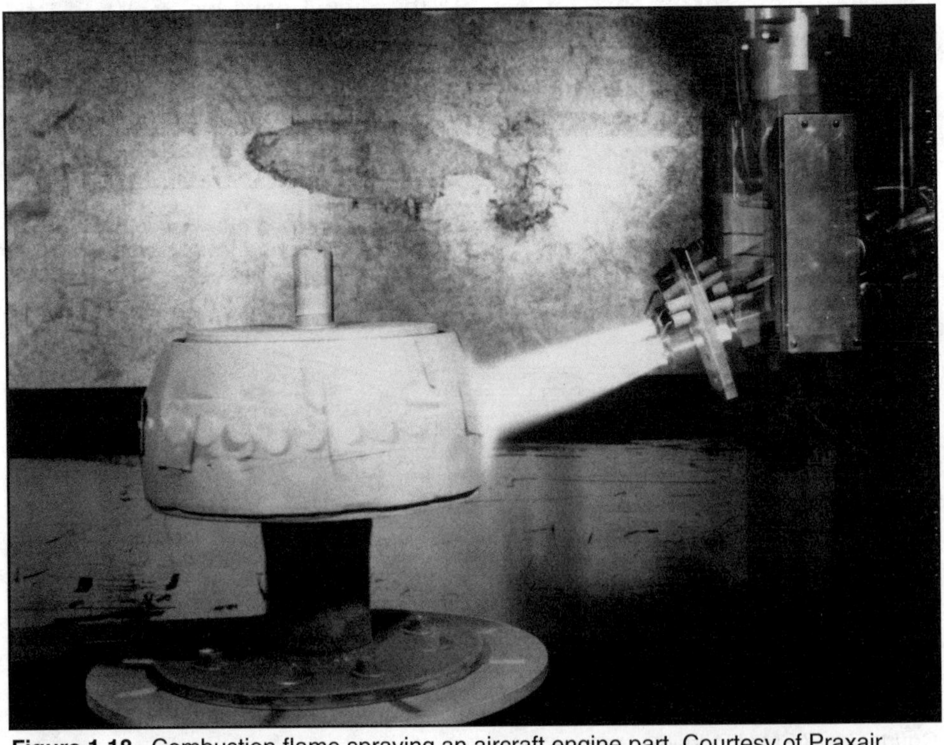

Figure 1.18. Combustion flame spraying an aircraft engine part. Courtesy of Praxair.

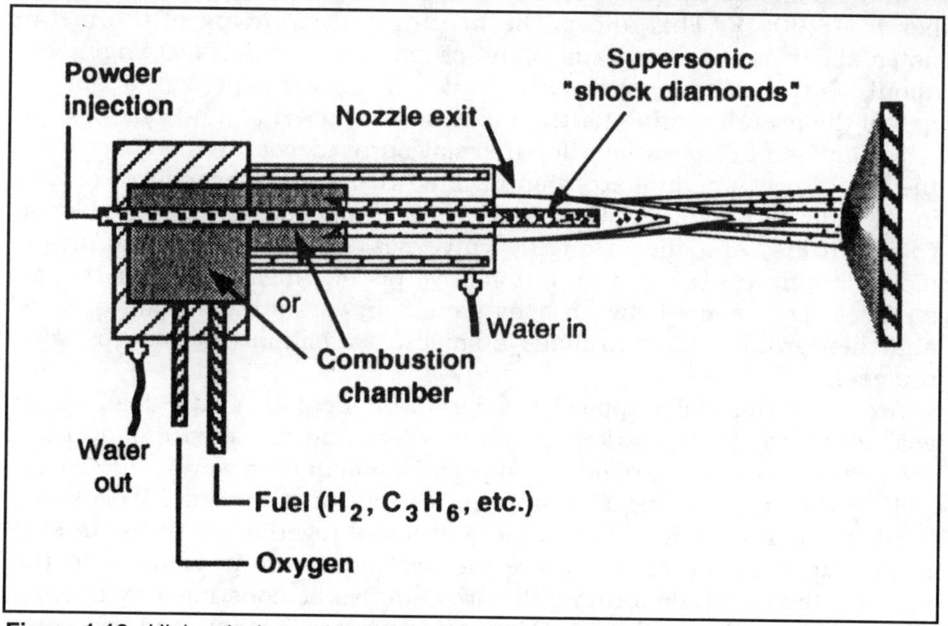

Figure 1.19. High velocity oxyfuel thermal spray gun. Courtesy of ASM Handbook, Volume 7 and Tafa Inc., a Praxair Surface Technologies Co.

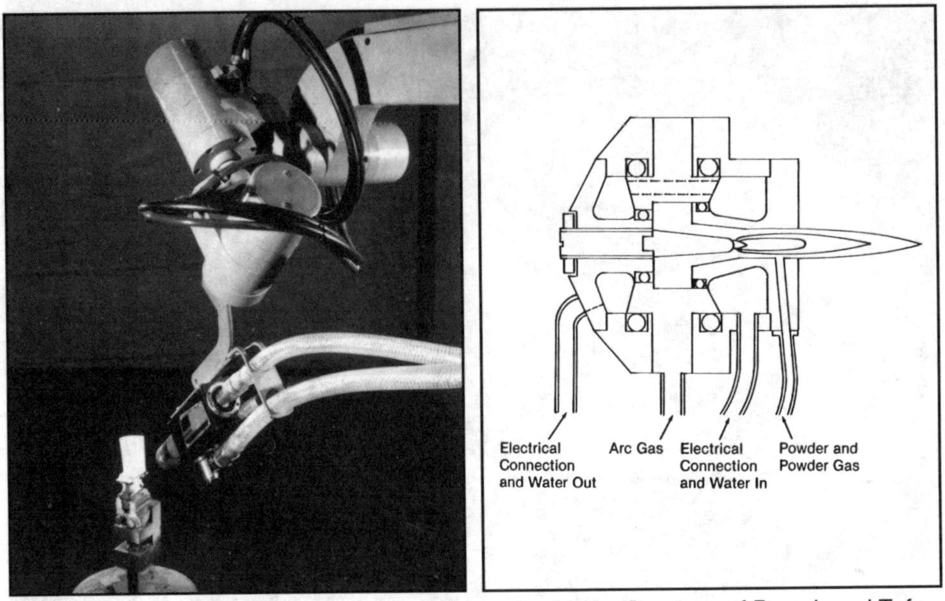

Figure 1.20. Plasma spray gun coating a jet engine blade. Courtesy of Praxair and Tafa.

In a plasma spray gun, an electrical discharge locally heats the chamber to 15,000°F. This allows the heating and spraying of the metals noted above as well as ceramics and chromium carbide. A jet engine has about 2000 plasma sprayed parts in it. Jet engines represent about one half of the overall market for thermally applied powders and services.

Examples of applications for thermal sprayed coatings are knives in the food processing industry that need corrosion resistance (for example in cutting up tomatoes, cobalt base), valves and pumps that handle corrosive fluids, and diesel engine valve seats. Significant amounts of molybdenum are sprayed on automotive piston rings and other bearing surfaces. The seams between body panels on cars are thermally sprayed and then ground down to make a smooth transition from one panel to the next.

Are these thermally applied powders a big deal or a little deal? In the year 2000, the U. S. market for the powders and the service of applying the powders is over $1 billion, and over $2 billion on a world-wide basis.

The term hard-facing also applies to a welding process. We usually think of welding as joining two piece of metal together by using an electric arc to melt the region where the two pieces are in contact. In this process, the electrode forming the arc can also be consumed by melting, and end up in the weld joint. Carry this one step further, and you could just build up a surface on a worn part by depositing the electrode's metal on it. A person working with a hand welder can deposit about 4 lb./hr. this way. The deposit is then ground back by a hand grinder to approximate the original surface. This deposit would be a very "hard face" if you had put the right metal alloys into the stick electrode. Welding stick electrodes can be cast, but they are also made from metal powders. The alloying elements for wear or corrosion resistance are first created in powder form. The powders are then filled into a tube which is subsequently drawn down through several dies until it is about 1/8 to 5/16 in. in diameter. It is cut into so-called tube electrodes. Although sophisticated alloys like cobalt based and nickel based alloys are done this way, the most common materials are iron based. This is a very common way to repair large equipment like an earth mover, even in the field, as it can be done with ordinary welding equipment. You can use an electric welding generator like you see on the back of many welders' trucks, or you can use a gas torch, e.g., oxy acetylene. With the gas torch, the welder heats both the surface being repaired and the stick of hard facing material, until it melts in place.

Friction Products

Friction materials are used in clutches and braking systems, for example on automobiles and aircraft. They slow the rotation of the wheels by pressing against the rotating iron drum or rotor. The friction materials are composites of several materials such as iron, copper, tin, molybdenum disulfide, graphite, coke silica and alumina. If they were made only from the metals like iron or copper, friction materials would squeal on contact with the iron drums or rotors, and they would overheat and seize. The non-metallic constituents like molybdenum disulfide and graphite provide a controlled degree of lubrication and friction. The objective is to have high friction so that there is considerable drag acting to slow the rotating wheel. In slowing the wheel, the mechanical energy of the wheel and the car is converted by friction into heat. P/M metallic friction materials can withstand considerable temperature (e.g, 600°F) and have a lower wear rate than organic brake pads. Figure 1.21 shows a semi-metallic disc brake pad for a pick-up truck. The photomicrograph shows what we found when we cut up this pad. You can see atomized and sponge iron particles and a conglomeration of everything else. It would be no picnic to reverse engineer this mix and try to make one like it! Is this application a big deal? The people who know the answer today are not talking, but in the 1980's about 8 million pounds of iron powder went into brakes each year. In 1994, the friction grades of powder sold for $.32 to $.39/lb.

The copper based mixes are often used in aircraft brakes. A large thin

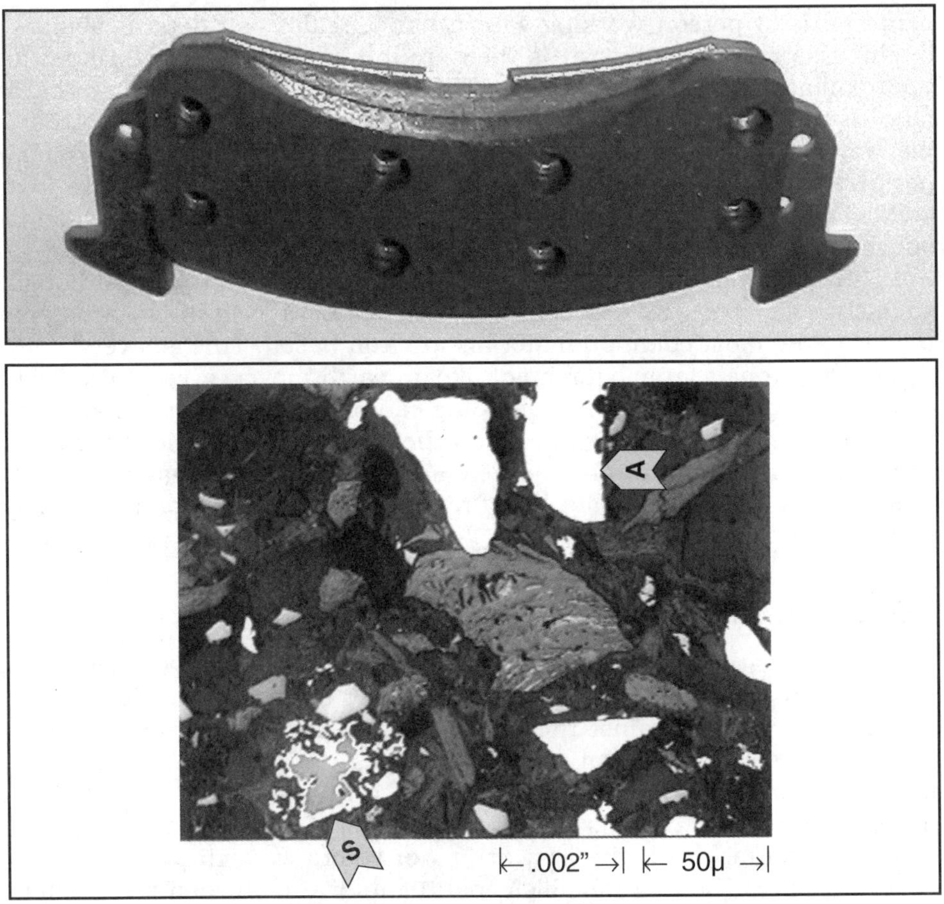

Figure 1.21. Semi-metallic disc pad brake for a truck, $3, at your local auto parts store. Photomicrograph shows iron powder and many other wonderful ingredients. 250x, A=atomized iron powder, s=sponge iron powder.

disc is cold pressed from the powder. This disc is set atop a copper plated steel backing plate. Then a whole stack of these plates is built up in a furnace chamber. Pressure is applied on the discs by a rod acting down through the furnace roof. Sintering can take two hours at 1200°F. The pressure helps bond the particles of copper together and also causes the friction material to bond to the backing plate.

Powder Rolled into Strip

Thin strips of metal are usually made by wrought metal techniques. The iron or copper alloy can be cast into large billets, or it can be continuously cast into strip about .3 in. thick or slabs. To make thin strip, the billets or strip cast materials must be reheated to soften them and then rolled while still hot. There is a limit to how much reduction in thickness can take place in a single pass through the rolls, i.e., about 30%. The strip must then be heated again and rolled again. This continued heating and rolling adds cost. There are also purity problems because of included slag. To avoid the problems of several rolling operations, it is possible to roll metal powders directly into strip. Figure 1.22 shows how the powders are carefully fed into the gap between two rolls. The rolled product is immediately passed into an adjacent furnace to bond the particles together. In this way, one can make strip about .125 in. thick in one pass. The process allows the direct rolling of much thinner strip: why don't they do that? (Lower pounds per hour rolled and the need for a very long adjacent sintering furnace!) The .125 in. strip can then be heated again and rolled to even thinner sections. One example is nickel strip, which is high in purity and high in electrical conductivity. This is used to make battery tabs which are used to connect terminals in multiple cell batteries. The nickel strip forms very well and the dies used to stamp out the tabs last a long time. In the U.S.A., the only manufacturer of powder rolled strip is Ametek, which atomizes the powders in

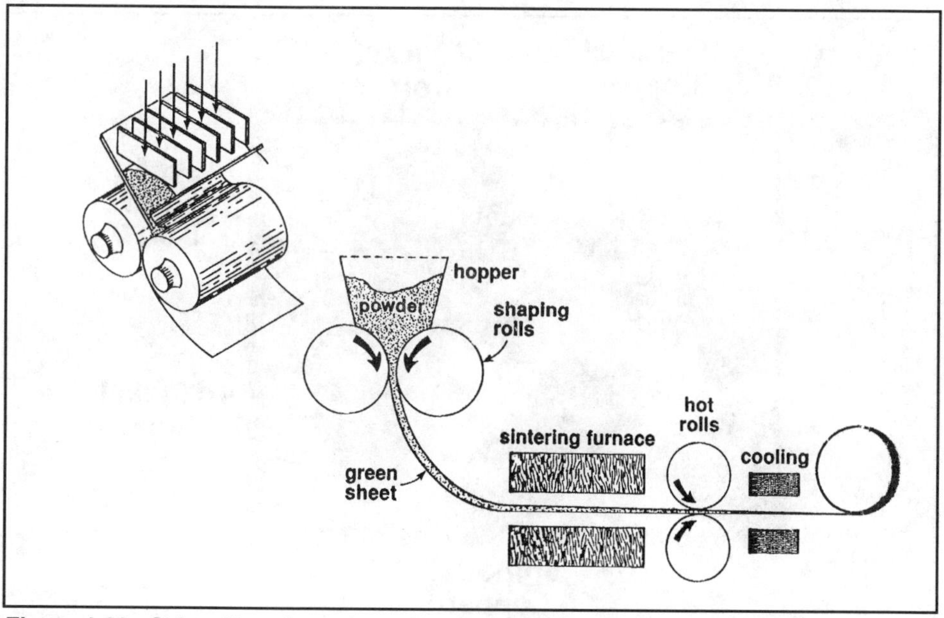

Figure 1.22. Strip rolling of metal powder. Courtesy of of R.M. German, Powder Metallurgy Science.

Pennsylvania and then rolls the strip at their Connecticut plant. We will spare you the photo of a battery, but Figure 1.23 does show some typical parts stamped from the strip.

Another powder rolling example is the nickel sheet stock used by the Canadian mint in forming their coins. Since 1961, Sherritt Gordon Mines in Alberta Canada, and its 1996 spin-off Westaim have been rolling nickel powder into strip. The nickel powder is the normal product of their hydrometallurgical refining process. It is carefully chosen for particle size and shape. The powders are fed into to two horizontal counter-rotating rolls, emerging as 80% dense strip. The strip then passes through a sintering furnace to further bond the particles of nickel. The strip is hot rolled to full density and then cold rolled to the desired thickness. Westaim has, in the past, gone on to form the coinage blanks, but now, the strip usually goes to the Canadian mint. At the mint, round blanks are stamped out of the strip. The blanks are re-struck to form the raised edge on the coins. (Coins need a raised edge so that a stack of coins sits on the nice flat edges, and not on the teetering embossed face of a beaver.) The coins are next annealed to soften them and then highly polished or burnished. The final step is the embossing or coining of the features in the middle of the coin. This process is used on the Canadian nickels, dimes, and quarters. For the one dollar coin, "the loonie", the

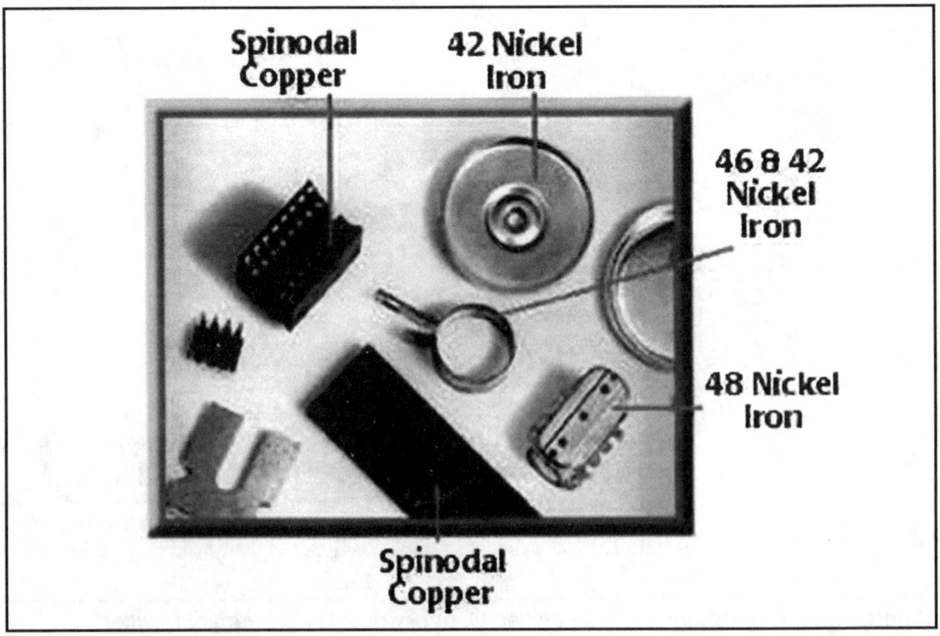

Figure 1.23. Examples of parts made from strip rolled powder. Courtesy of Ametek, Inc.

nickel blank is copper-tin plated to give its bronze color. The two dollar coin, "the twoonie" features an outer shell of nickel and an inner medallion which is an aluminum nickel bronze alloy. As our cousins on PEI astutely observed on first seeing this marvelous coin, it features a likeness of the Queen on the front, with a bear behind! Just as we provided no photo of a battery, modesty forbids showing you the twoonie!

Superalloys

Modern superalloys are used in the engines of jet powered aircraft because they have excellent strength and oxidation resistance at the high temperatures found in these engines. The alloys are modifications of the early Nichrome material, 80% nickel and 20% chromium, long used as heating elements in appliances like toasters. This early alloy was then strengthened by the addition of titanium and aluminum to form compounds with the nickel. High temperature resisting metals such as tungsten, molybdenum, and niobium were added and then some cobalt was substituted for some of the nickel. This resulted in modern superalloys like René 95, low carbon Astroloy and Alloy 720. The René 95, for example, has 61% Ni, 14% Cr, 8% Co, 3.5% Mo, 3.5% W, 3.5% Nb, 2.5% Ti, 3.5% Al, 0.16% C, 0.01% B and 0.05% Zr. (The "partridge in the pear tree" was omitted, but chickens simulate bird-strikes and are used to test jet engine blade strength!) The powder sells for $15-35/lb.

Why did the jet engine manufacturers choose to make their engine components starting from powder, rather than a casting or forging? Today, the powders used are generally smaller than .002 in., and this limits the maximum size of an inclusion or gas bubble that can end up in the part. At this small size, the inclusion can not start a crack that will move across the part. Superalloys are very difficult to forge because they are very strong at high temperature and they are prone to cracking during forging. P/M avoids most of the forging steps. Cast superalloy billets must be trimmed and machined to avoid surface contamination and casting cavities called "pipe". P/M avoids this waste. Finally, the P/M powders are so small that there is no measurable alloy segregation.

The clean superalloy powders are consolidated to full density in large billets or in shapes close to the desired final size. This is done by filling the powder into an oversize metal can of the desired shape. The can is evacuated of all air, heated, and sealed off. The can is next densified by Hot Isostatic Pressing (HIP'ing). HIP'ing consists of heating the can in a high pressure chamber filled with argon gas. The heat softens the can and the powder, and the gas pressure forces the can onto the powder, pushing the powder to near fully density. When cooled, the can is machined or dissolved off the part. If the final application is rotating discs in jet engines, the HIP'd material is forged prior to finish machining the disc. As of 1999, Crucible Compaction Metals had produced over 80,000 superalloy P/M high pressure turbine discs and blade retainers. They are currently in service in commercial and military airplanes and helicopters. Some of these parts weigh up to 1,500 lbs. Other superalloy powders are canned and then extruded into rods and billets. These are machined to size and do not need a subsequent forging operation. Some HIP'd superalloy parts ready for final machining are shown in Figure 1.24. Such parts sell for \$20 to \$40/lb.

In 1997, the U. S. Census Bureau reported that the value of superalloy powder sold in the U.S was \$106 million. We suppose that by the time it was canned, HIP'd, forged, machined, and inspected, that the

Figure 1.24. Aircraft engine parts made from superalloy powders, René 95, low carbon Astrology, and alloy 720. These are sonic shapes ready for final machining a turbine discs and blade retainers. Courtesy of of Crucible Compaction Metals.

value added was considerable! From 1999 to 2009 the group Resource Strategies forecast that about 800 civilian jet airplanes would be built each year. In addition, there would be more rapid growth in the non-aerospace applications for superalloys. These include power generation, marine engineering, environmental control, chemical and oil and gas industries. By 2009, they forecast a world-wide market for all the forms of superalloys at 330 million pounds. However, many of these applications are based on cast or forged superalloys, for applications that are less critical than aircraft engines.

Molybdenum and Tungsten

These two elements are formed into useful objects via powder metallurgy because they have very high melting points: (Mo = 4730° F and W = 6170°F). and are difficult to cast. (By comparison, cast iron melts at 2100°F.) When casting of tungsten was finally achieved, the product was brittle. Powder metallurgy is the preferred route to making tungsten components. Tungsten ore is first converted to an ammonium compound to get rid of the impurities. The ammonium paratungstate is oxidized in air, and finally reduced back to tungsten powder by heating it in hydrogen.

In 1910, Coolidge developed a way to take this tungsten powder and form it into electric light bulb filaments, solving a problem that had existed since Edison invented the bulb in 1879. (The prior filaments were based on carbon and had been very brittle.) Today, tungsten powder is pressed into rods 18 in. long by .8 in. square. This is done in special dies 18 x 8 in. in a hydraulic press. The parts are not ejected from the tools, but, instead, the dies are taken apart and the fragile tungsten bar carefully removed. It goes onto a tungsten plate and then into a presintering furnace for preliminary heating in hydrogen at 2200°F to begin bonding the tungsten. After the rods have been cooled back down to room temperature, they are clamped between two electrodes, one of which is free to move. Electric current is run through the rod, to heat it by its own resistance, and to sinter or bond the particles at 5400°F, in a protective atmosphere. (This is like one humongous light bulb!) Once cooled, the loose rods are reheated in protective atmosphere to 2000°F - 2200°F and then swaged in air, to a smaller diameter. (In swaging, the rod is fed horizontally into a machine that has tapered dies that repeatedly slam together on the rod, forcing it into a smaller diameter, while gradually elongating the rod.) After swaging, the rods are drawn hot through diamond dies, gradually reducing the tungsten rod to the fineness of a lamp filament. Prof. Randall German of The Pennsylvania State University reports that 150,000 filaments for 40 watt bulbs come from one pound of tungsten wire. The raw material as powder sells for $8/lb.

Although it seems like ancient history, the rocket nozzles on the little lander that descended to the moon's surface in 1970 were made from tungsten powder. The nozzles were cold pressed isostatically, sintered and then forged. Tungsten, even when mixed with small amounts of nickel and iron, has a very high density, about 18 g/cm^3. This high density makes it useful in counterweights, gyroscopes and additive weights in golf club heads to change their center of gravity or percussion. In a friendly game of darts, the tungsten alloy in the middle of the dart gives it better balance, while in a less friendly mode, tungsten penetrators, as part of an ammunition shell, have enough momentum to penetrate an enemy's tank. Figure 1.25 shows molybdenum heating elements. (We thought it a good idea not to show a picture of a lightbulb.) In 1997, the value of tungsten base powders was $184 million and molybdenum was $14 million, while in 1999, in North America, about 3 million pounds of

Figure 1.25. Molybdenum metal used inside of a hot isostatic press. The network of horizontal rods are the electrical heating elements. The sheet of material just outside the elements is a molybdenum heat shield. Such a heating system inside of the pressure vessel can operate at 2,500°F. Courtesy of Bodycote IMT.

tungsten powder were produced.

Parts made from molybdenum have good strength at high temperature, as long as they are not exposed to oxygen above 1100°F. (Molybdenum forms an oxide that evaporates above 1100°F.) It is excellent for the hardware that is used inside furnaces or hot isostatic presses, as shown in Figure 1.25, operating at 2400°F. That includes heat shields and the heating elements. It is also used as the heating element in pusher type sintering furnaces, but as long as the furnace is above 1100°F protective atmosphere must be present. Alloys of molybdenum strengthened with carbides (TZM) are used to make the dies that hot form ("Gatorize") superalloy powders into turbine discs for jet engines. The die and the superalloy material are all heated together, and under pressure, the die gradually forms the turbine disc.

Because molybdenum conducts heat well, it is used to make heat sinks. These are made with mixtures of molybdenum and copper to carry the heat away from computer chips and other solid state power devices. In X-ray tubes, molybdenum forms the X-ray target and the support structure for the target. Molybdenum powder and its alloys are thermally sprayed on the piston rings in an automobile engine for better wear resistance. Finally, molybdenum electrodes are used to increase the heat input to furnaces that melt glass. The glass is partly heated by burning gas, but once the glass becomes molten, it becomes electrically conductive and the molybdenum electrodes can send electric current through the glass to heat it even faster. The glass protects the molybdenum form oxidizing.

Food Supplements

Did you ever wonder what makes corn flakes crunchy? Could it be the iron powder that goes into all enriched flour and food products that contain an iron supplement? One of us, not WW, actually prepared a polished "metallographic" mount of an airline pretzel, looking for the iron additive. In retrospect, the .002 in. iron particles probably dissolved in the pretzel batter just as the fine iron would dissolve in the acid juices of your stomach. A more logical path of investigation would have been to look at some King Arthur Flour, right out of the bag. Putting deed to word, using the scanning electron microscope on some of our bread flour, we found the little iron particle shown in Figure 1.26. It looks like it was beaten around by the milling process on the flour or by the sponge iron powder producer. The flour contains about 30 parts per million of iron, so your 10 lb. bag has about .0005 oz of reduced iron powder. In concrete terms, that would be about 26,000 particles like the one shown in Figure 1.26.

Figure 1.26. Elongated Fe powder particle, .002" long in the middle of a field of flour (bread flour).

When you read the fine print on the iron addition, it refers to "reduced iron", and you will have to be patient until Chapter 2 to find out what "reduced" means. Are food supplements a big deal or a little deal? The reduced iron powder sells for $1.09/lb. in lots of 7,900 lbs., and about 2 million pounds are sold each year in the US. If you want to "take your iron straight", at the drugstore, you can buy 50 mg caplets of carbonyl iron, another kind of iron, and a term defined in Chapter 3. The 30 caplets, containing 30 x 50 = 1500 mg = 1.5 g of iron sell for $5. With 454g in a pound, you will quickly see that these folks are charging $1,513/lb. for what the manufacturer of carbonyl iron sells for $4.50/lb.! Surplus iron powders are also added to fertilizers to help make your lawn green. Fortunately, the fertilizer companies have not yet figured out how to get $1,513/lb.!

Tantalum Powder in Electrical Capacitors

R. German reports that tantalum powder is made from enriched ore by dissolving it in hydrofluoric acid and precipitating out potassium tantalum fluorine (K_2TaF_7). This is reacted with molten sodium and releases the tantalum as metal powder. The sodium salts, along with potassium and fluorine, are washed out with water. The resulting tantalum cake is ground up into fine powder. The powder is purified by electron beam melting, hydriding and recrushing. The principle use for purified tantalum powder is in electrical capacitors, with 60% of the powder being used in this application, shown in Figure 1.27. This figure shows a capacitor with electrical leads, but 80% of capacitors are smaller and have no leads, They go directly onto computer chips. Capacitors temporarily store electric charge and then give it back to the circuit, allowing the circuit to tune, filter, or time electrical signals. Tantalum powder is used inside the capacitor for the anode which stores the charge. Tantalum sheet and wire made from powder are used in the case and the leads. Tantalum capacitor usage grows at the rate of 20% per year, but folks keep making smaller capacitors that are more efficient, and so tantalum powder usage does not grow that fast. Tantalum powder sells for $200/lb and about 1,700,000 pounds were sold on a world-wide

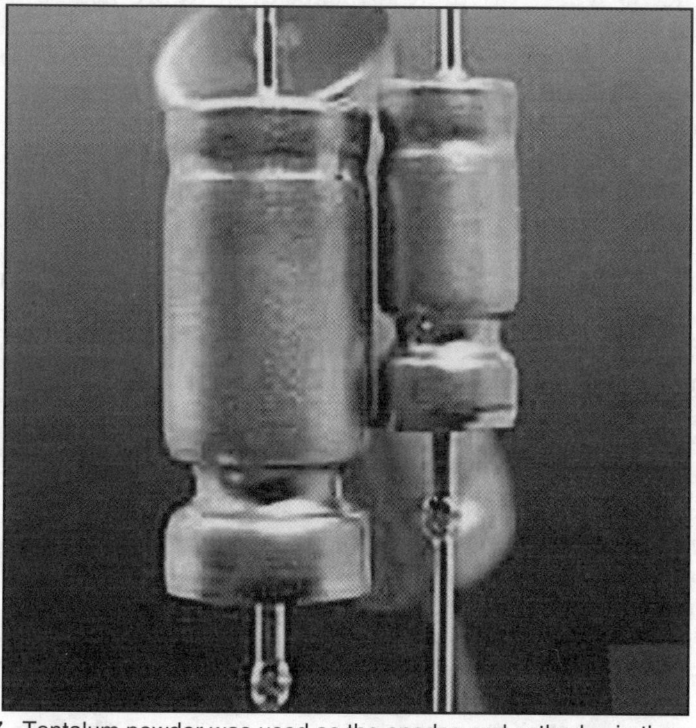

Figure 1.27. Tantalum powder was used as the anodes and cathodes in these tantalum electrolytic capacitors.

basis in 1995. Because it melts at 5,425°F, tantalum sheet stock is usually produced from metal powders, although it can also be produced by electron beam melting. The sheet forms of tantalum are used for heating elements, heat shields and chemical process vessels that require extreme corrosion resistance. The tantalum powder with the usual oxygen levels must be sintered near 5425°F to reach high density. This means that the sintering furnaces have short life spans! Tantalum powders under development will sinter at only 2550°F, and will lower the cost of producing sheet, bar stock and wire, as well as improve capacitor anode efficiency.

Pyrotechnics, Aluminum and Magnesium Powders

The oxides of aluminum and magnesium, Al_2O_3 and MgO, are very stable compounds. They became very stable because, when the aluminum or the magnesium reacted with oxygen to form them, a lot of heat was liberated. To put the rather clinical term "liberated heat" in perspective, when the space shuttle is boosted into orbit by those huge rockets, the hot gases driving the rocket upward are heated by reacting (read "burning") aluminum powder with oxygen! One shuttle launch uses between 160,000 and 500,000 lbs. of aluminum! The aluminum powder sells for $1.05/lb. while the metal in ingot form is worth about $.80-.85/lb. The coarse aluminum powder is mixed with an oxidizing agent like ammonium nitrate and binder so that it can be cast into semi-solid fuel. When the oxygen in the ammonium nitrate (NH_3 NO_3) reacts with the aluminum to form aluminum oxide, the heat of the reaction heats the gases and forces them out of the rocket engine.

Total aluminum powder produced for all purposes in 1999 was 106 million pounds. Of that, 2.7 million pounds were used to make P/M parts such as cam cap bearings in automobile engines, in Figure 1.28. Development is ongoing for mirror mount brackets, shock absorber parts, gerotors, pumps, and connecting rods.

Aluminum powder has many uses other than just pressed and sintered parts. Tom Gurganus of Alcoa (Adv, Mat. and Proc. 1995, vol.148, #2, p.57) reports that aluminum powder is used as follows:

1. Production of Cr and Fe-Nb
2. Chemical industry as catalyst, reducing agents and fillers.
3. Aluminum chlorhydroxide is used as a deodorant
4. Fillers in epoxy resin
5. Ball milled flake is used as a pigment or colorant in paints for cars and trucks

As a young undergraduate in 1955, LP heard from the "big boys" how they used to use a little "thermite" to weld street cars to their metal

Figure 1.28. P/M aluminum bearing cam cap made by GKN Sinter Metals, Conover for Chrysler Corporation automotive engines.

tracks in Cambridge and Boston! Aluminum plus iron oxide = thermite, which releases great heat when it reacts!

Frank Beaumont, retired from Ampal, Inc, tells us that the big boys now have a more socially acceptable use for the thermite. If you wanted to join together two iron rails at their butt ends, position them end to end, about 1/2 to 1 in. apart. Construct a little ceramic mold that will enclose the sides of the space between the rails and fill the space with thermite. Top it off with a child's sparkler and ignite. The reaction reaches 3500°F and iron melts at 2800°F. It then flows down between the rails and gives an excellent weld. If you mix a little ferro-vanadium in with the thermite, you create a nice alloy steel weld.

About 5% of the 106 million pounds goes into military explosive items. One example is the stun grenade, which is a mixture of fine aluminum and an oxidizer. On ignition, it creates an incredibly bright flash, accompanied by a very loud bang. Anyone near this is momentarily stunned and disoriented, allowing the person using the grenade to seize control of a situation.

Magneto-Rheological Applications of Iron Powder

Earlier, we conveyed the idea that iron could be magnetized, and so you will not be surprised to learn that iron powder can also be magnetized. One novel use of magnetized powder is in a clutch. Clutches are used to start and stop mechanical motion, hopefully smoothly. For instance, when the face of a rotating plate is forced against a non-moving plate, after a little slipping and chatter, the second plate will come to rotate at the same speed as the first one. That is a simple clutch. But, if the two plates are in an enclosed box containing some iron powder and oil, magnetized iron powder can be used to connect the two plates so that one drives the other. A magnetic field is created between the two plates and the iron powder hops into place between the plates, creating a friction or linkage. Turn off the magnetic field and the powder returns to its former loose state, and the two plates are no longer linked together. Rheology describes the viscosity and flow character of the oil/iron mix. The magnetic field makes the mixture behave like a thick viscous fluid. Phrased another way, what cold mornings do to molasses, a magnetic field does to the iron powder and oil mix.

Delphi Automotive has recently developed the shock absorber or strut shown in Figures 1.29a and b. When the shock absorber gets compressed as the car hits a bump, an oil and iron powder mixture has to pass through a small internal hole for the shock absorber to shorten. If the material in the absorber were just oil, it would always flow through the hole with the same degree of resistance. But, with the oil and iron powder mixture, a sophisticated computer program can sense what the car is doing, and make the mixture either thinner or thicker in 1/1000th of a second, as needed to better control the car and the bump. It changes the mixture's thickness or viscosity by applying a magnetic field though it. The iron powder consists of $5/lb. carbonyl iron about 5 micrometers in diameter and a coarser iron, presumably the less expensive atomized material. This device called Magneride will be on some General Motors cars in 2003.

Rapidly Solidified Powders (RSP)

If metals or alloys are cooled very rapidly from the melt, 2 million to 2 billion °F/sec., the atoms do not have time to arrange themselves in their usual neat orderly rows. Instead, they are highly disordered or jumbled. The usual forms of crystals that we expect, do not form, but instead, the material is a kind of solution of all the various metals present (called an amorphous or glassy structure). Powders in this form can be very hard because atoms of different size are forced together, straining the structure. The materials are unstable in the sense that if the rapidly solidified powders are heated at low temperature, the atoms quickly re-arrange themselves in their customarily neat rows (crystallinity). This reversion to the normal structure destroys most of the interesting properties the RSP material had.

Well, what interesting properties does RSP have, and how on earth do you ever convert the hard powder into a useful part if it is unstable and

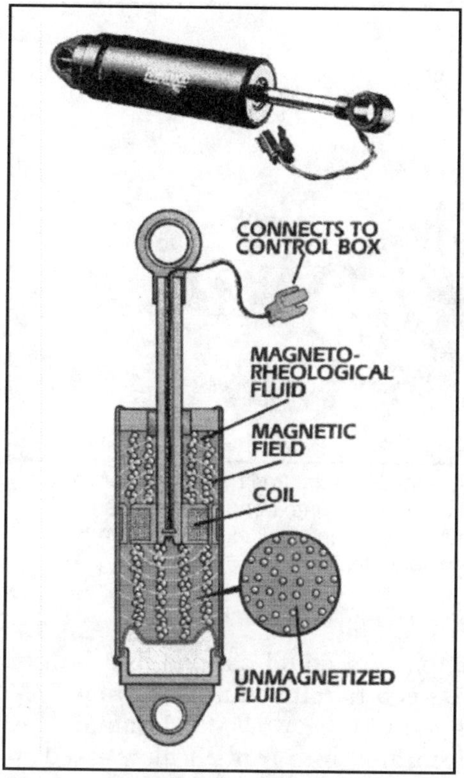

Figure 1.29a. Magneto rheological fluids shown schematically in a shock absorber, courtesy of Popular Mechanics and Lord Corp.

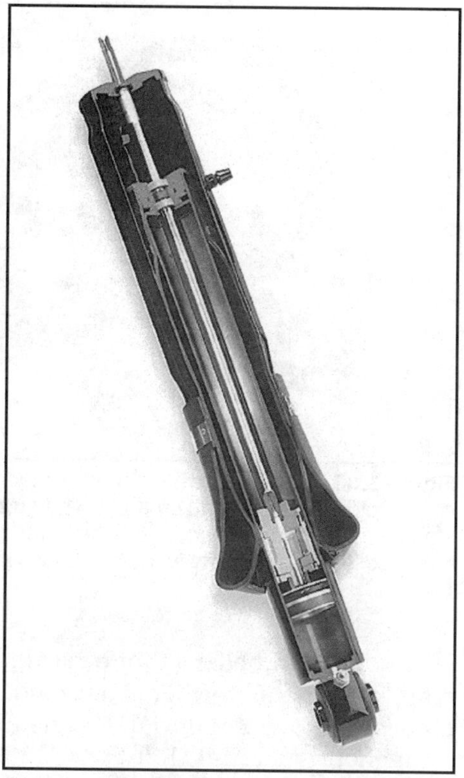

Figure 1.29b. Magneto-rheological effects at work! Magnetic iron powder is suspended in oil in this shock absorber. When a magnetic field is applied, the mixture stiffens in proportion to the field. This allows the continuous and instantaneous control of the stiffness of the shock absorber and a better ride. Courtesy of Delphi Automotive.

you can't heat it? Prof. German, "Powder Metallurgy Science" p.149 cites the hard magnetic material $Fe_{14}Nd_2B$ used in automotive starting motors and the superalloy IN 100 used in jet engines as examples. The IN-100 does not remain truly amorphous or glassy, but develops a very fine crystal structure during hot extruding of the powder. This process increases the tensile strength by 70% and the ductility by over 500%, compared to the cast alloy. Figure 1.30 shows some RSP parts. The $Fe_{14}Nd_2B$, familiarly known as neodymium iron boron, is solidified as a thin sheet. The sheet is crushed and ground to 3 micrometer particles. The powders are pressed in a magnetic field and sintered. The resulting parts can then be highly magnetized, making the best permanent magnets.

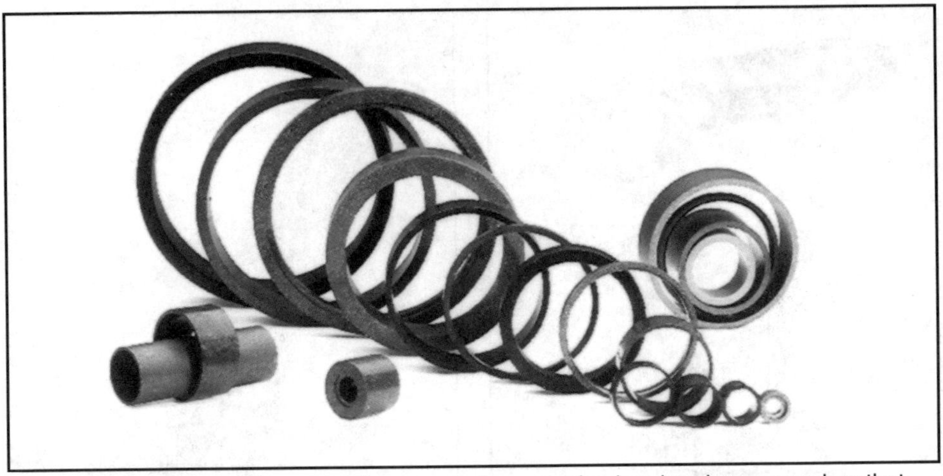

Figure 1.30. Permanent magnet rings made from neodymium iron boron powders that have rapidly solidified and crushed. They were then formed into bonded magnets by MagneQuench International Inc. These magnets are used in small electric motors automobiles and in industrial applications like power drills.

It is hard to believe that anything important could happen if it did not start with metal powder. That said, we need to tell you about a non-P/M amorphous strip material, Metglas made by Honeywell-Allied Signal. It is a magnetic material that can be incorporated into transformers used to distribute electricity. It wastes very little energy compared to the normal iron laminations. The Metglas costs more than the iron, but pays for itself in energy savings. Vacuumschmelze in Germany makes 500 miles of thin amorphous strip each year. It goes into the security tags on retail items in stores. It has the advantage that, when demagnetized by the cashier, it does not trigger false alarms at the exit. In the future, such tags will carry product information such that you could wheel a shopping cart through check-out, and each product would be automatically identified and the cost tallied. No more laser scanners.

Osprey Process

A stream of liquid metal can be poured through a nozzle. The stream can be struck by inert gas and broken into droplets. The droplets can be directed to land on a base or substrate, cooled and frozen very quickly, thus building up a rough form. For instance, if the droplets hit a rotating shaft, a roughly shaped tube can be formed around the shaft, and subsequently removed. If the droplets are directed at a round pedestal, a long round billet can be built up. The process is best applied to expensive alloys that are difficult to cast or to form into shapes like tubes. The P/M part has all the advantages of uniform chemistry and small crystal size that do not happen when one pours a large ingot. Tubes, rings and rolls up to 30 in. in diameter by 26 feet long can be made. The metal is deposited at rates of 20 to 220 lbs./minute.

At Peak in Germany, about 3,000 ton/yr. of high alloy aluminum with 25% silicon, 4% copper, plus magnesium, are being produced in billets 12 in. in diameter by 9 feet long, deposited at 33 lbs./minute. The billets are extruded into long tubes, swaged, and cut off to make cylinder liners for the Mercedes Benz engines. The aluminum engine blocks are cast around these cylinder liners. The liners have much better friction and wear properties than the ordinary aluminum alloy that makes up the rest of the engine block.

Howmet and Pratt and Whitney have a joint venture called Sprayform Technologies International which they created to sprayform superalloy rings and cases for jet engines. The rings can be 30 in. in diameter by up to 32 in. long and have a wall thickness of 3 in. The rings are deposited at 80 to 90 lbs./min. with a deposition yield of 80% (i.e. 80% of the sprayed powder hits the ring and sticks to it). A hot spray-formed ring is shown in Figure 1.31. With three separate spray forming units, the plant has the capacity of 3,000,000 lbs./yr. In one unit, the vacuum melting furnace can melt 880 lbs. of metal for the spraying run. The spray formed rings are nearly fully dense, but they are further processed to reach full density. This includes HIP'ing or HIP'ing plus some form of hot ring rolling. They are then finish machined. The process works well on alloys like IN 718 and difficult to work alloys like Waspalloy. The P/M advantages are fine uniform grain size, no coarse carbides and no macro-segregation of alloying elements as found in a large casting.

Osprey Metals is spray forming silicon with 30 to 50% aluminum in billet form. The material has a low and controlled thermal expansion when it is heated. It is used for electronic packages and heat sinks. It has good thermal conductivity to carry the heat away from the high speed electronic chips in computers. Computer chips can be brazed to this material because its thermal expansion matches that of the chips. If the expansions are not the same, the chips could be stressed or fractured during cooling from the brazing cycle, or in service. Billets 24 in. in diameter are spray formed at up to 440 lbs./minute. The density of the billets is 97 to 99.5% of full density.

Laser Free-Form Fabrication of Powders

If a thin stream of metal powder is allowed to fall into the focal point of a laser which is directed at a base or substrate, the particles will melt

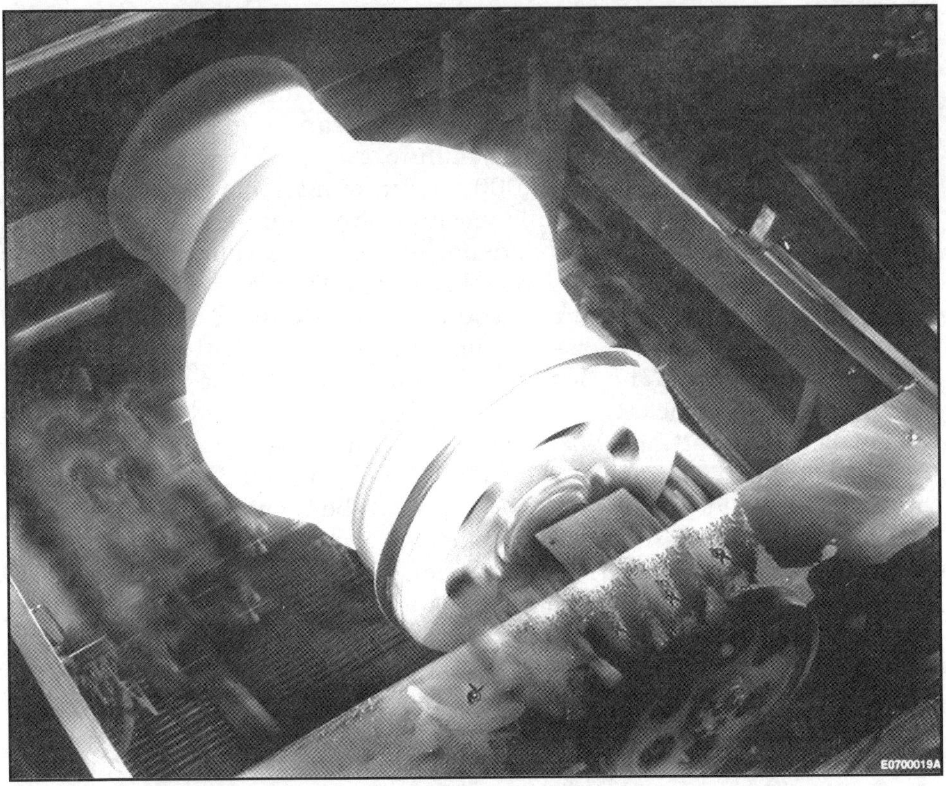

Figure 1.31. Spraycast-X®, process for forming superalloy rings such as turbine support housings, 12 inches in diameter. The melting of the alloy and the spraying all take place in a 7,800 cu. ft. vacuum chamber. Courtesy of Greg A. Johnson, Howmet Corp.

and attach themselves to the region which is being struck by the laser, shown in Figure 1.32. The powders are typically 10 to 180 micrometers (.0004 in. to .007 in.) in diameter. The work is carried out in an argon filled chamber, essentially free of oxygen. If the base or substrate can be moved in a controlled way, in one pass, a layer about .005 in. thick of fused metal can be built on the base. If additional fused layers are placed on the first layer, a fully dense three dimensional object can be built up. The area heated by the laser, where the powders melt and attach, is small, about .030 in. in diameter. For this reason, a part with close dimensions and decent surface finish can be built up. This laser free-form fabrication requires no die or tooling to form the metal part. The technique was demonstrated by Gary Lewis of Los Alamos National Laboratory and David Keicher of Sandia National Laboratory in the mid 90's. David Keicher then moved on to be one of the principals in Optomec

Design in Albuquerque, NM, building machines to do the laser forming.

By 1998, the technique was being used to make injection molding tooling in H13 tool steel. The injection molding tools could be produced in a few days, so that the development and testing of molded plastic parts could be vastly speeded up (P/M industry shoots self in foot?) Of course the same laser formed tools could be used to make Metal Injection Molded, MIM, parts, starting with metal powders. Because the tooling is built up layer by layer, it is possible to leave cooling passages in exactly the right places in the tools. These passages conduct cooling water during the injection molding process and allow the molded parts to be quickly cooled and removed from the tooling, speeding up production. There is no other known way to place these water cooling channels in the best possible location. In late 1998, the state of the art for producing such tooling included powder deposition rates of .5 to 2.0 in^3/hr., and a tolerance of ±.002 in. over a distance of 6 in. in the horizontal plane. The vertical tolerance was ±.015 in..

In a related technique developed by DTM in Austin, TX, the starting material was steel coated with a thin plastic coating. The coated powder was spread out in a layer. A movable laser mounted above the powder bed traced out a heated pattern on the powder. Where the laser struck, the plastic melted and the coated particles fused together. Another layer

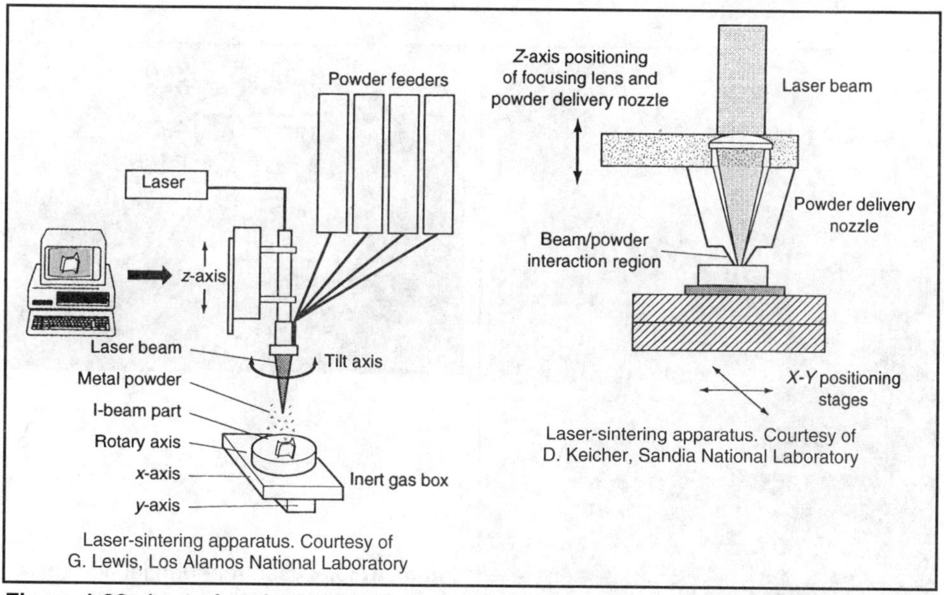

Figure 1.32. Laser free-form deposition of metal powders. Courtesy of ASM Handbook, Volume 7.

about .003 to .010 in. thick was spread on top of the first layer. This was traced over by the laser, fusing the particles to themselves and to the layer underneath. This allowed them to build up a three dimensional object, such as injection molding tooling. The plastic was then burned off in a furnace, and the temperature raised to above the melting point of copper (greater than 1981°F). A block of copper or bronze placed near the tool, melts and runs into the tool, filling all the void spaces created when the plastic burned out. In 1998, more than 20 companies had the laser sintering station and software needed to make these tools. The tolerances on these tools are: ±.010 in. and a tool 6 in. x 6 in. x 12 in. can be produced in 48 hours. The 48 hours is a combination of the laser forming time and the time to process and infiltrate the iron/plastic skeleton.

Work done at The Pennsylvania State University under a DARPA contract demonstrated laser forming of titanium powder with properties equal to the ASTM cast equivalent. The technology has been carried further by Aeromet of Eden Prairie, MN under Frank Arcella, who was the Johns Hopkins program manager responsible for the work at Penn State. Figure 1.33 shows some of Aeromet's early laser formed titanium pieces. This is laser forming on a large scale, aimed at replacing the large titanium forgings used in aircraft, like the bulkhead of an F-16 fighter plane, 9 ft. by 3 ft. by 3 in. thick. The titanium and alloy powders are placed in a bed that can be up to 10 ft by 10 ft. An 18 kw laser beam strikes the

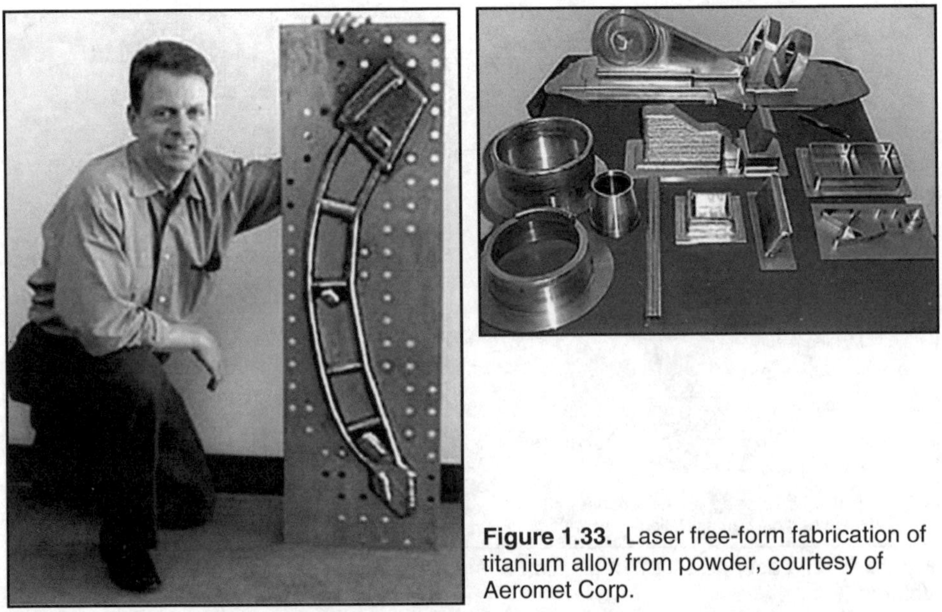

Figure 1.33. Laser free-form fabrication of titanium alloy from powder, courtesy of Aeromet Corp.

bed of powder and melts a spot about ½ in. diameter. The laser is mounted above the powder bed on a gantry so that it can move in any direction and fuse a layer about .040 in. thick. After one layer is created, a new .040 in. thick layer of titanium and alloy powder is spread on the first layer. The laser then fuses the second layer to the one under it. A three dimensional object can be built up one layer at a time. The deposition rate is 2-10 lbs./hr. so that a 330 lb. bulkhead might be built in 60 hours. Because the laser spot size is large, the laser formed parts will have a rough surface and must be machined all over. The alternative to the laser forming of the titanium powder is the production of conventionally forged parts that can require 12 months. Part of this time is what is required to build forging tools and then there is the waiting queue for forging, machining and heat treating. The original forgings result in an extraordinary waste of 5700 lbs. in reaching the final machined part which weighs 330 lbs. By comparison, the laser forming of titanium powder wastes very little!

Beryllium

Beryllium is a very light metal, 1.85 g/cm^3, being 25% as dense as iron and steel and 62% stiffer. It is used in aerospace structural applications like mirror mounts and gyroscope hardware. Cast beryllium has very little ductility so it is most often processed from powder. For the best ductility in the final parts, the powders are hot isostatically pressed, HIP'd. A metal can shaped like the final part, but 18 to 20% larger in all dimensions is filled with gas atomized beryllium powder. Under high temperature and pressure, the can is forced inward and presses against the powder, removing all the porosity. The can is machined off or dissolved in acid, and the part machined to shape. This process gives such good properties because the consolidating temperature is relatively low and the metal grains do not grow. Powder is also vacuum hot pressed to near full density in shapes up to 72 in. in diameter at 1,000 psi and about 2000°F. Machining of beryllium parts removes much of the initial stock, with parts weighing only about 30% of the raw material. The beryllium is sold at $300 to $400/lb. in rod or bar form, for subsequent machining. The P/M products are not brittle and will stretch 5% before failure, with tensile strength of 50,000 psi (hot pressed) to 120,000 psi for extruded rod. Brush Wellman is the only U.S. company that starts with ore, and converts it to powder and then to rod and billet form. They do not sell the beryllium powder. Oxides of beryllium are toxic, and great care must be exercised in handling them.

Beryllium is mainly used where extreme lightness and stiffness are required. Applications include parts for fission and fusion reactors

The light weight of this beryllium internal structure for a commercial satellite is obvious in this picture. The structure is an assembly of adhesive bonded and riveted beryllium sheets (Courtesy, Electrofusion Corporation).

CIP/Sinter/HIP

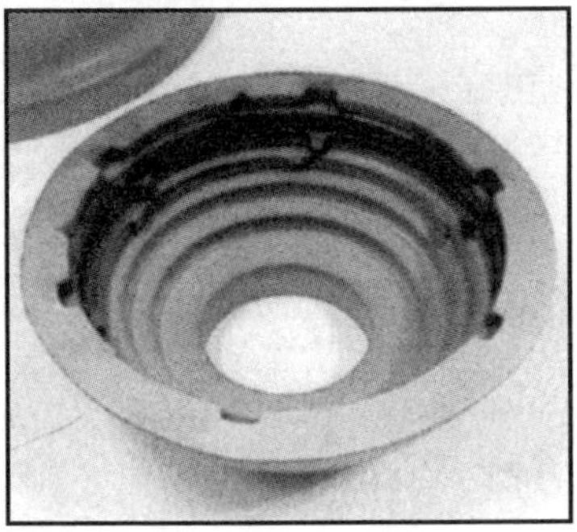

Near-net-shape beryllium blank for a missile guidance component was produced by the CIP/sinter/HIP process. This process provides a cost-effective way to produce complex shapes with fine details.

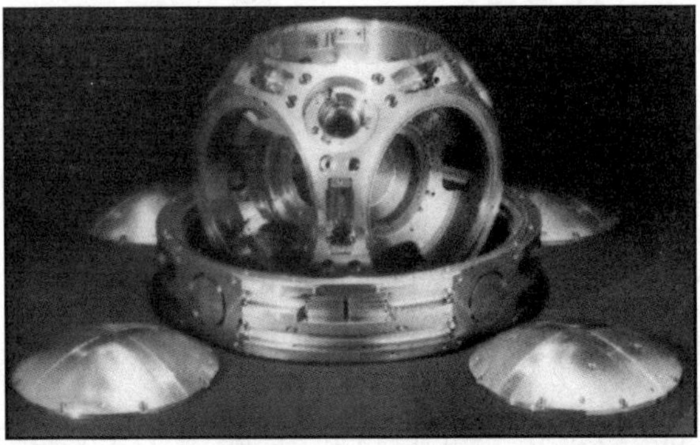

Figure 1.34. Beryllium parts made from metal powders by hot isostatic pressing or hot pressing of metal powders. Courtesy of Brush-Wellman.

(beryllium can control neutrons in a reactor), inertial guidance, space telescopes, satellite structures, computer disc drives, and windows for X-ray tubes. Figure 1.34 shows parts formed from beryllium powder.

Electrical Contacts

We noted above that tungsten has a very high melting point, 6160°F. This feature allows it to resist the high temperatures caused by sparking, when it is made into an electrical contact. In an electrical contact,

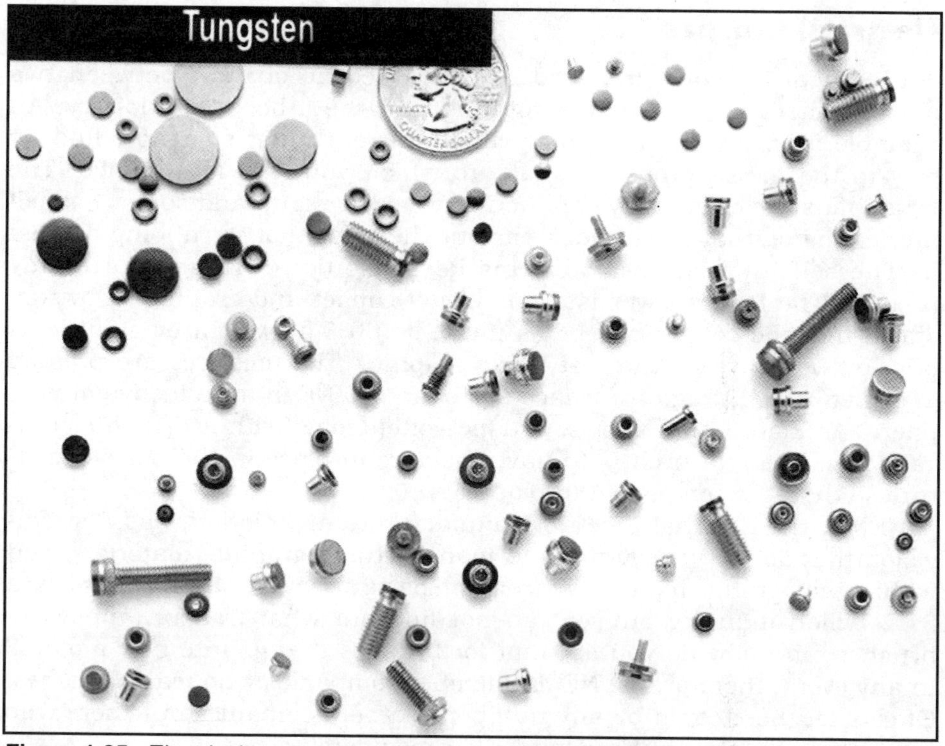

Figure 1.35. Electrical contacts made from tungsten powder by pressing and infiltrating with copper and silver. Courtesy of Norstan, Inc. of Waukegan, IL.

every time it is opened to stop the passage of electrical current, the current sparks and heats the surface of the contact, tending to boil away some of the contact material. The tungsten resists the high heat and also helps resist the tendency of the contacts to weld together under the heat of the electric sparks or arcs. The P/M contacts are made by pressing the tungsten powder into a little button, and then heating it in a furnace. A small piece of copper or silver placed on or near the tungsten will melt at 1981°F and then run into the spaces between the tungsten particles, (infiltration). With the spaces between the tungsten filled with copper or silver, the tungsten is bonded together. In the contact, in service, the heat from the electric spark will be carried away by the copper or silver while the tungsten resists boiling away. Figure 1.35 shows the many forms of electrical contacts.

Electrical Brushes

Electrical brushes are used to carry electric current between two objects where one surface is rotating and the other is stationary. An example is in an electric motor where current must somehow find its way to the inner rotor and shaft, to which your fan is attached. The brush must be electrically conductive, wear resistant, and somewhat self lubricating so that it will not wear out the rotor that is rubbing against it. The self lubrication also helps keep it quiet during this rubbing process. The P/M answer is a mixture of copper and graphite. Dowson, states that the copper can be as much as 80% by weight or as little as 30% by weight, with the rest being graphite. The mixtures are pressed and then run through a furnace at about 1800°F in a protective atmosphere. As Figure 1.36 shows, brushes often come with attached electrical wires. The real trick is to have the wire in place in the powder mixture so that it is compacted in place!

NEMA, the National Electric Manufacturers Association, has a subdivision that deals with carbon and manufactured graphite materials, and within that group, motor replacement parts (www.NEMA.org) That is a $1.2 billion industry but we can not find out what fraction applies to brushes, and how do you account for brushes that go into new motors? In any event, there are 12 NEMA member companies who make brushes. Of course, there could be any number of other companies out there who are making brushes but are not NEMA members!

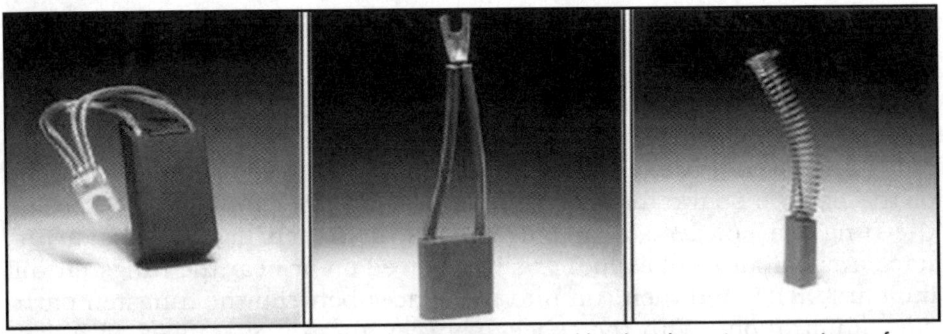

Figure 1.36. Graphite and metal powder brushes used in electric motors, courtesy of Repco, Cherry Hill, N.J.

CHAPTER 2

Visual Basics – A Quick Tour of Powder Metallurgy

We all pick up a text with the best of intentions. But, if all that you ever read are these first two chapters, we want you to have a clear idea of the powder metallurgy, P/M, process. To be sure that you can visualize the fine details that are occurring when metal powder is formed into useful parts, we have selected a series of images photographed through the scanning electron microscope (SEM). They have an unusual three dimensional quality to them that makes the powders look real. The scanning electron microscope can trace the process from powder to finished part in such detail that you will easily be able to follow the more technical descriptions of the later chapters. We think of the photographs as mental pegs. This series of images assumes that the reader is interested in following the conversion of metal powder into useful things like gears or bearings.

If you have never seen a high magnification image of metal powder, you might imagine it to look like fine BB shot or ultra miniature ball bearings. Not wanting to disappoint, Figure 2.1a shows gas atomized powders, .0003-.0010 in. in diameter. How were the particles created and why are they such perfect spheres? The metal alloy was first melted at about 2000- 3000°F and then poured into a ceramic vessel with a small plugged hole in the bottom. When the plug is removed, the molten metal falls as a stream about the diameter of one's little finger. Jets of high pressure inert gas such as argon or nitrogen strike the liquid metal stream at an angle and disintegrate it into fine droplets. Surface tension causes the liquid droplets to take a spherical form. The chamber surrounding the stream is large enough that the metal droplets have time to solidify before reaching the chamber walls. The particles are like perfect little castings. Figure 2.1b shows some rotating electrode steel powder, AISI 1018, (REP) which is even more spherical. If you look closely at the surface of the particles you will see fine structures that are individual metal grains. A good lesson to learn early: metal particles can have many thousands of grains in each particle. In an individual metal grain, the atoms are arranged in orderly rows and stacked neatly up in planes, sort of like a deck of cards. In one of the fine 75 micrometer (.003 in.) grains visible in Figure 2.1b, there might be about 9 x 10^{15} (9,000,000,000,000,000) atoms. Looked at another way, two iron atoms in steel fit nicely in a cube about 11 billionths of an inch on a side, .000,000,011 in. or .29 nanometers, nm. To give you a physical idea about powder particle size, think about a new sewing needle and its fine sharp point. Figure 2.1c shows a needle point against a background of one of the usual water atomized iron powders. Powder particles are about the same size as the tip of the needle, or perhaps a little larger.

That's a lot of information about spherical powder that almost never ends up in a car or a washing machine transmission! Spherical powders

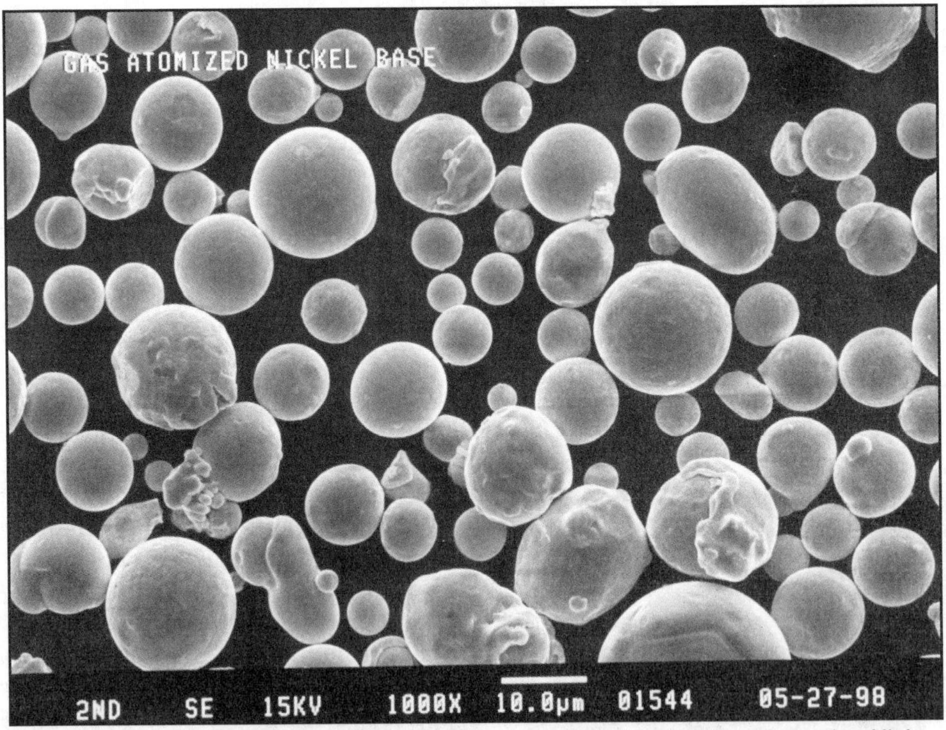

Figure 2.1a. Scanning Electron Microscope image of gas atomized metal powder. High pressure gas hits a stream of molten metal and breaks it into droplets. Surface tension of the liquid creates perfect little spheres that freeze before they reach the walls of the atomizing chamber. A few particles collided with each other while molten, and are not quite such perfect spheres. The scale at the bottom of the photo shows a line 10 microns long. Use it like a little ruler to measure the particles. (.001" ≈ 25.4 microns, or, properly, ≈ 25.4 micrometers.

have their uses, but compacting them into ordinary P/M parts is not one of them. The spherical powders certainly pour well into a die, but the usual compaction pressures of 40,000 to 100,000 pounds per square inch, psi, do not cause them to stick together. They pack so uniformly that they are probably only touching other particles at six points, like oranges in a wooden case. The high pressures of compaction rub one particle against the other, and a small metallic bond forms. We think of it like cold welding or the phenomenon that happens in a bearing and shaft that runs out of oil, and the bare metal of the shaft freezes up or sticks to the bearing. When the pressure on the powder is released, the compressed powder tries to spring back a little, to its uncompressed form. If there are not enough cold welded particle bonds, the existing bonds get broken, and you are then faced with the unhappy sight of a pile of loose powder or a very fragile part.

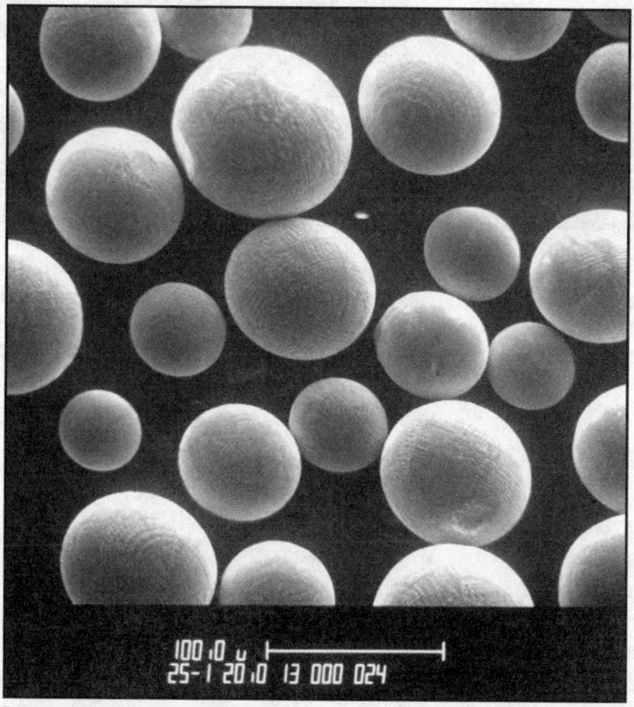

Figure 2.1 b. Scanning electron microscope image of rotating electrode powder, 1028 steel . The particles are all 50 to 100 micrometers in diameter (.002 to .004"). The fine structures on the particle surfaces are grain boundaries. This shows thousands of grains in any one particle. In the REP process at Starmet, a steel electrode in a vacuum chamber is rotated at high speed. When an electric current , like a welding arc, hits the end of the bar, the surface melts and little particles are flung off. They spheroidize from surface tension and freeze before they reach the vacuum chamber wall.

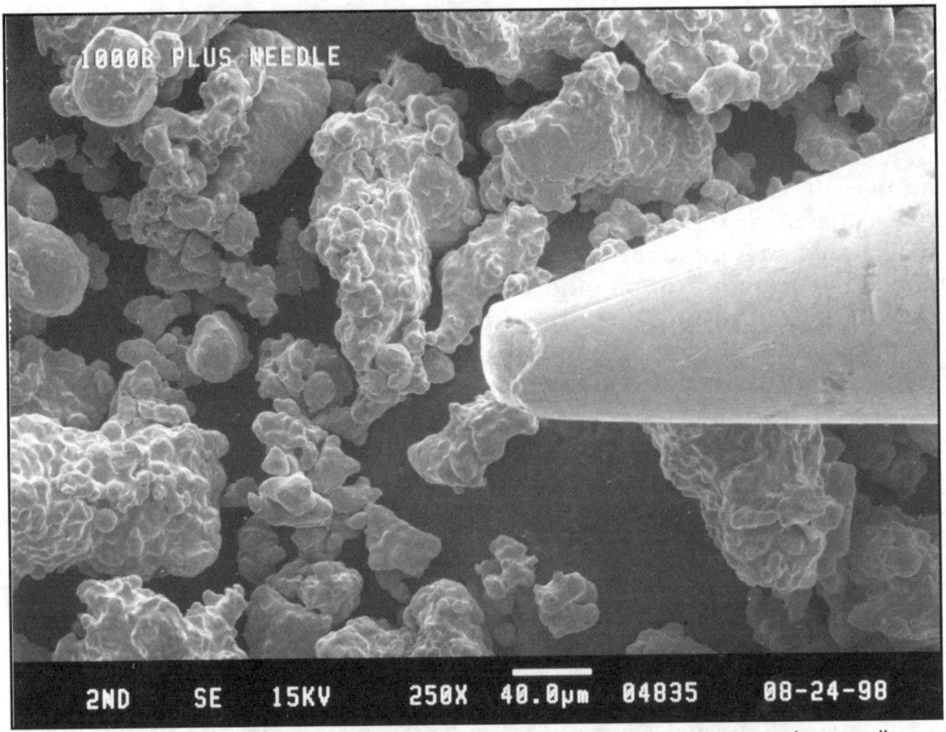

Figure 2.1c. Scanning Electron Microscope image of a very fine sharp sewing needle lying on a bed of water atomized powder, Ancorsteel 1000B. This is intended to give you a physical idea of the size of typical iron particles. The lower touching particle has 7×10^{14} atoms. It also tells you something about how needle points are formed, apparently by swaging or drawing operations that flow the metal toward the tip. Also, needles have a round point, perhaps to avoid splitting threads? Courtesy of Drew Kilius Analytical Answers, Woburn, M.A.

You may be wondering why we P/M folks don't solve this problem by making irregularly shaped powders. Then, with lots of arms, legs and protrusions sticking out of each particle, there would be many more bonds formed during pressing. Furthermore, these irregularly shaped powder particles would tend to mechanically interlock, sort of arm in arm. Well, we did think of that idea, and it was practiced long before gas atomized powders were ever made. Figure 2.2 shows sponge iron powder, made from mill scale. It is the ultimate in irregularity, with multiple arms, legs and protuberances. A moment's study of the photo and the reason for the name "sponge iron" powder will be apparent, but what is mill scale? In large steel mills, the 10 ton billets are heated in air and then hot worked into the desired shape. During heating and working, the oxygen in the air reacts with the iron to form a black scale of iron oxide.

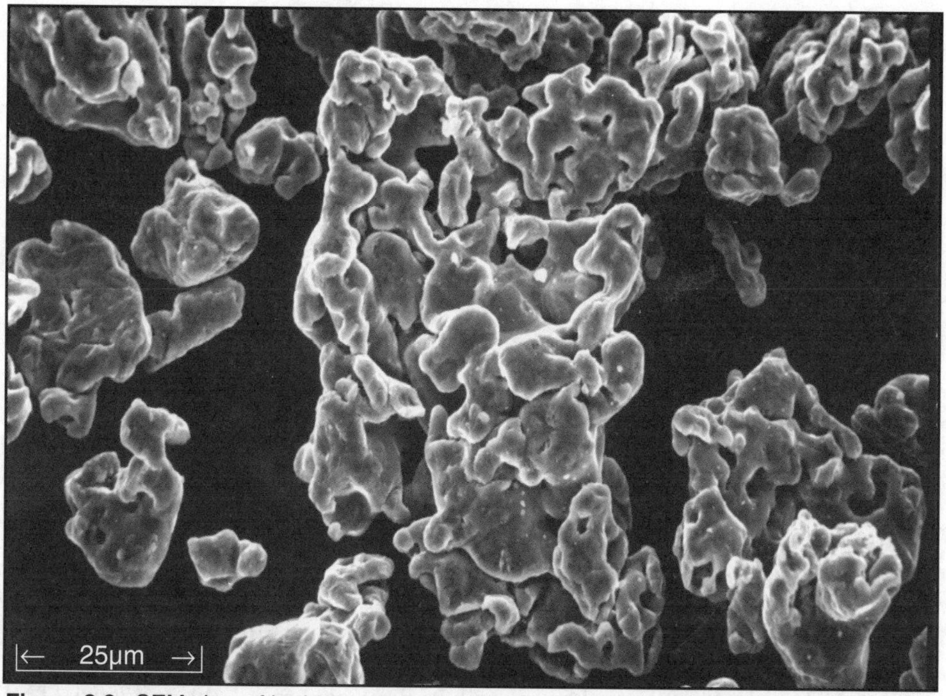

Figure 2.2. SEM view of hydrogen reduced sponge iron. The hydrogen reacted with the oxygen in mill scale (Fe_30_4) and carried it away. This left a series of fine pores throughout the iron particle and the spongy appearance. The irregular shape provides multiple points of contact at compaction. This leads to particle interlocking and good green strength.

The scale thickens and eventually flakes off to land on the floor. The mill scale flakes can then be collected and converted into iron powder. (LFP's wife's great uncle Percy Moody from New Hampshire once said that the Pennsylvania Dutch can live off of what the Scottish throw away!)

Pyron Corporation located in Niagara Falls, N.Y. purchases the mill scale and deposits it as a 2-3 inch thick layer on a belt of stainless steel sheet which passes through a furnace at about 1700°F. The company is happily situated close to an inexpensive source of hydrogen which forms the atmosphere inside the furnace. The hydrogen reacts with the oxygen in the mill scale to form water vapor, and the water vapor then exits the furnace. After a few hours, the oxygen has been pretty well removed. The original mill scale contains more oxygen atoms than iron atoms (4 to 3). When all that oxygen is removed, it leaves a lot of space behind, in the form of pores that run throughout the iron particle. In Figure 2.2, you can see right down inside the particle. Of course, oxygen atoms are only .00,000,008 ins. across and the pores are much larger, .0001 ins., so the tiny vacancies created by the oxygen atoms' leaving must have coalesced together to make the pores in Figure 2.2. The 1700°F furnace conditions makes the iron atoms quite mobile. They adjust their positions to minimize the surface area and surface energy of the particle. It is just like the atomized liquid iron particle turning into a low surface area sphere instead of a thin disc or needle. Of course, the unmelted sponge iron particle is a lot more viscous, and surface tension takes a long time to round out the pores and the particle. Figure 2.2 does show that many of the curves and pores are nicely rounded. You will soon see that minimization of surface area and surface energy is the fundamental driving force in sintering particles together. But we are getting ahead of our story!

Mill scale is not the only source of sponge iron powder, and a much more common form is the carbon reduced product produced by Hoeganaes in the U.S. and its former parent, Hoganas, in Sweden. The Swedes have some high purity iron ore bodies. That means the iron oxide (Fe_2O_3) they dig out of the ground does not have many other minerals mixed in with it. It would be natural to assume that they mimic the Pyron folks and pull out the oxygen by reacting it with hydrogen. They don't. Instead, they take crushed ore, coke, and limestone and mix it together. The coke is a source of carbon. The limestone is calcium carbonate. The mixture goes into a tall hollow ceramic tube (sagger). The sagger is about one foot in inner diameter and 6 ½ ft. high. A great many saggers are loaded on an open freight car and the freight cars go right into the furnace (tunnel kiln). The trip through the furnace takes about 2 ½ days. In the furnace, at 2200°F, the carbon in the coke reacts with oxygen in the iron ore to form carbon monoxide and carbon dioxide. That removes the oxygen from the iron ore and leaves sponge iron behind. It also forms a protective furnace atmosphere that keeps the iron from re-oxidizing. The limestone reacts with any sulfur present in the coke and converts it to calcium sulphate. Because the exposure time is long and the temperature is high, the conditions are right to allow the surface area of the particles to be minimized. The process forms coarser pores than in the reduced mill scale (Figure 2.3). At the end of the furnace, the saggers are emptied, the lumpy contents are crushed into fine powder, and the iron is magnetically separated from the non-magnetic oxides, sulphates and other debris. In magnetic separation, a powerful magnet attracts the iron particles and leaves the non-magnetic particles behind.

So, we now have two ways to make spongy iron powder with highly irregular surfaces and great green strength. When the powder is pressed at 20 to 50 tons/in.^{2}, the parts ejected from the tools hold together very well. So why is the usage of sponge iron powder declining? You have now seen both ends of the spectrum of particle shape. The spherical gas

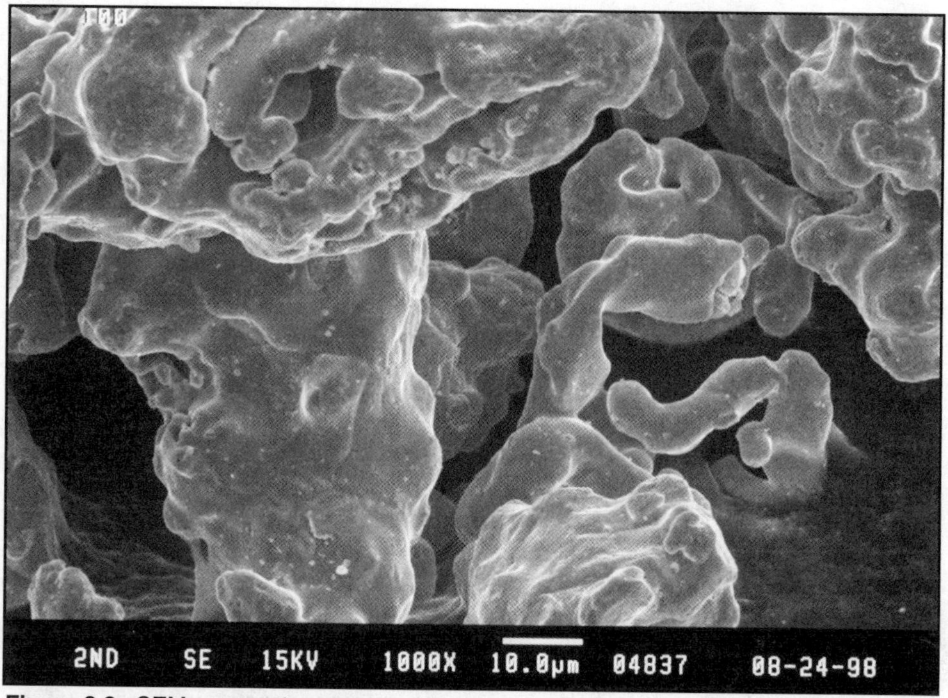

Figure 2.3. SEM, sponge iron made by carbon reduction of Swedish iron ore. The carbon reacts with the oxygen in the Fe_2O_3 and carries it away as carbon dioxide and carbon monoxide. The removal of the oxygen results in the spongy appearance. The higher temperature and longer time of the process gives coarser pores than for the hydrogen reduced mill scale of Figure 2.2.

atomized powders will quickly flow into a die and they will pack together very closely. But, after pressing and ejection, the parts will fall apart. The sponge iron powders are highly irregular. On pouring into a die, the particles rub their rough surfaces against each other and flow more slowly. Once they settle in the die, all the arms and legs and protuberances keep them from packing really closely. To make a part of a given weight, the die will have to be much deeper to get enough powder down inside. Finally, when you are ready to press the sponge irons, you find that they don't compress very easily. The same 30 tons/in^2. pressure that leaves only 12% air space left in a compressible iron, results in about 19% air space in the sponge irons. The good news is that sponge irons hold together well in the green state and, at a given level of air space (porosity), they sinter to a higher strength than the compressible irons. For all their warts, they were the mainstay of the P/M industry from the late 1950's until the mid 1960's.

So what, exactly, is this nice compressible iron, to which we have been comparing the sponge iron powder? It is water atomized iron powder, first introduced by the A.O. Smith Company in the U.S. in the mid 1960's. If you think about the melting and gas atomizing described earlier, try to imagine substituting high pressure water jets as the means to break up the molten stream of iron. At 2000 psi, the molten iron stream is disintegrated and the iron particles become highly irregular in shape. They freeze solid before surface tension has time to minimize their surfaces into little spheres. They freeze in the same tortured positions created by the high pressure water. They are irregular on the surface while being quite solid in the interior. Enough words! Look at Figures 2.1c and 2.4. You can not see down inside these particles, as you did in the sponge irons. Figure 2.4 is the powder you purchase, not the product that leaves the atomizer. As water atomized, the powder has a thin iron oxide layer on it and contains a fair number of fine spherical particles. You would not like those little spheres! They flow down between punch and die and tend to bind the tools, they lower green strength, and they are dusty. So, the product of the water atomizer is dried, and then thermally treated, much like the operation done on the mill scale at Pyron. The powder is fed onto a stainless steel sheet or belt, and run through a furnace at 1400 to 1800°F in a protective hydrogen containing atmosphere. The hydrogen reacts with the thin surface oxide layer and carries it away. This leaves a fine spongy surface, temporarily high in surface energy. Surface tension has its way, minimizing surface area, and those unwanted super fine particles end up bonded or sintered to the larger particles. As you might guess, this serves to create the irregular particle surfaces which will later be compressed into parts with good green strength. (We once photographed some agglomerated water atomized powder, which looks like a caricature humorously making this point-see Figure 2.5!) So, those clever powder folks have made a raw material with surface roughness in between the gas atomized and the sponge. The water atomized iron powders are very pure, which means they are soft and compress easily and they do not have the convoluted interiors of the sponge irons with their myriad fine pores. Coarse pores are easier to collapse than fine pores, so water atomized powders are easier to compress. As you would expect, their flow rates into dies and their ability to pack well in the dies are also midway between the spheres and the sponges. Life is full of compromises!

We have been speaking of compaction of metal powders as if we had actually given you some details, and have, therefore been depending on your imagination or prior knowledge. Figure 2.6 shows you the three basic steps of filling the powder into a die, compacting it, and ejecting the part from the die. The part shown is a simple cylinder. "The genius is in the details!" (And no one person actually knows all the details!)

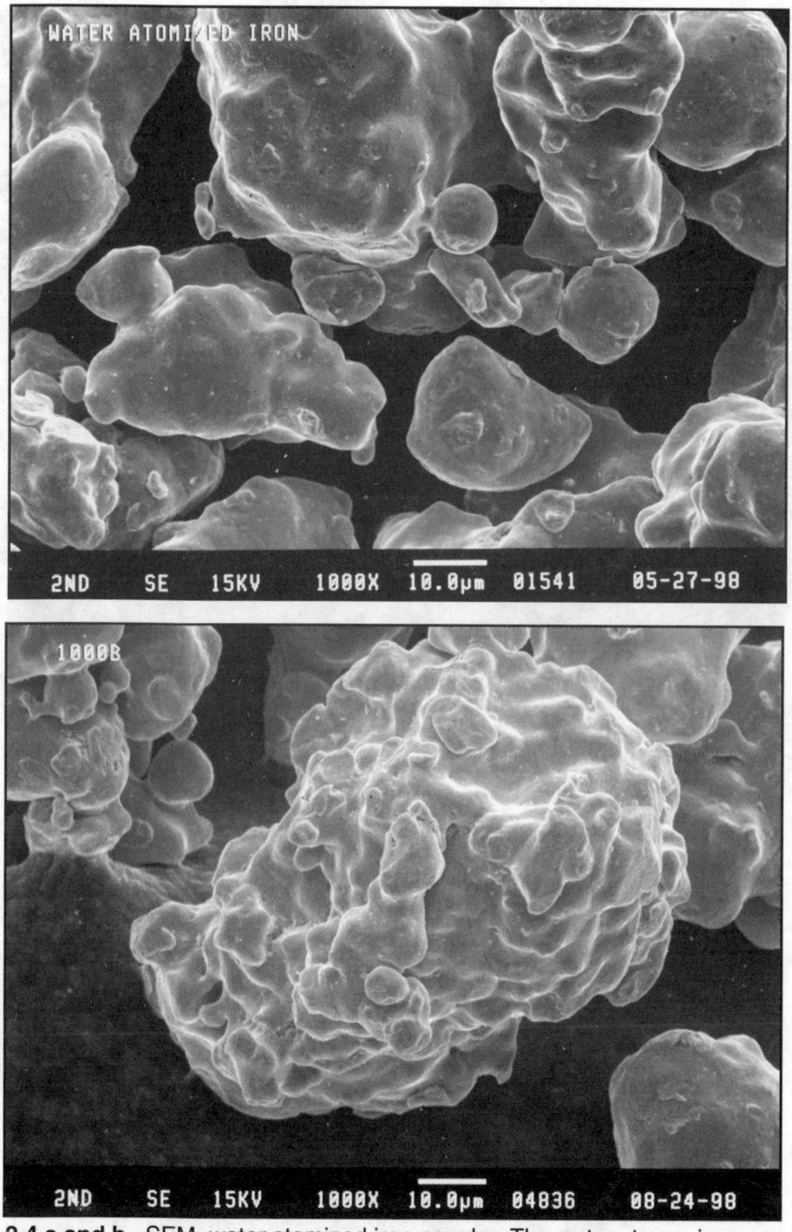

Figure 2.4 a and b. SEM, water atomized iron powder. The water stream is more violent than the gas stream of Figure 1a and 1b and forces the droplets into irregular shapes. The water rapidly freezes the particles in their distorted shape. The product from the atomizer was then annealed in a belt furnace, which joined small particles to large and created some of the agglomerated particles shown here. The surface roughness aid green strength during compaction.

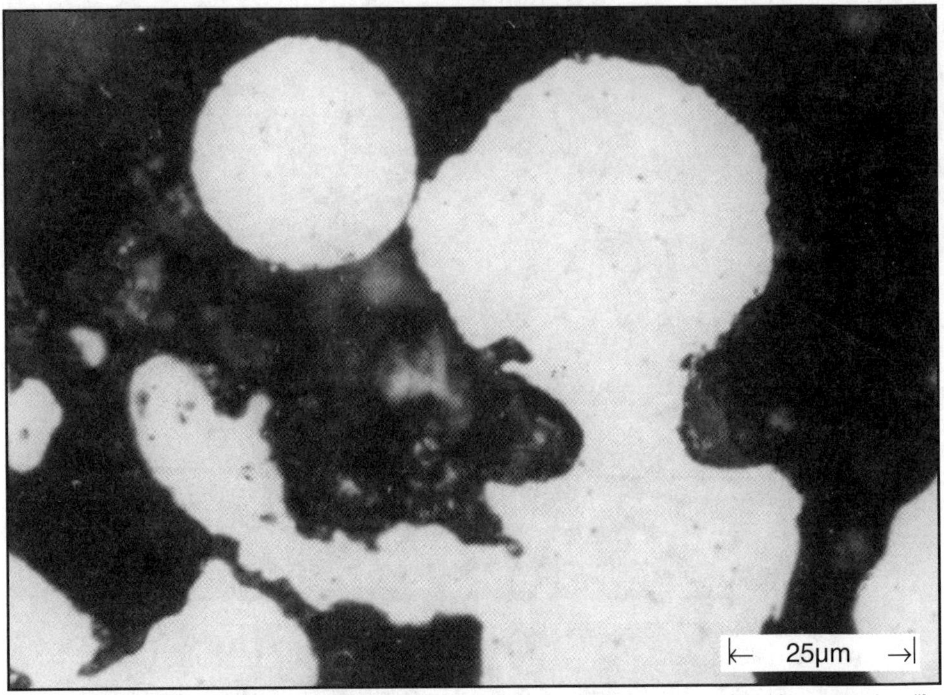

Figure 2.5. Optical microscope, polished cross section of water atomized iron powder: "I assure you, sir, there is no information available on methods to increase the green strength of water atomized powders!"

The powder flows down through the feed shoe and fills the die to a depth L. The shoe then moves away from the die opening, while leveling the powder surface even with the top of the die. In this way, the die now contains enough loose powder so that after compaction, it will have the desired weight and volume. The loose powder in the die has a density which we define as the weight of one cubic centimeter (one cm^3) of loose powder. In principle, you could imagine taking a little one cubic centimeter cup, scooping up some loose powder to fill it up, and weighing the number of grams in the cup. For water atomized iron powder, that would be about 3g in 1 cm^3, i.e. $3g/cm^3$.

In the real world, the density of loose powder is called the "apparent density" and is measured by MPIF Standard Test Methods 04. All of the MPIF, Metal Powder Industries Federation, test methods are in one handy little volume which insiders call the "Bumblebee Book" because of its distinctive black and yellow jacket. More formal users call it "MPIF Standard Test Methods for Metal Powders and Powder Metallurgy Products" and it is available from MPIF, Princeton, New Jersey. Similar test methods, written by essentially the same group of clever P/M folks,

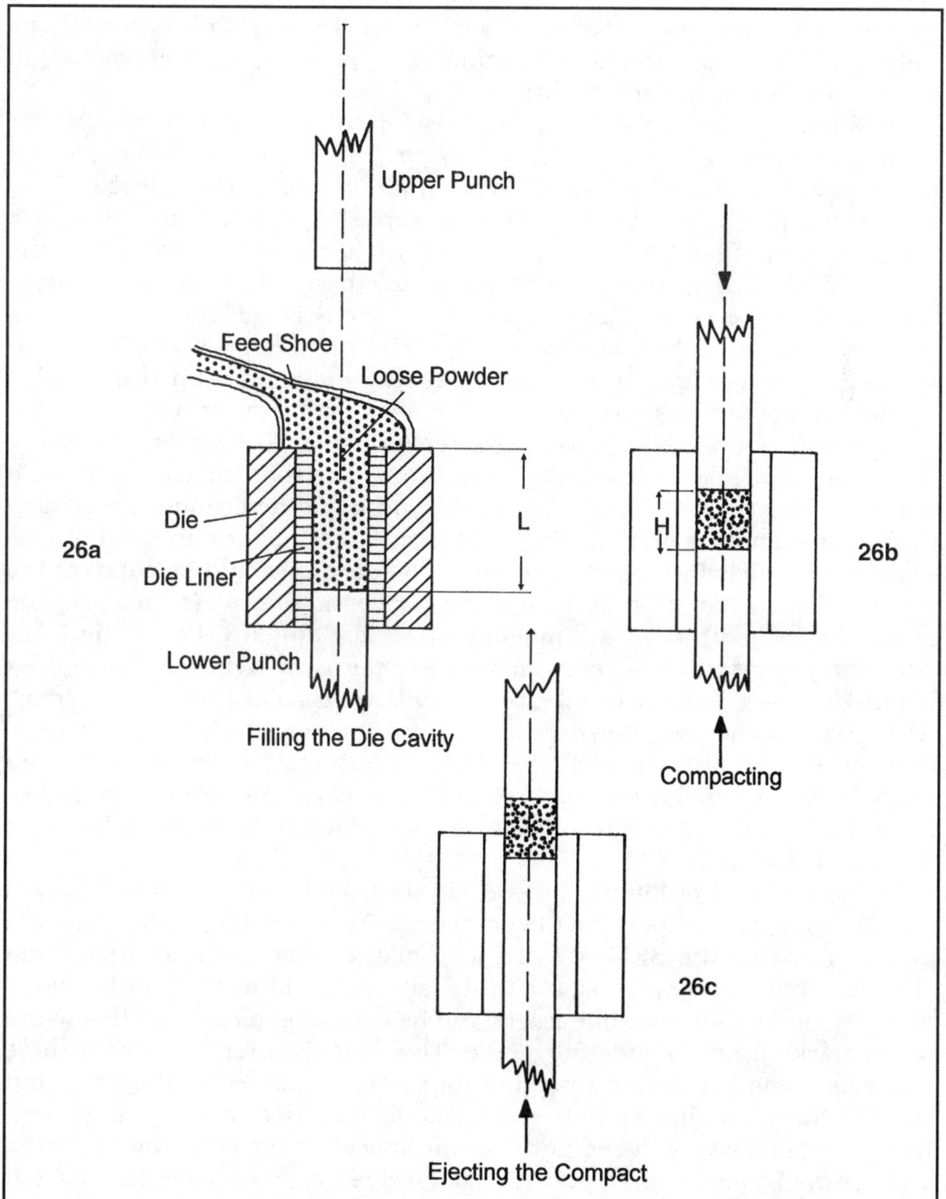

Figure 2.6. Basic steps in compacting metal powders. 2.6a, metal powders flow through feedshoe into die and as the shoe moves aside, the die is filled level with the top. 2.6b, the top and bottom punches move toward each other, compressing the loose powder of height L to a compact of height H. Compression ratio = L/H. 2.6c: the lower punch moves up through the die to eject the compact. Courtesy of "Powder Metallurgy, the Process and its Products", published by Adam Hilger, 1990.

are available from the American Society for Testing and Materials, as Volume 02.05 or as individual test methods, at West Conshohocken, (no we are not kidding!), Pennsylvania.

Note that the length of the loose powder column, L, is much greater than the compacted part height, H. Figure 2.6 shows schematically that particles are farther apart in the loose state than after the punches have compacted it. The compaction process rapidly increases the density of the powder confined in the die, stopping when the distance between the punches is H. The confined powder has had its apparent density increased by the ratio of L/H. Thus, if L were 3 in. and H were 1.5 in., the compression ratio would be 3/1.5 = 2. If the powder started out at 3 grams in each cubic centimeter, its density would end up being doubled by the compaction to 6 grams in each cubic centimeter or 6 g/cm^3. Even though there is some air space still remaining in the compact, we now drop the term "apparent density" and simply say that the compact has a density of 6 g/cm^3. Again, in principle, you could determine the density of the compact by machining out a little one centimeter cube and weighing it on a scale which reads in grams. The density would be equal to the weight of the little cube (Almost nobody does it that way, they use the more elegant method of Archimedes in MPIF Standard 42). If the compact at 6 g/cm^3 still has some air space or porosity in it, how dense does it have to be to get rid of all the air and equal solid iron, 7.87 g/cm^3, which we call the pore-free density of iron? Try not to call it the "theoretical density" because the ASTM B09.01 Subcommittee on Nomenclature can't agree on a definition for "theoretical density." However, if you have such a definition, please feel free to pass it along to LFP who will take it to B09.01, for you!

In the last part of Figure 2.6, we see that the lower punch has moved upward and the compact has been ejected from the die. In Figures 2.7a and 2.7b, using the SEM, we can look at the sides of the compact and see just what the compression and densification did to the powder particles. The lines that look like fine scratches running almost vertically are score marks from the side of the die. This is a vivid reminder that there must be some friction or dragging forces on the sides of the part that contact the die walls. The other notable feature is that the part is anything but solid. We still see pores or air spaces in between the particles. The arrows in Figure 2.7b point to cold welded regions between adjacent particles. If you think of the force on the particles being in the direction of the score marks, you can see that it is concentrated in the small areas where the particles make contact. This concentration of the force is a locally high pressure which deforms the metal particles and slides one against the other. This sliding action breaks up any surface films and allows the atoms in one particle to contact those in the next particle to

Figure 2.7a and 2.7b. SEM image of the die wall contacted surface of an unsintered compact. The marks that are slightly off vertical are score marks from the die surfaces and are evidence of friction against the die. Many pores or sir spaces are still evident between the particles. In the lower photo, the arrows show where particles have cold welded to each other and provided green strength.

create a small metallic bond. This bonding is the basis for the green strength that holds the unsintered compact together.

In the sintering process, we strengthen the part by increasing the amount of bonding between the particles. If we heat the iron part, the atoms become more mobile. If, at the same time, we first remove any surface films on the particles, and then prevent formation of such films, the iron atoms will be free to move about on the surface and within the particles. This motion happens without the particles melting. The motion increases in intensity at higher temperatures. Once the iron atoms can migrate, they move in such a way as to lower the surface energy of the compact. This atom migration lowering of surface energy is the fundamental driver for sintering. They lower the surface energy by lowering the surface area. It is not in our everyday experience to think of surfaces as having an associated energy, after all, they just sit there! One analogy is the beading up of water on a polished surface: the water does not stay in a thin (high surface area) film, it pulls together into rounded or spherical beads of low surface area. Water is pretty weak compared to iron, and its surface tension quickly spheroidizes it into low surface droplets. The surface tension easily moves the water molecules to form the beads. Iron, even at elevated temperature, is pretty viscous, and it takes a long time for the atoms to move about and round out the surface.

Well, Virginia, atoms do move, and the Scanning Electron Microscope at 5000X shows the result in Figure 2.8. This iron part was sintered less than 10 minutes in the hot zone at 2050°F and then deliberately broken to show the fracture surface. The atoms have migrated over the surface toward the point where the particles were originally cold welded. They are building up a neck , as shown by the right arrow. When iron particles are touching in the green state, the pores between particles have very thin edges near the points where the particles are touching. The atoms move into those thin regions and round them out, as shown by the right arrow. In doing this pore rounding, the surface area of the pore is lowered. An analogy is a round balloon. You can feel the surface energy or tension in a balloon. If you twist it out of shape by stretching the rubber surface, you have added surface area and surface energy. Release the balloon, and the surface tension quickly draws it back to its original shape. The molecules of air inside the balloon are free to move easily. Think of the pores in a green part as being an irregular stretched surface, like the twisted balloon. At room temperature, the iron atoms on the surface of the pores would like to move around and get rid of the high surface. But the atoms are not very mobile at room temperature. When we heat iron to1500 to 2050°F, atoms can easily move and respond to the surface tension forces. The result is filling in the sharp spaces between particles as in Figure 2.8. Of course, when the thin spaces where the particles were touching get filled in with iron atoms, a strong bond or neck develops between the particles. The bonds are the basis for the strength of sintered parts.

Figure 2.9 shows the surface of the sintered part. This was the surface that was contacted by the upper punch. The external particle surfaces are just like the balloon: give the iron atoms some mobility and they will round out an originally sharp particle surface. This degree of sintering causes the pores to appear somewhat rounded. The grain boundaries inside the iron particles are shown by the arrow. Figure 2.9 looks rounded, and Figure 2.8 shows neck and bond development, but how do we know that the iron atoms have actually bonded particles together in a way that makes the material strong and ductile, like pore-free materials? SEM Figure 2.10 at 5,000X is a view of the fractured surface of a sintered iron part, where we have deliberately broken the bonds. We see a series of ductile cups and cones where the iron has been pulled apart. It

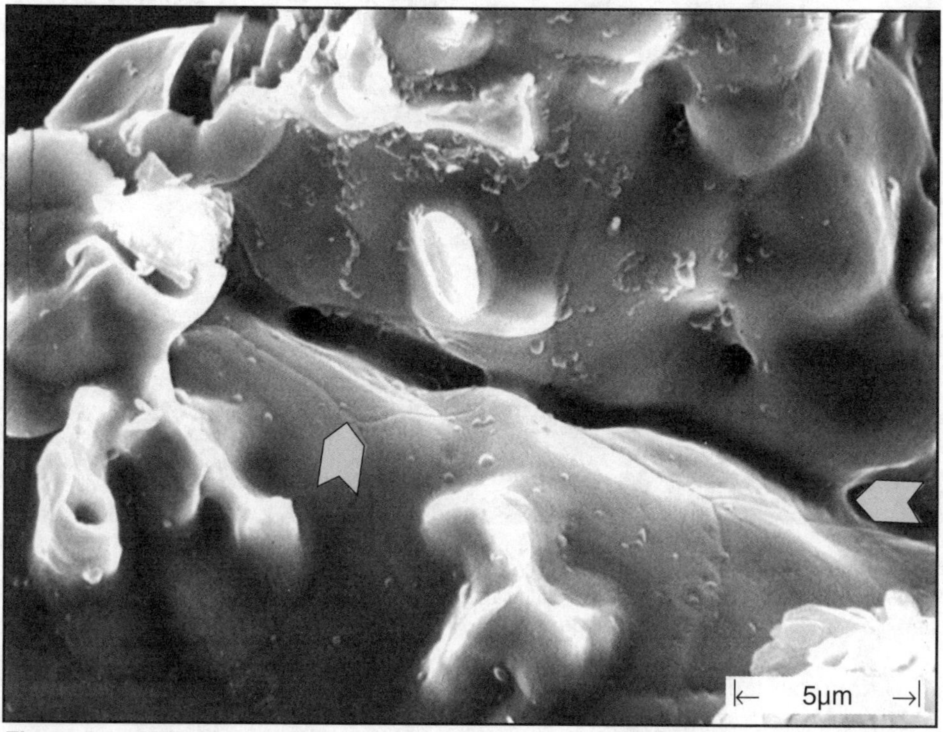

Figure 2.8. SEM view of fracture surface of iron powder compact that was sintered for 10 minutes at 2050°F. The atoms have migrated over the surface to form a bond or neck between the particles, see right arrow. The left arrow shows a grain boundary on the surface of an iron particle. The little round dots on the surface of the iron are probably unreduced oxides.

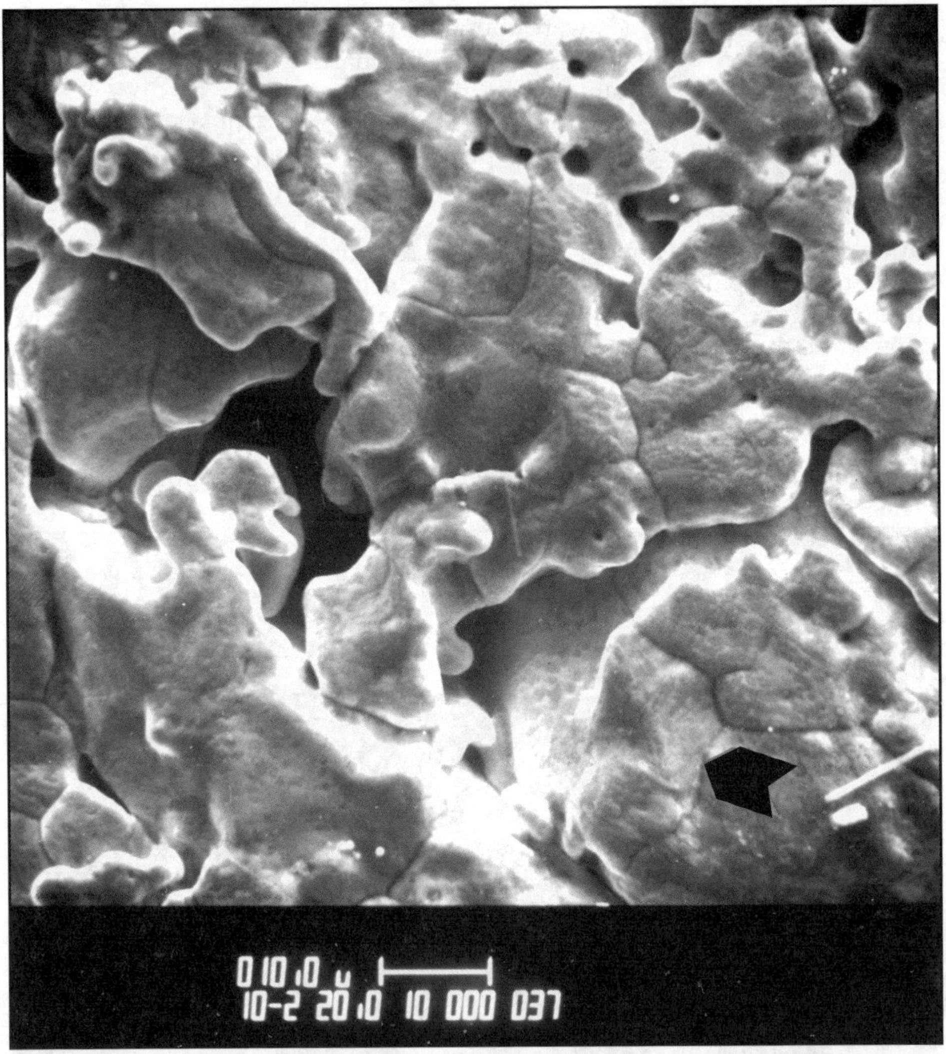

Figure 2.9. SEM view of the top surface of a sintered compact. Atomic mobility has allowed the sharp edges of the particle to become rounded. Compare with figure 2.7a. The arrow shows a grain boundary exiting the surface of a particle. This shows how one particle can contain several metal grains.

is reminiscent of a taffy pull. Where the particle bonding has actually occurred, the fracture appearance is the same as it would be in wrought or pore-free iron. The sintered iron stretched and deformed quite a lot, before it finally let go. The rub is that this ductility only occurs where there is particle bonding. Figure 2.11 is a lower magnification view and we can see a particle in which there are local areas of bonding on its sur-

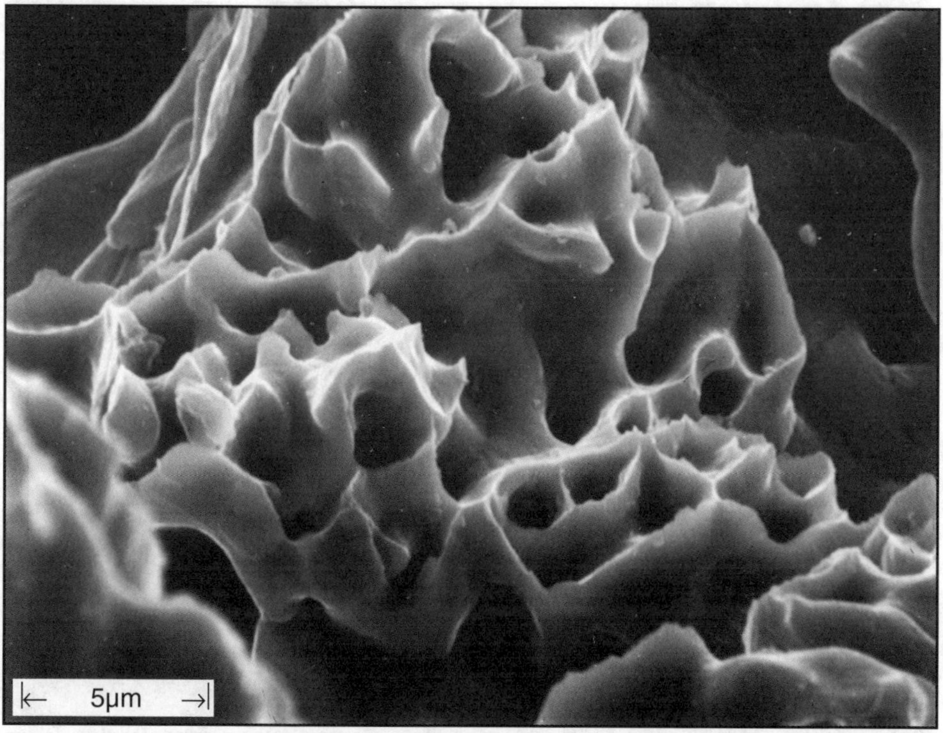

Figure 2.10. SEM view of the fracture surface of a sintered iron part. Sintering was in the normal range of 20-30 minutes at 2050° F. On fracturing, the region where one particle was bonded to another as pulled apart ductility creating the cups and cones seen here (Like a taffy pull!). When a piece of solid iron is fractured, its surface looks just like this. So, where there are bonds in a sintered part, the bonds are just as strong and ductile as those in a pore-free material.

face, but the whole surface is not covered with bonds. Why not? For one thing, we still have quite a large volume of porosity, and where the iron is "attached" to the air in a pore, we will not see any bonding! Also, the development of bonds goes on gradually with time. Give the part more time in the hot zone, or raise the temperature and speed up the iron atoms, and you get more bond area.

Sometimes a crack can occur in a P/M part during ejection from the tooling. So that you can learn to avoid them, the folks at the Carl Gunnard Johnson P/M Material Research Center (H. Cai et al) at Worcester Polytechnic Institute have filled a whole book with examples and causes, (see references). After the fact, you might want to know whether a crack happened in the green state, or after it was sintered. You know from Figure 2.11 that a crack or fracture that occurs in a sintered part shows lots of ductile cups and cones (formerly bonded regions)

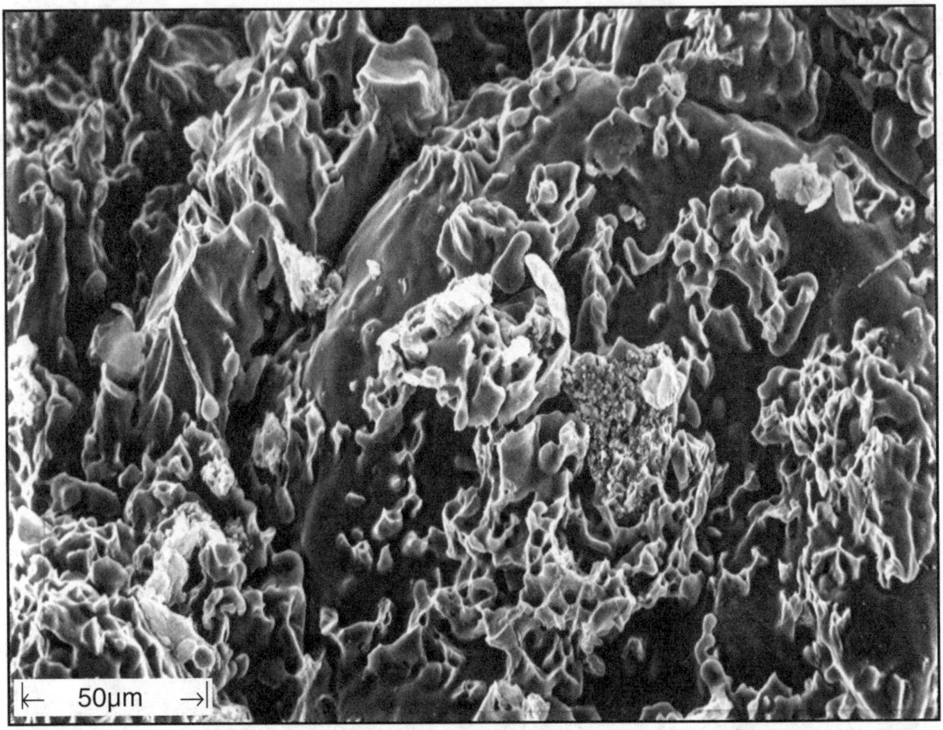

Figure 2.11. SEM view of the fracture surface of an iron part given a normal sinter. This is a lower magnification view than Figure 2.10, but the cups and cones of the ductile fracture are evident as are the original particle surfaces which are now the borders of the pores. The greater the area of bonding, the stronger and more ductile the part. A longer sintering time and higher temperature create more bonds and larger bonds, giving better strength.

on its surface. By comparison, a part that cracks in the green state, and then is sintered, will not show these ductile regions. You will mainly see the rounded original particle surfaces, Figure 2.12. What is even more fascinating about this photo is that we can see the rounded indentations where the particles were pressed against each other, before the crack separated them. Now you want to know about the little white specks scattered over the particle surfaces. We could claim they are oxides, but we really do not know. In a normal P/M steel part, they definitely should not be there. At the time the photo was taken, we could have turned on the EDAX (energy dispersive analysis of X-rays), dropped the electron beam on one spot, and obtained a quantitative analysis of most chemical elements. EDAX is a trade name. The generic term is EDS (Energy Dispersive Spectroscopy).

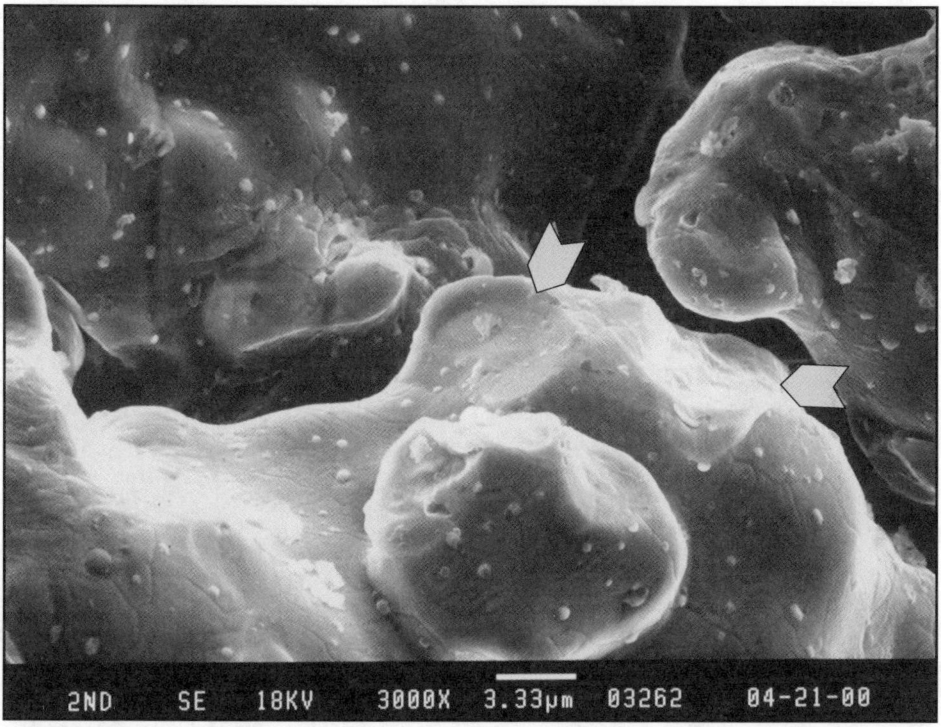

Figure 2.12. Fracture in a green part, that was then sintered. Rounded particle surfaces show there was no bonding, as was seen in Figure 2.11. Arrows show a shallow depression formed by the adjacent particle having been pressed onto it. White specks on surface are abnormal and of unknown origin.

Well, there is quite a bit more to powder metallurgy than we have presented here! However, the images from this chapter will help you to understand the details in our other chapters, and in the other excellent advanced texts that cover the science of powder metallurgy.

CHAPTER 3

Metal Powder Properties and Production

In the last chapter, via the visual tour, we hope you developed some idea of what metal powders look like. You visited gas atomized spheres, irregular sponge iron, and the industry work horse, water atomized iron. The P/M world is a lot wider than iron powder. As nearly as we can tell, all metals, save mercury, are available in powder form. Presumably one could make useful objects out of them by pressing and sintering. And at least one of our colleagues does: Sherri Bingert of the Los Alamos National Laboratory says she processes everything from aluminum to zirconium! Well, this little text does not have enough pages to cover all the metals. What makes sense is to talk about the general properties of metal powders and how the more common ones are made.

Particle Size and Shape

Why are size and shape of any importance? If we are thinking about using tooling to press the powder to form a gear, the powder better fit in the die. Well, powder particles the size of marbles will fit in a die, but they won't compact to form a useful P/M part. These very large particles will not form the strong green bonds described in Chapter 2. Perhaps that is why the ASTM B243 glossary limits powders to being less than one millimeter in diameter (.040 in.). As a practical matter, for materials as strong as iron, the particles need to be smaller than .010 in. in diameter to have enough green strength in the compacted form, and to provide a reasonably smooth surface finish. Green strength refers to the force needed to break a compacted part before it is sintered. If green strength is too low, e.g. below 800 psi, parts may crack during ejection from the tools, or in handling, on the way to the sintering furnace. As we saw in chapter 2, the best way to impart green strength is to create a particle with a highly irregular shape.

If fine particles make for a fine surface finish on the parts, why not make the particles much smaller than .010 in.? There really is an optimum size range for metal particles that are to be compacted. For iron, it is a mixture of sizes from .010 in. down to about .001 in.. You could learn that by measuring the particles in Figures 2.2-2.4 in Chapter 2, using the little micron rulers. (Remembering that 10 microns = .0004 in.) Why not make all the particles .001 in. (25 microns) in diameter? There are a number of problems. First, the manufacturing process, water atomization, economically produces the particle size range from .001 to .010 inches. It would cost considerably more, to separate out the .001 in. particles and throw away or recycle all the larger ones. Second, the fine particles would not flow well into the compaction tools, so the presses would run more slowly and with greater variation in part weight. Third, the fine particles would tend to be less irregular in shape and would have lower green strength. Fourth, some of the fine particles would run down between punches and die, and tend to jam the compaction tools. Are there any benefits to all fine particles? Yes, the surface finish would be smooth, because the pores would be small, and the fine particles, with their high surface area, would sinter rapidly, producing a higher density part. Of course, rapid shrinkage leads to problems with dimensional control. In metal injection molding of mixtures of fine particles (.0005 in.) and wax and thermoplastic, use is made of the fine particles. Once the organic binders have been driven off, the part is about 60% dense. With such fine particles, at 2400°F in an hour, the part has shrunk so much that it is about 97% dense.

How do you measure particle size? The obvious way is to use the little micron rulers and check the largest dimensions of each of the particles in photos like Figures 2.1-2.4 in Chapter 2. After you had measured a few hundred, you could calculate an average size. You could even group the results by size, indicating how many were between 0 and 10 microns, and 11 and 20 microns, etc. This would create a size distribution based on the number or population of particles of each size. You could imagine that if the particles were mainly between 0 and 10 microns that the mix would press and sinter differently than if they were mostly from 90 to 100 microns. For this reason, powder metallurgists do not pay as much attention to the average particle size as they do to the distribution of sizes, when they are trying to control a powder.

If you are now sufficiently tired of measuring particles with the small ruler, you must be thinking there must be an easier way. There is, and indeed there are many ways to measure particle size. The problem is that they all give different results, and none are quite as satisfying or easy to visualize as using the micron ruler on individual particles. Of course, even the ruler and the SEM photos are not foolproof. What if the particle is being viewed at an angle? Are we really measuring its longest dimension? It finally comes down to, "Who cares? Give me a cheap and easy way to characterize size so that my powders will always press the same way or always shrink the same amount during sintering!" The problem is that characteristics like dimensional change on sintering depend on more than just the particle size. So, controlling particle size is just one of many characteristics that must be managed. Do we know what all the factors are that must be controlled? No. But we know quite a few, and in the real world, the P/M industry makes a lot of parts with closely controlled dimensions and strength.

Historically, and largely at present, powder manufacturers have measured the size of particles using sieves made from finely woven wire cloth. The size of the holes in the cloth or screen is controlled by the size of the wires and the number of wires/in. The sieves that are used for compacting grades of metal powders are shown in Table 3.I.

TABLE 3.1

Sieve Designation Micrometers*	Sieve Designation Mesh, Wires/inch	Sieve Opening Inches	Wire Diameter Inches
250	60	.0098	.0063
180	80	.0070	.0049
150	100	.0059	.0039
125	120	.0049	.0035
106	140	.0041	.0028
90	170	.0035	.0025
75	200	.0029	.0020
63	230	.0025	.0018
53	270	.0021	.0014
45	325	.0017	.0013

*The designation is the sieve opening, expressed in micrometers.

Many compacting grades of iron powder are nominally minus 80 mesh. That means that the particles will pass through the 80 mesh screen and all the particles are smaller than .0070 in. or 180 micrometers. A powder with closely controlled particle size might be described as - 80 +100 meaning the particles are smaller than .0070 in. and larger than .0059 inches. Because the sieves are woven from wires, the openings are nearly square. But when you look closely, as in Figure 3.1, you see that in three dimensions, the bending of the wires creates a shape that is not exactly square, or even in one plane. Figure 3.1 shows a particle trying to squirm through the opening, but ending up getting captured or retained. You can imagine that a long thin particle like a piece of very thin wire or a cat's whisker could work its way through the opening. So the 80 mesh sieve only tells you that anything that went through had a cross section less than .0070 in. in diameter, but it tells you nothing about its maximum length. That's kind of an academic point for most iron and copper powder since all three of the dimensions of a particle are roughly the same (equiaxed). But it is probably a life and death matter for folks who sell steel wool fibers to make brake linings! Wire mesh sieves are not woven much finer than 325 mesh because, according to Dr. Lenel, retired Professor from RPI, the wires smaller than .0013 in. were not readily available. For comparison , in the ASTM E 11 Sieve Standard, the coarsest mesh opening is 5 in., a fact that probably is only of interest to the coal companies in Dagus Mines, PA!

How do you use the sieves to measure the particle sizes in a metal powder sample? This is a trick question like "How to you make a rabbit stew?" (First catch the rabbit!) What you don't do is take a hand scoop to the top layer of a barrel of powder, and call that the sample. The sample needs to represent the whole barrel. ASTM B215 tells you to insert a thief all the way down into the barrel of powder. A thief is a hollow pipe with little shuttered widows at different points along the length, as in Figure 3.2. With the thief down inside the barrel, twist the thief to open the shutters and in flows the powder from the different locations. But you're not done yet. ASTM B 214 says to test only 100 grams and the thief gave you many hundreds of grams. So you put the powder in a small blender for a minute or so. Then subdivide the mix using a splitter, Figure 3.3, to get approximately 100 grams for the sample. You then accurately weigh the sample and you are ready to begin.

You assemble a series of sieves such as 80, 100, 140, 200, and 325 in a stack with the 80 on top, and a metal pan under the 325, as in Figure 3.4. This goes in a special sieve shaker for 15 noisy minutes of rotation and tapping, Figure 3.5, at which point you declare the test over. The powder retained on each screen is weighed to .1 gram. The report will show the percentage of the overall powder sample that was retained on each screen. An example from a recent lot of atomized iron was as follows:

TABLE 3.2. TYPICAL SIEVE ANALYSIS FOR WATER ATOMIZED IRON

Mesh Size	Weight % Retained on Mesh
60	0.0
80	
100	11.0
120	
140	
170	
200	
230	
270	
325	70.3
Pan	18.7

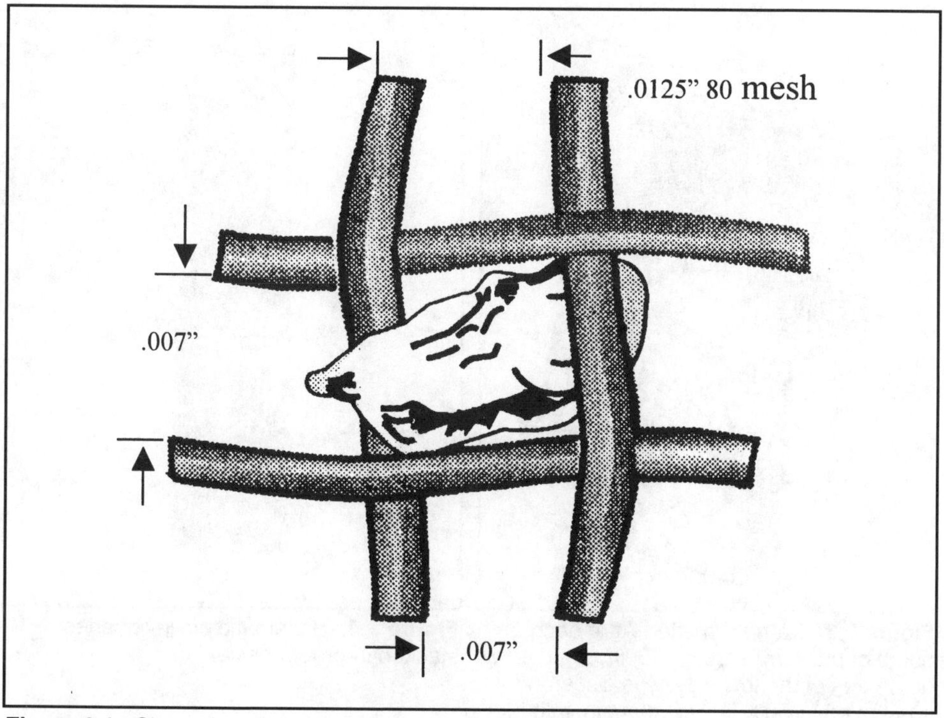

Figure 3.1. Sieve size of an irregularly shaped particle at 80 mesh.

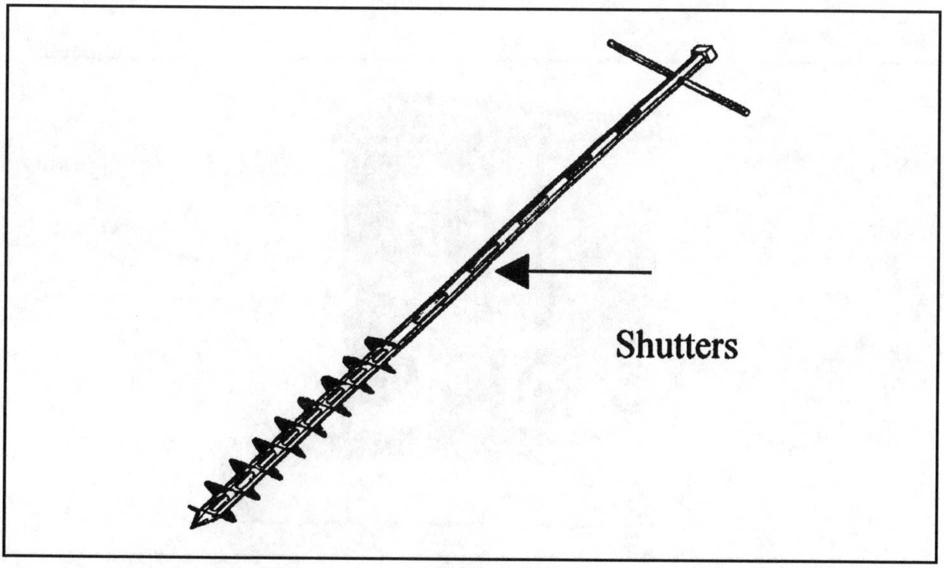

Figue 3.2. Sample thief approximately 3 ½ feet long. For accurate sampling, this "augers" well!

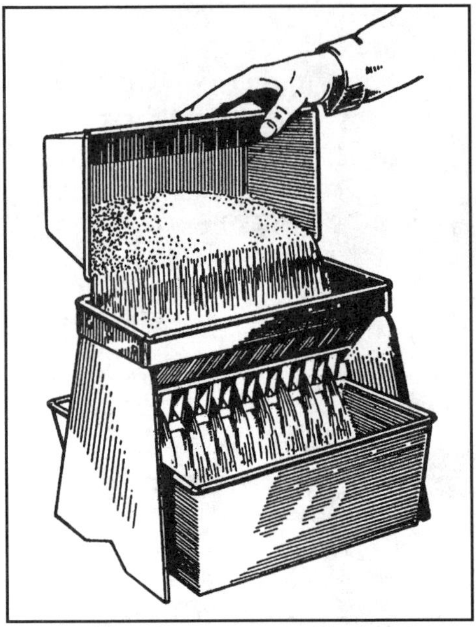

Figure 3.3. Sample splitter. After each dump of powder in the top, 50% of the powder in one of the lower pans is discarded and the other 50% is cycled again, into the top. After a few cycles, the powder in one of the pans is about the right amount to test.

Figure 3.4. Schematic of sieve series stacked in order of size.

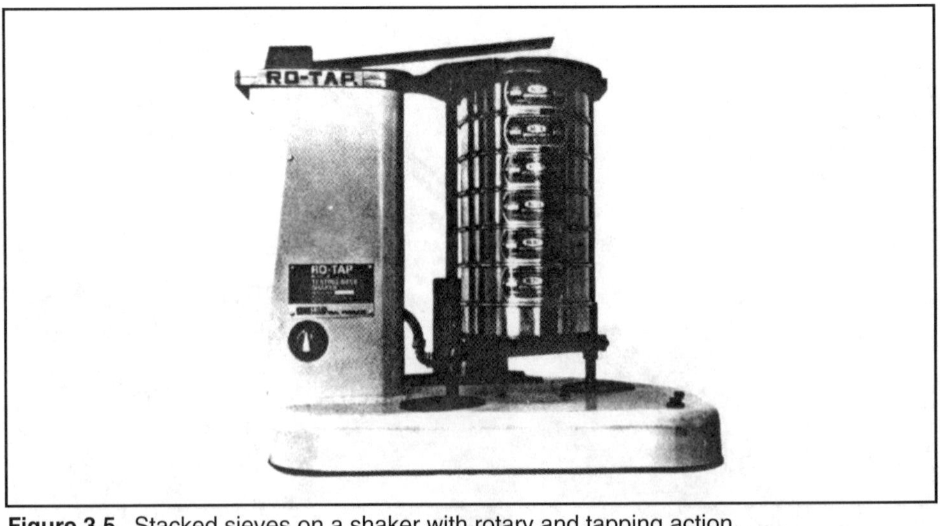

Figure 3.5. Stacked sieves on a shaker with rotary and tapping action.

The powder manufacturer did this analysis by selecting only three sieves: 60, 100, and 325, and a collecting pan at the bottom of the stack. It surely does save on sieves! Presumably, folks who buy powder negotiate the % ranges for each of the sieve sizes, and possibly the sieve sizes that are checked.

One neat point of difference between this analysis and the one you were supposed to do with the micron ruler on the SEM photos in Chapter 2 is that in the earlier measurement, you counted the number of particles in each size range. Using the sieves, you measured the weight of particles falling between two different sizes. You can imagine that there were a lot more particles in 10 grams of the fine mesh sizes than in 10 grams of the coarsest size. So what do we really need to know: the *number* of particles between 45 and 53 micrometers, or the weight % of particles between 45 and 53 micrometers? Either measure could control the process, so throw away your micron ruler and do it the easy way with the 15 minute sieve test!

The use of particle size in terms of the number (or percentage of population) of particles in a certain size range has another practical problem. Suppose you had two heaps of powder and someone told you that in pile A, 10% of the particles were between 45 and 53 micrometers and for pile B, there were 20% in that range. Suppose you wanted to make a pile with something in the middle, say 15% of the particles between 45 and 53 micrometers. How would you actually do it? What would happen if you just weighed out equal grams of A and B? The problem is that you do not know how many particles there are in a gram of Powder A and if Powder B has the same number of particles in a gram. Unless there are the same total number of particles in a gram of each powder, you can't depend on weighing equal amounts to end up with the 15% by number in the 45 to 53 micrometer range. That is why real people prefer particle size distributions by weight! (Then you just mix equal weights in the example above.) If you really are determined to make the 15% mix by number of particles, you will have to figure out how many particles there are in a gram, and that is a story for another day or an interesting student exercise.

Particle Size by Laser Light Scattering

After this exercise, we assume you will go through life looking only at particle sizing devices that give the weight percent of each size! One really useful one is based on the use of laser light, shining through a flowing stream of particles. An often used device is the Microtrac instrument. The layout for the instrument is shown schematically in Figure 3.6. The powder is dispersed in water plus wetting agent, and flows in front of the laser beam. The result seems sort of baffling: small particles push the beam further to the side than large particles, while large particles result in a much stronger light intensity (the intensity is proportional to the square of the particle size, e.g. double the particle size means 4 times

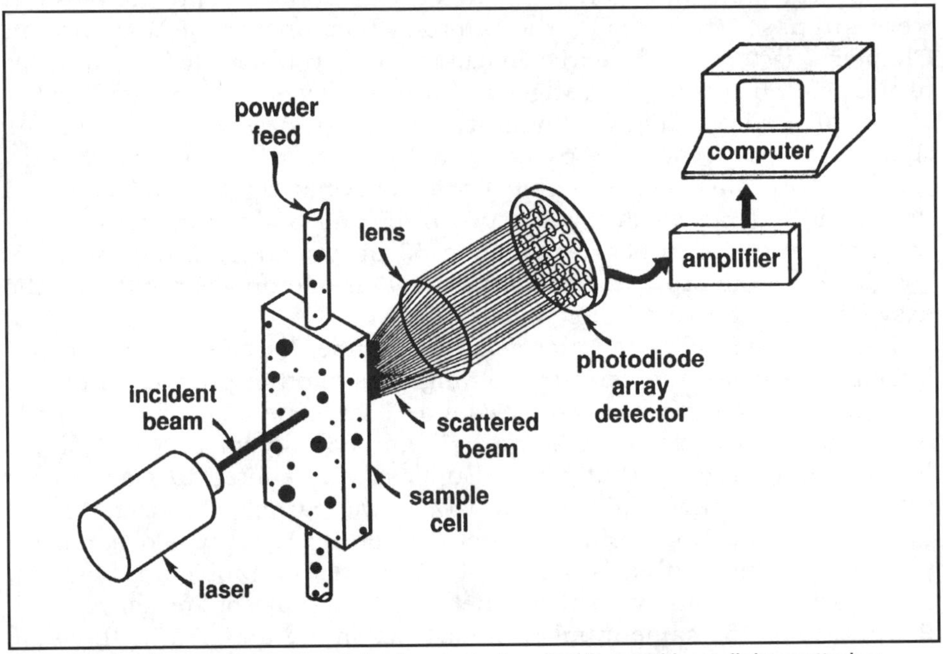

Figure 3.6. The principle of particle size analysis using forward laser light scattering. The powder is dispersed and fed into the sample cell where scattering takes place. The detector system measures the angular intensity for subsequent computer calculation of the particle size distribution.

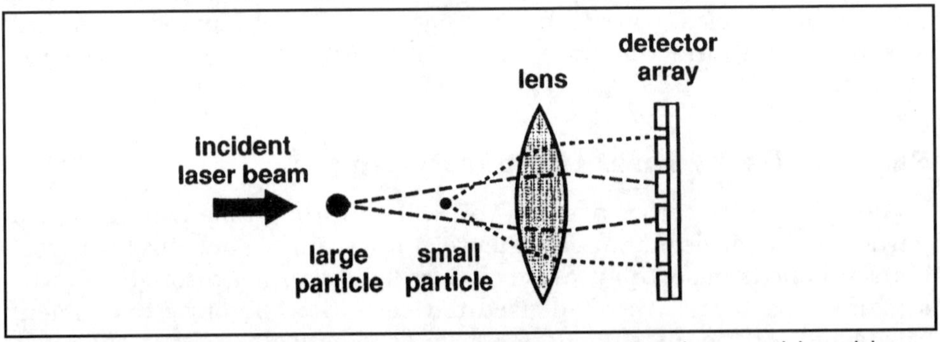

Figure 3.7. Small particles provide a wider scattering angle than large particles, giving a basis for particle size analysis based on laser light scattering.

the intensity of the light). The detector picks up the strength of the light at many different locations, as shown in Figure 3.7. Then, in a calculating *tour de force,* the computer solves equations based on Fraunhofer and Mie scattering of light (ref. ASM Vol. 7 page 250-252) and gives us size distribution by weight %.

Usually, when we see this data, it is not broken down to look like our sieve analysis in Table 3.2. Instead, they plot the *cumulative* weight % smaller than a given particle size. Lets give it a try based on the water atomized iron in Table 3.2.

TABLE 3.3

Mesh Size	Particle Size Micrometers	Weight % Retained on Mesh	Cumulative Weight % Smaller Than
60	250	0.0	100.0
80			
100	150	11.0	89.0
120			
140			
170			
200			
230			
270			
325	45	70.3	18.7
Pan	0 to 45	18.7	not applicable

The cumulative weight % column is just the total of all the percentages from the finer sieve sizes. You can see at 60 mesh and 250 micrometers, 100% of the particles are smaller than 250 micrometers. Figure 3.8 shows how this would be plotted, and Figure 3.9 shows some actual data from a laser analysis of water atomized iron powder, courtesy of Hoeganaes. The laser light scattering allows measurement of particles from .05 microns to 5000 microns (5 mm, and not far from the size of marbles which we never press!) .05 Microns is small! How small? About 1/1500 of the tip of the needle shown in Figure 2.1c of Chapter 2, or about 200 atoms of iron standing shoulder to shoulder!

Fisher Sub-sieve Analyzer

This is an older device, predating the use of laser light scattering, which attempted to characterize small particles, .2 to 50 microns in size. The sub-sieve idea came from the fact that the particles were smaller than the 45 micron sieve openings. The principle is based on work published by Carman in 1937 (ref. see #18, page 98 in Lenel's text). You would not be surprised to hear that in flowing air through a tube packed

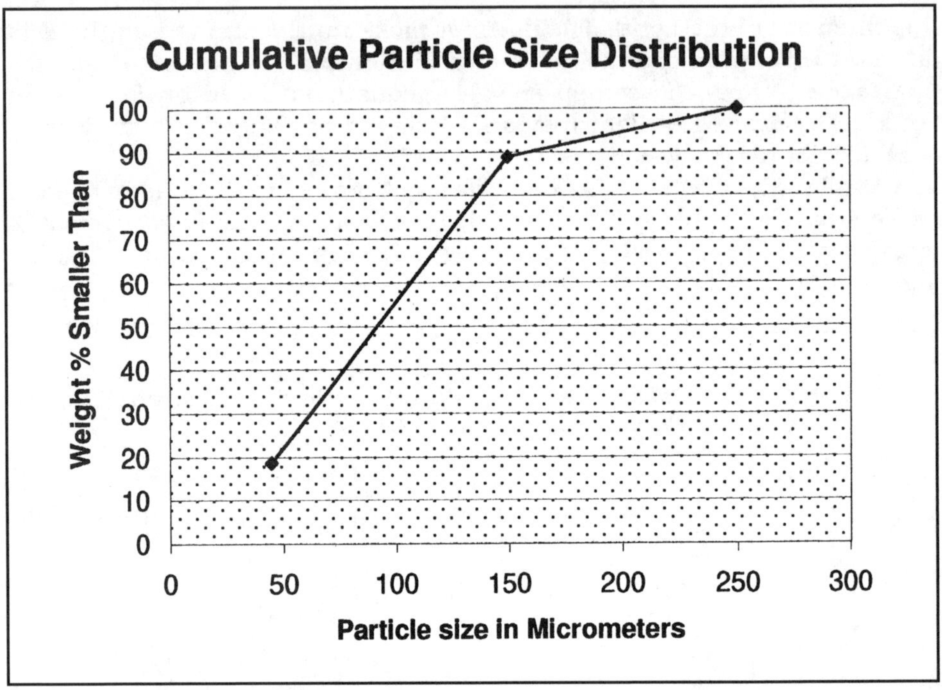

Figure 3.8. Cumulative particle size distribution. Illustrates the data in Table III for atomized iron. Note that about 50% by weight of the particles are smaller than 90 micrometers.

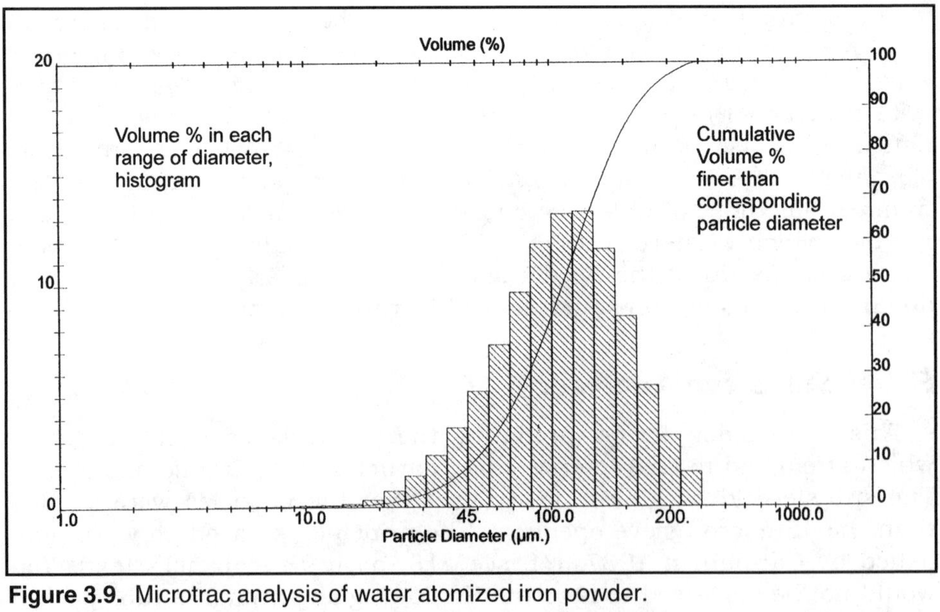

Figure 3.9. Microtrac analysis of water atomized iron powder.

with metal powder, the friction of the particles reduces the air flow or pressure. Carman showed that this friction depends on the surface area of the powder particles. It is logical that smooth spheres with low surface area wouldn't impede the air flow as much as particles with rougher surfaces. Refresh your memory by looking at Figure 2.1 (smooth) and Figure 2.2 (rough) surfaces in Chapter 2; you will be convinced. With Carman's equation, if you measure the flow rate of the air and the loss of pressure as the air flows through the powder, as well as how tightly the powder is packed (fraction of full density like 35% or .35), the length of the powder bed and you look up the viscosity of air, you can calculate the surface area in square cm. for one gram of the powder.

The device that is used to make this measurement, the Fisher Subsieve Analyzer is standardized in test method ASTM B 330. Strangely enough, that device was not set up to tell you the surface area in cm^2 for 1 gram! Instead, it reads out the Fisher "size" for the powder. What it does is assume that your powder is made up of perfect spheres, all one size, and then tell you what their diameter should be. That might work pretty well for the little gas atomized spheres in Figure 2.1 of Chapter 2, but for any ordinary irregularly shaped powder with twisted surfaces, it would read out a number that would be too high. This leads folks to refer to the Fisher sub-sieve data as "Fisher numbers", while keeping one's fingers crossed. The instrument serves well in process control in making fine powders like tungsten or molybdenum. If all you had for control were sieves, and all the tungsten particles were less than 45 microns, you would not know much about your powder by doing a sieve analysis. The Fisher number was useful on these fine powders because it is easy to do and the instrument is not expensive ($4,825) Figure 3.10 shows a schematic of the operation and Figure 3.11 shows a photo of the instrument.

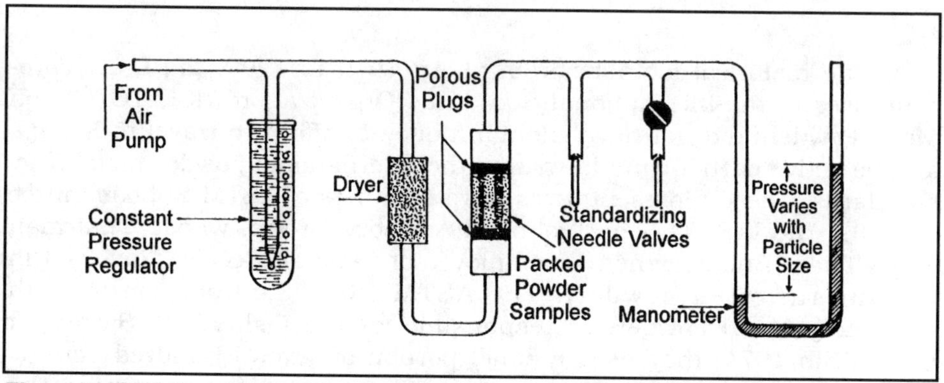

Figure 3.10. Schematic of Fisher Subsieve Sizer operation. Courtesy of Lenel.

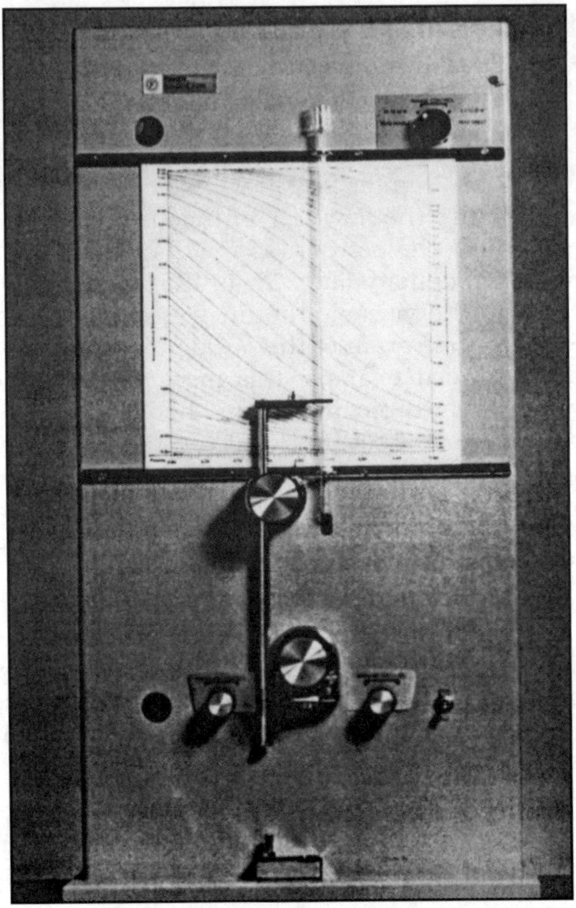

Figure 3.11. Front view of Fisher Sub-Sieve Sizer (Courtesy: Fisher Scientific Co.). Courtesy of Lenel.

A little historical footnote: from about 1970 to 1990, the U.S. representatives to the International Standards Organization , ISO TC-119 on Metal Powders did not have enough money to afford to travel to Europe and be active participants in writing the international powder metallurgy standards. These representatives were members of ASTM Subcommittee B-09.09, the U.S. TAG or Technical Advisory Group, whose Chairman was William Buerkel, then an employee of General Electric, (involved in making tungsten powder). The ASTM B09.03 Subcommittee on Refractory Metal Powders is responsible for the Fisher Sub-Sieve test method. In 1974, they used a small portion of several hundred pounds of donated tungsten powder in a round robin test to determine the precision of the test method. The remainder of the tungsten powder was carefully stored in Argon, in double plastic bags, and became a reference standard for calibrating the Fisher apparatus. Folks who were interested in a standard for their Fisher device paid a modest fee ($50/lb.) to ASTM to get additional samples of this particular lot. Over time, these fees were used to fund the air travel expenses of the B-09.09 TAG members, so that the U.S. could play an active participatory role in writing the ISO powder metallurgy standards. (You did what you had to do!) In recent years, as the original tungsten powder supply has dwindled, MPIF, the Metal Powder Industries Federation has begun funding these travel expenses. Since 1998, some forward thinking U.S. and Canadian companies have also contributed money.

Having offered a brief insight into the real world of P/M standards development, we need to say that P/M test methods and material standards definitely respond to suggestions from individuals, such as, for instance, people who might be reading this fine little text. If you have a comment or a problem in using a standard or test method, contact MPIF in Princeton, NJ and/or ASTM and let them know your thoughts. In 1978, MPIF launched a whole revision to its mechanical properties standard, number 35, based on one question raised by an engineer from a company that builds tractors. Anyone with interest in P/M standards can exercise his view point by joining ASTM, Committee B-9, Metal Powders and Metal Powder Products. Contact ASTM, in how could you have forgotten already, West Conshohcken, PA. (We almost never have our committee meetings there!). If your company makes P/M parts, and is an MPIF member, you could ask to join the Powder Metallurgy Parts Association Standards Committee.

And Now, a Few Other Ways to Measure the Size of Particles

Well, we've been through, sieving, laser light scattering, and the Fisher "number"; how many more ways can there be? You're about to be saved by Prof. German who prepared an excellent listing of the various methods, as shown in Table 3.4. The table shows you the particle size for which the method works well, how quickly the test can be run, and whether the size distribution is based upon sample weight or number of particles (population). Now you know which ones to avoid! For those of you who are determined to learn all about these methods, a good starting place is Prof. German's text and Volume 7 of the ASM Metals Handbook . Unsolicited testimonial: ASM, Volume 7 is a terrific reference for all areas of powder metallurgy, and so are Prof. German's many texts. Beyond that, to learn any more, you will have to take up playing with metal powders. *Deus ex machina!*

TABLE 3.4. COMPARISON OF PARTICLE SIZE ANALYSIS APPROACHES

Technique	Size Range, μm	Dynamic Ratio	Sample Size, g	Relative Speed*	Basis+
microscopy					
optical	0.8 and up	30	<1	S	P
electron	0.001 - 400	30	<1	S	P
sieving (wet and dry)					
wire mesh	38 and up	20	100	I	W
electroform	5 - 120	20	>5	S	W
sedimentation					
gravity	0.2 - 100	50	5	I	W
centrifugal	0.02 - 10	50	1	S	W
light scattering					
Fraunhofer	1 - 800	<200	<5	F	W
Mie	0.1 - 3	30	1	F	W
velocity	0.5 - 200	400	1	F	P
Brownian	0.005 - 5	1000	<1	I	P
electrical zone sensing	0.4 - 1000	30	5	I	W
light blocking	1 - 600	45	3	I	P
x-ray					
broadening	0.01 - 0.2	-	1	S	P
small angle	0.001 - 0.05	-	1	S	P

* S = slow (1 hour or more), I = intermediate (~1/2 hour), F = fast (1/4 hour or less)
+ P = population basis, W = weight basis

Particle Shape

We all remember the advice about the value of a picture versus 1000 words. Well Figures 3.12 through 3.20 taken from the original ISO 3252 Powder metallurgy - Vocabulary will give you a good tour of powder particle shape. After looking them all over, you may remark that none of them seem to be similar to our favorite water atomized iron, which we presented in Chapter 2, Figure 2.4, or to the two sponge irons in Figures 2.2 and 2.3. We assume that the ISO TC-119, SC-1 Terminology made the same observation, and so, in the 1996 edition, they changed the example of irregular powder in our Figure 3.18 to look a lot more like water atomized iron, (see our Figure 3.21). (If you are a fan of reading the obsolete versions of the ISO 3252 glossary, you may wonder whatever happened to the powder shape known familiarly as "botryoidal". The U.S. representatives on the committee couldn't think of many examples of powders whose shapes reminded you of a bunch of grapes, and we happily recommended its deletion.)

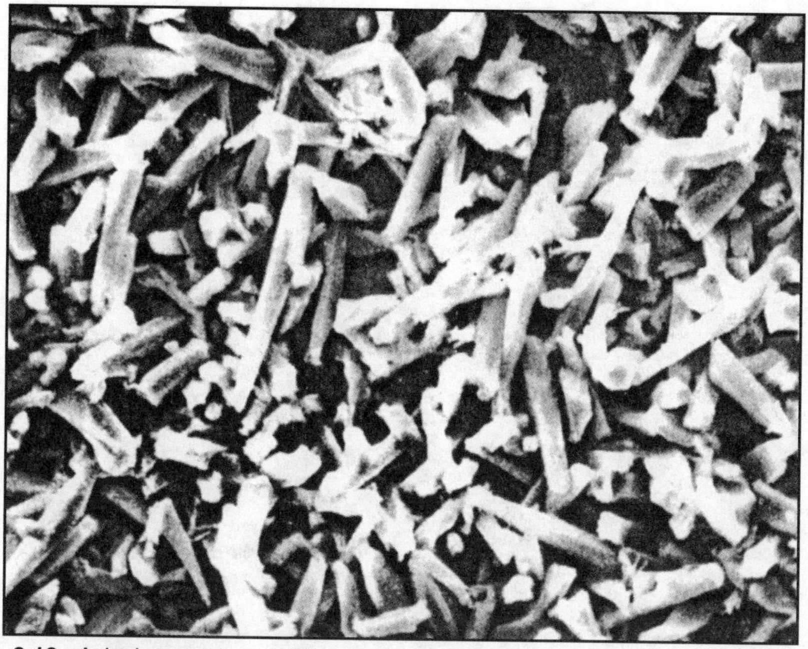

Figure 3.12. Acicular powder particles.

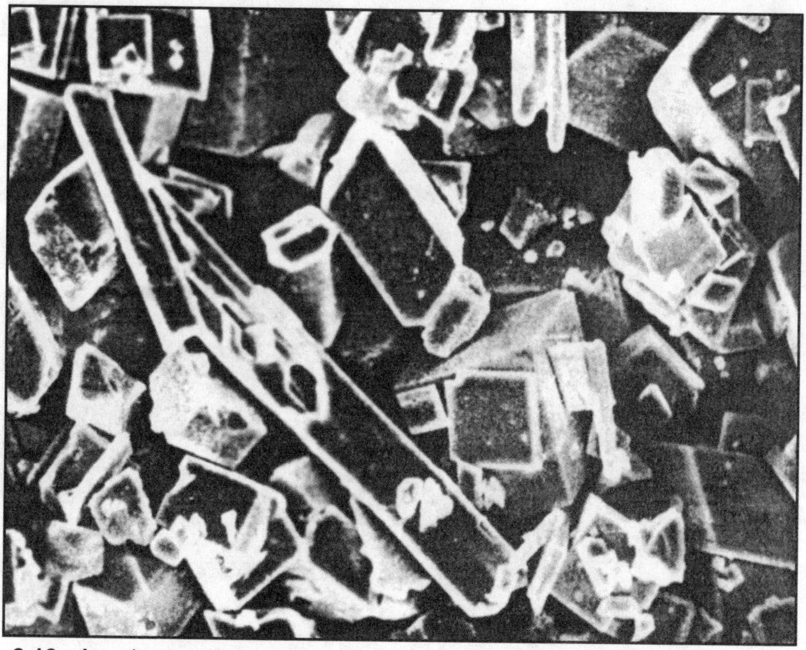

Figure 3.13. Angular powder particles.

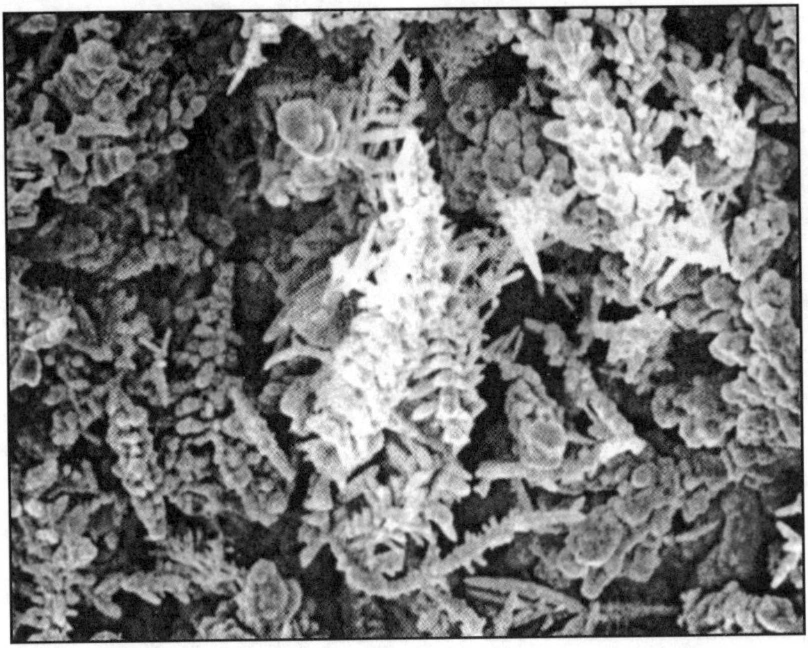

Figure 3.14. Dendritic powder particles.

Figure 3.15. Fibrous powder particles.

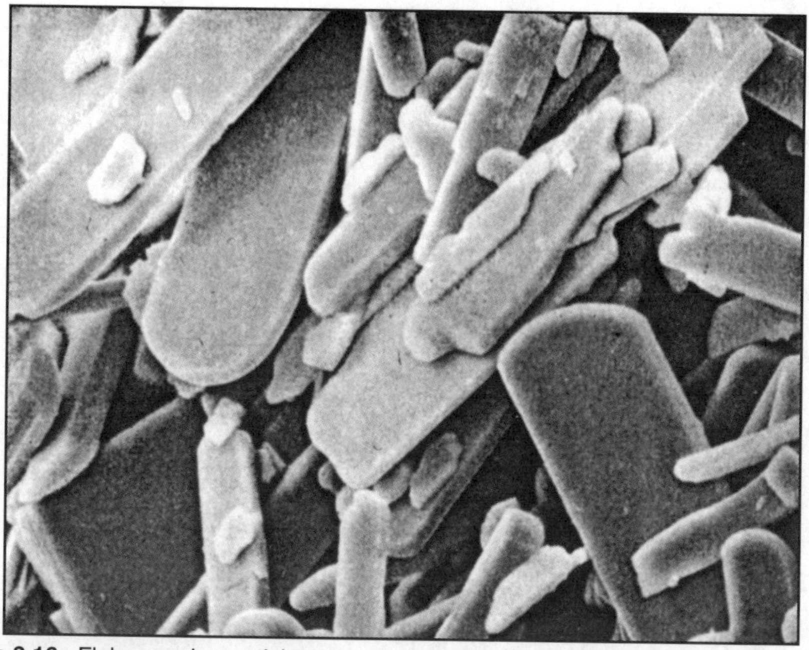

Figure 3.16. Flaky powder particles.

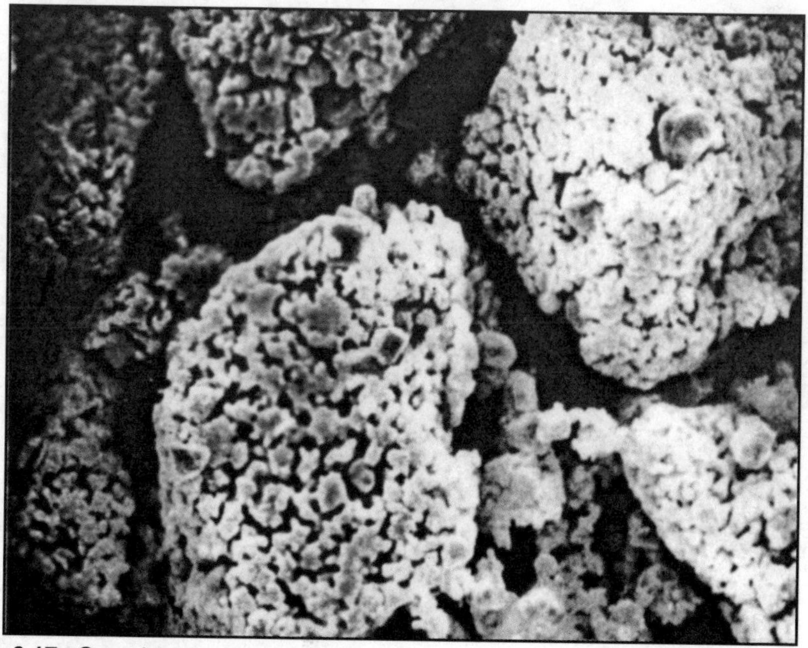

Figure 3.17. Granular powder particles.

Figure 3.18. Irregular powder particles.

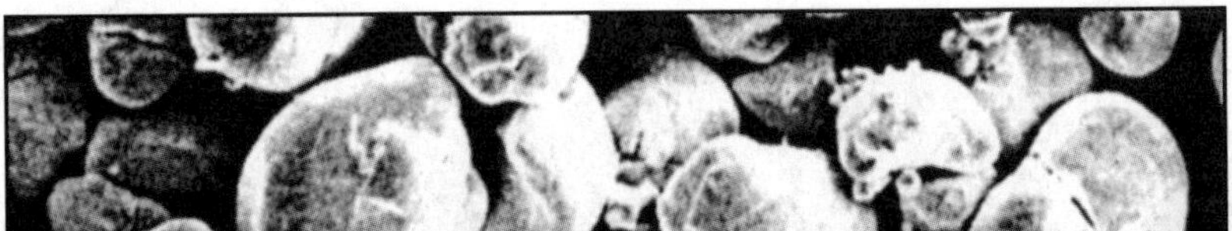

Figure 3.19. Nodular powder particles.

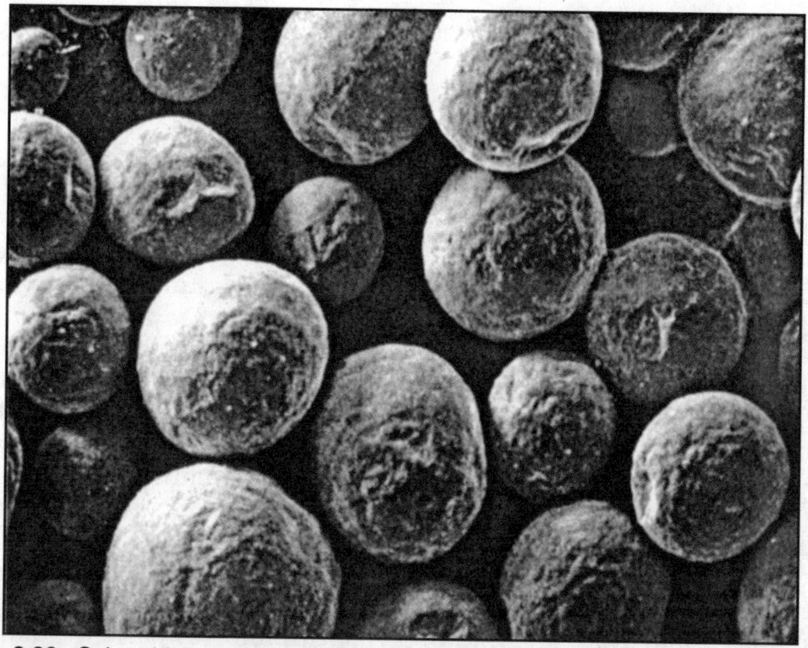

Figure 3.20. Spheroidal powder particles. Courtesy of Lenel.

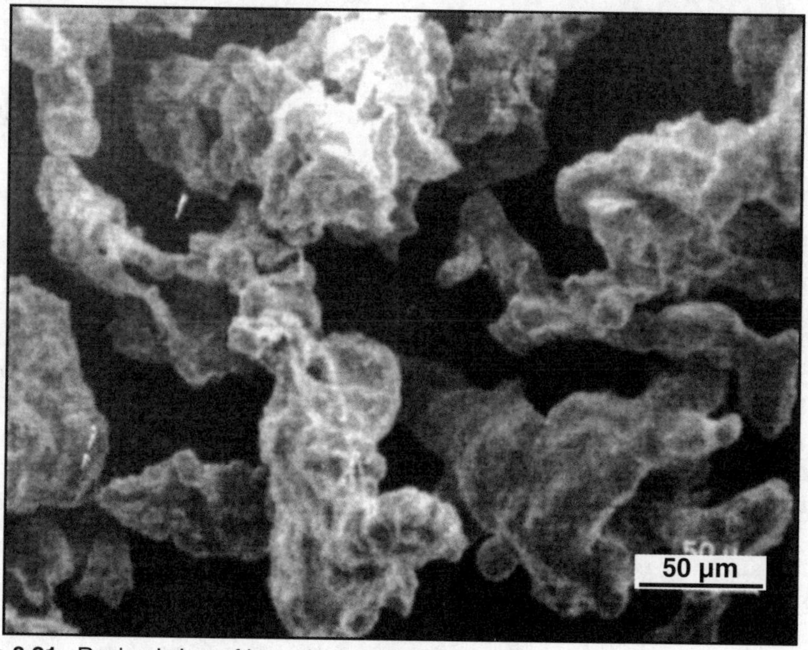

Figure 3.21. Revised view of irregular powder from 3252.

The good news is that it is possible to attach a mathematical number to the shape of a particle. The questions are, "Is this an important thing to do?" and "Does anybody do it?" We all believe that particle shape affects how fast particles flow, how easily they compact into parts, how strong the green parts are, and how easily the particles in those parts sinter together to form strong bonds. We will shortly describe how to measure how fast powder will flow into a die. You will learn that for a decent iron powder, 50g will flow through a small hole in 27 seconds. Does anyone ever measure the shape of the particles and then try to predict this flow rate? Not from what we have been able to learn! For instance, Prof. Lenel said, "Although quantitative shape parameters for powder particles have been defined, they have been little used in the characterization of powders." Former U. Wisconsin Prof. Joel Hirschhorn reported, "Many attempts have been made to place shape characterization on a quantitative basis by the use of various mathematical descriptions. These are difficult to apply and not particularly enlightening or accurate." And more recently, Prof. German indicates, "Because it is difficult to quantify particle shape, qualitative descriptors are frequently used to convey a sense of shape." Now that you have been warned, but are still determined to look into applying numbers to particle shape, we recommend Volume 7 of the ASM Metals Handbook, pages 263-273. The part that is easiest to follow was done by B. H. Kaye, pages 263-266, but we believe that the definitions in their Figure 12 are mislabeled.

If you can imagine that a particle can be thought of as having a length, width and thickness as in Figure 3.21, then the aspect ratio would be the length/width. Long thin particles would have a high number, particles with length = width would have a 1 for a ratio. Chunkiness is width divided by length. If the width is about equal to the length, the ratio is 1 and chunkiness is 1. Chunkiness equal to 1 is as chunky as it gets. If you ever wondered if flakiness could be measured, it is width/thickness. In staring at Figure 3.21, you may have wondered how one could measure all three dimensions: length, width and thickness. We don't see any easy way either, so folks look at the shadow outline of the particle, which puts the shape in only two dimensions. Then you imagine a rectangle enclosing the particle. Its sides are called length and width, chosen so that the rectangle has the minimum area. This gives you the length and width that you need to calculate the aspect ratio (length/width) or the chunkiness (width/length). If you wanted to go further, it would seem rather tedious to turn all the little particles on their "sides" so that you could measure thickness. So we conclude that although flakiness is rather apparent when you encounter it, in Figure 3.16 it is not all that easy to measure!

If you were thinking that it would be a long job to fit rectangles around all the particle shadow outlines, you were right. That is what computers and image analyzers are for. A TV camera picks up the particle image and turns it into the pixels that you see on your TV monitor. The computer can then be programmed to do calculations, including trying all the possible surrounding rectangles, until it finds the one with the least area. That gives the length and width needed to get at chunkiness etc. The same programs can count all the pixels (TV monitor dots) inside the particles' shadow which will equal a cross sectional area of a particle. It can count all the pixels on the perimeter. You can imagine that highly irregular and rough particles will have a long perimeter. If the same particle area were shown as a circular area, its circumference would be much less. The image analyzers readily calculate this ruggedness effect by taking the ratio of the equivalent circle's circumference to the rather large perimeter of the actual particle. You can see that for a very rough particle you would get a low number for this ratio. If the particles were like spheres and their shadows like circles, you would get a ruggedness ratio near 1. (Frankly, we prefer to calculate the ratio the other way: long perimeter of the rough particle divided by the small circumference of the equivalent round circle. This means that rough and rugged particles get characterized by high numbers. Seems more natural that ruggedness should be "big").

Surface Area of Particles

The surface area of a gram of particles can be very large, if the particles are small and have very irregular surfaces: for instance a gram of atomized iron has surface area of 525cm^2 (84in^2) while a reduced (spongy) iron might have 5160cm^2. Each particle is small and its area is small, but there are plenty of particles in one gram of iron. Particles with lots of surface area sinter more rapidly, so it is sometimes worth measuring the surface area. It is done by the BET (Brunauer Emmett and Teller, 1938) method. The sample of metal powder is first cleaned by evaporating off any adsorbed films in a vacuum, with some added heat. In this atomically clean state, it is cooled to the temperature of liquid nitrogen. When low pressure nitrogen is passed over the surface of the particles, the nitrogen molecules stick to (adsorb onto) every nook and cranny in the particles. First measure the weight of nitrogen picked up on the particles surfaces, then the BET formula is used to calculate the weight of nitrogen that would form just one layer of molecules, packed as closely as possible on the particle surfaces. Since we know what a nitrogen molecule weighs, ($28/6.02 \times 10^{23}$ g) we can calculate how many molecules have been adsorbed on the surface. Then, knowing that nitrogen molecules occupy an area of 1.62×10^{-19}m^2, we can easily calculate how many square meters or inches have been covered, and that is the surface area of the powder. If you want to do this yourself, Micromeritics will sell you the instrument, or, you can do what we do, and send the powder out to be tested.

FLOW RATE

Once you own a compacting press and a set of tools, the only thing standing between you and producing lots of parts is how fast the powder is going to flow down into the die. If you want to test your powder to see if it flows well, the easiest way is to use the Hall Flow Meter funnel shown in Figure 3.22. Devised by Dr. E. J. Hall of Metals Disintegrating, now ACuPowder Int.,and standardized in 1946. It is simplicity itself. Weigh out 50g of powder, place one finger under the funnel bottom, dump in the powder, remove your finger, and start your stop watch. When the last bit exits the funnel, stop the timing: Ah! 28.3 seconds for 50g, not bad! We hope you won't ask why a flow rate is expressed as a flow time for 50g when you expected to get an answer like 50/28.3 = 1.77g/second. That is life in the P/M world. Anyway, the lower the flow time, the faster you will fill the die.

This method is covered in good detail in MPIF 03 or in ASTM B 213.

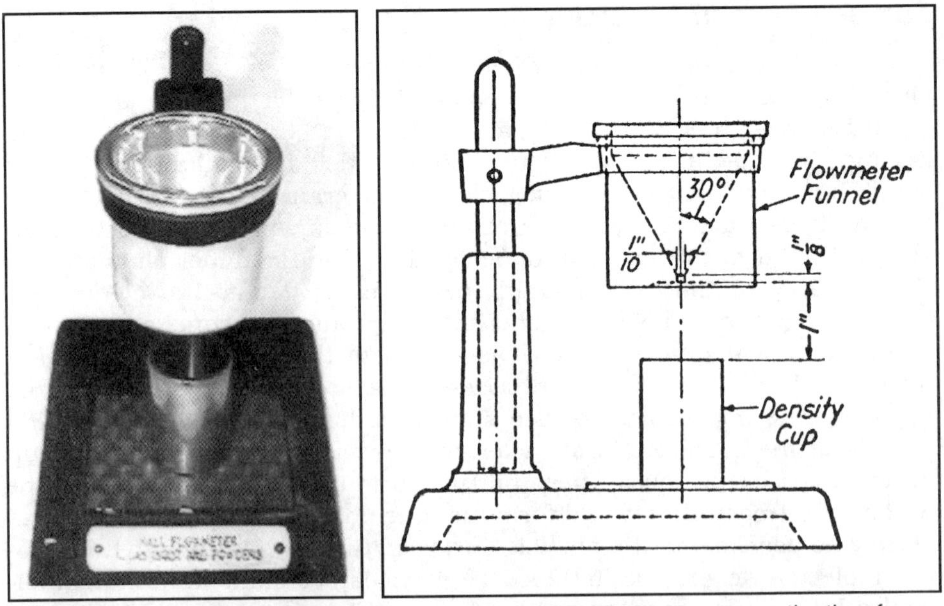

Figure 3.22. Hall Flow Meter used in test method ASTM B213 to measure the time for 50g of metal powder to flow through the funnel.

The method is applicable to free flowing powders. What does that mean? On a bad day while doing this test, you pull your finger away from the funnel and the powder does not flow. Or it starts, and then unaccountably stops. What do you do then? Resist the temptation to strike the funnel! Resign yourself to recording the mix as non-flowing.

What factors affect flow rate? Particle shape is the most important. BB shot and little ball bearings flow well (a.k.a. gas atomized and rotating electrode powders). On the other hand, fibers of steel wool and flaky powders flow slowly, if at all. The people who describe particle shape will eventually learn a way to characterize particle shapes with a number that relates to our measured flow rates.

The density of the parent metal affects the flow rate of the powder. Iron has a density of 7.87 g/cm^3, so that 50 grams of iron occupies 50/7.87 = 6.4 cm^3. Aluminum at 2.64 g/cm^3 would occupy 50 g/2.64 g/cm^3 = 18.9 cm^3. If you picture the 6.4 cm^3 of iron in the form of a pile of bb's and the 18.9 cm^3 of aluminum in a larger pile, you can imagine that the smaller iron pile will easily slip through the funnel very quickly, compared to the aluminum. Even when we compare powders of the same parent metal (iron), the powder that packs most densely flows the fastest.

APPARENT DENSITY

Since we slipped in the concept of packs most densely, we had better explain. First, density is simply weight (mass) divided by volume. If we had a little 25cm^3 cup and filled it with powder, we would certainly know the volume of the powder. Dump the powder on a balance and we learn the weight, e.g. 75g. Density is weight (mass) ÷ volume = 75g/25cm^3 = 3g/cm^3. That is, each little cubic centimeter of powder would weigh 3 grams. (Maybe we could use a little 1cm cup and save the division step! Don't be silly, you say! Well the Scott Volume Meter, ASTM B 329, carefully dumps tungsten powder into a 1 cubic inch cup. Voila! grams/in^3 No division!)

Why did we title this section "Apparent Density" and then show you how to measure the density of a pile of powder? Another peculiarity of the world of powder metallurgy. Metallurgists think of density in terms of solid objects: a 1 cubic centimeter block of steel weighs 7.87 grams. If you somehow disintegrate the steel into a pile of loose powder, dump it into our 1 cm^3 cup and weigh it, you might get 3.0 grams in the cup with the rest of the 7.87g having flowed over the edge of the cup. The steel now has a density of only 3g/cm^3. Lets call it an apparent density , since it is now less than the solid steel's 7.87g/cm^3. Is that the "weigh" it happened? Who knows? That's the way it is: measure the density (weight divided by volume) on loose powder and call it an apparent density. Now for some fun. Take a little 1 cm^3 ceramic cup full of iron powder, run it through a sintering furnace and bond the particles together. Dump the sintered and bonded particles out of the cup and you have a nice little 1 cm cube of stuff that weighs 3 grams. Its density is now 3 g/cm?, but since it is no longer loose powder, we call it density, not apparent density ! If tilting at wind mills is your game join ASTM Subcommittee B09.01 Nomenclature, and write this wrong!

Well, if filling little 1 cm^3 cups with powder doesn't make much sense, how would you fill the 25 cm cup? Dr. Hall cleverly provided a 25 cm^3 brass cup to sit under his funnel and catch the falling powder. So after the 50g goes through and completes the flow test, you add about 50% more, and it will continue to fill the cup until it overflows. Then level off the powder in the cup by carefully scraping with a spatula and weigh the powder contents. The apparent density will be the powder weight (mass) ÷ 25 cm^3. All this is covered in MPIF 04 and ASTM B 212, apparent density of free-flowing metal powders. You can see why we told you not to strike the Hall funnel: if you are collecting powder down below for apparent density, jostling it will cause the powder to settle a little bit, giving an erroneously high apparent density.

How do you get an apparent density if the powder won't flow well? Get a funnel with a bigger hole! The orifice for Dr. Hall's funnel is only .10 in. in diameter whereas Mr. Carney's funnel orifice is .20 in. As you can imagine, reluctant powders will pass through the Carney funnel and land in the Hall cup, there to be weighed, to give an apparent density. The Carney funnel method is covered in MPIF Standard 28 and ASTM B 417. Why don't they measure the flow rate of those powders that are reluctant at the Hall funnel, using Mr. Carney's more open funnel? We have no idea! Nor do we have any idea who Mr. Carney was!

Why is apparent density important? First, high apparent density powders flow quickly into the press tools. What if low apparent density powders flowed well, would we like them? Probably not. You can imagine that to make a given length of P/M part, we would need much longer tools if the powder were light and fluffy. Who wants to pay for all the extra tool steel? Also, presses have limited ranges of up and down motion. Who would want to pay for a very tall press with capability to move punches up and down great distances, just to accommodate fluffy powder?

For some sets of tools, it is important to control the apparent density quite closely. If a die has a step in it (Chapter 4, Figure 4.5) the density of the part in the region formed against the step depends upon the apparent density of the powder. Remember that the upper punch will always enter the die and stop a fixed distance above the step, in order to give the correct part thickness. If the apparent density of the powder falls, there will be less powder above the step. After compaction there will be less powder in that region of the part, and it will be a low density region. Our colleague, Norbert Arnold, retired from Keystone Carbon Co., wondered if the apparent density of the powder sitting down inside the die was the same as that measured with the Hall cup. To simulate the rapid dropping action of the powder, he devised a simple apparatus which bears his name: (ASTM B 703, MPIF 43). Figure 3.23 shows the Arnold Meter. In essence, a small brass bushing holds some loose powder, and simulates a press feed shoe. It is passed over a 1.2466 in. diameter hole in a steel block. The hole in the block has a volume of 20 cm^3. By weighing the loose powder that falls into the hole in the block, we can calculate the apparent density as the weight (mass) ÷ volume of the cavity. The method is fast and easy to carry out. It relates to the Hall density and for iron and bronze molding grades is generally, 15 to .26 g/cm^3 higher. Mr. Arnold also likes a volumetric flow rate: just pass the powder obtained in measuring the Arnold density through the Hall funnel. This will give a flow time for 20cm^3 of powder.

What factors influence apparent density? Particle shape is the most important. Round BB's pack very well. Perfect packing of the same size spherical powder would be 64% of full density, with 36% air space. This is true regardless of the size of the spheres, as long as they are all one diameter. Very irregular powders could have apparent densities of 10 to 20% of the parent metal. Water atomized iron powders are about 40% of

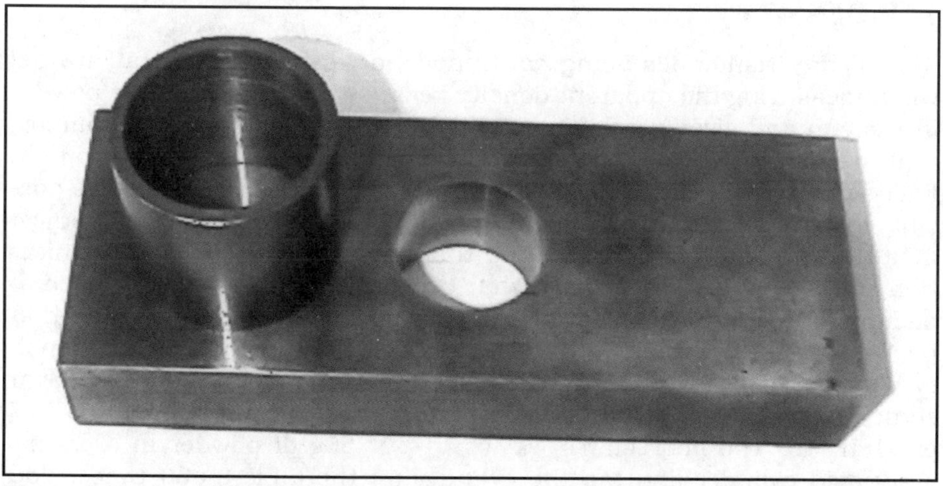

Figure 3.23. Arnold Meter for measuring the apparent density of powders. It simulates the density that the powder would have after it has fallen into the compacting die. The brass ring is filled to a specified level with powder, and then slid over the opening in the block. The weight of the powder that falls into the opening is divided by the volume of the cavity in the block, 20 cm^3, to give the powder's apparent density.

the density of steel. Most iron powders are mixed with a lubricant like Acrawax or zinc stearate (ZnSt, a colloquial abbreviation). The mixing takes anywhere from a few minutes to an hour. During mixing, the apparent density and flow rates improve, i.e. the apparent density goes up and the flow time for 50g goes down. This is attributed to a gradual rounding of the particles so that they slide more easily over each other. Also, SEM photos show how the lubricant particles smear and coat the metal particle surfaces. The mixing effect should not be over done - the better flow rate and apparent density could come at the cost of lowered green strength. When you are knocking the little arms off the powder particles to make the powder flow well, you are depriving it of the arm interlocking mechanism that creates green strength.

Particle size affects apparent density. Very fine particles, even rounded ones, sort of stick together and do not flow well (Van der Waals attraction). Particle size distribution affects apparent density. If you have some coarse powder you can mix in about 27% by weight of a powder that is 1⁄7 the size. The small particles fit in between the big ones. With spheres, the maximum particle density could go from 64% to 86%. We did this with some tungsten powder as follows: On the first try we used 85% by weight coarse tungsten with all particles passing 100 mesh at apparent density 8.77 g/cm^3 and 15% fine tungsten, 4.1 Fisher micrometers at 6.66 g/cm^3. After mixing, the apparent density was about 11.7 g/cm^3.

TAP DENSITY

You may remember being cautioned not to wiggle the Hall funnel when measuring the apparent density because it would settle the powder in the cup and give an erroneously high density. Sometimes it is important to know just how much powder will settle when we tap or vibrate it. Folks who do cold isostatic pressing have to get as much powder as possible into their urethane molds (Figure 12.12). The MIM/PIM folks inject mixtures of metal powder, wax and thermoplastic into molds. The injection pressures are lower than what are used for the usual P/M compaction. For that reason they want to have a high apparent density that packs as well as possible into the molds. It is easy to measure tap density which is the apparent density of powder that has been tapped or settled under certain conditions. The method is covered by ASTM B 527 or MPIF 46. The poor man's way is to put 50g of powder in a 25cm^3 graduated cylinder and tap the cylinder on the table 3,000 times. You know the weight of the powder and you can read the final volume of the powder in the graduated cylinder. By now, given mass and volume, apparent density is easy to calculate. On the other hand, the device shown in Figure 3.24 sells for about $3,353. dollars and does those

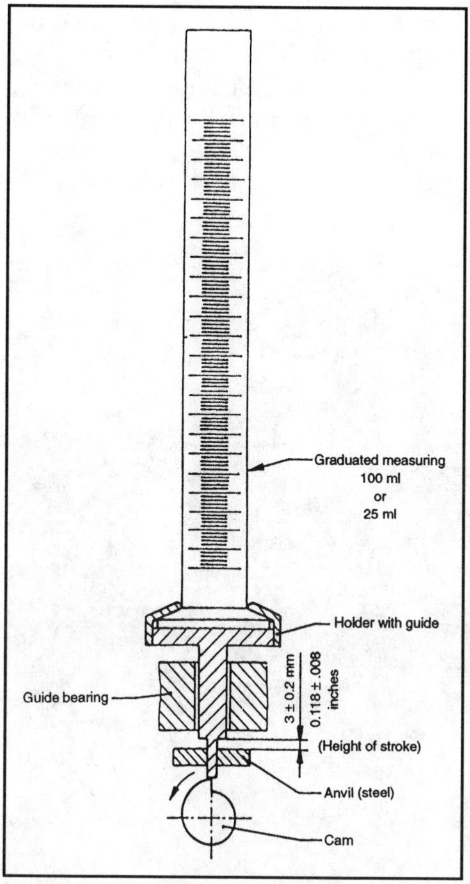

Figure 3.24a. Example of tapping apparatus.

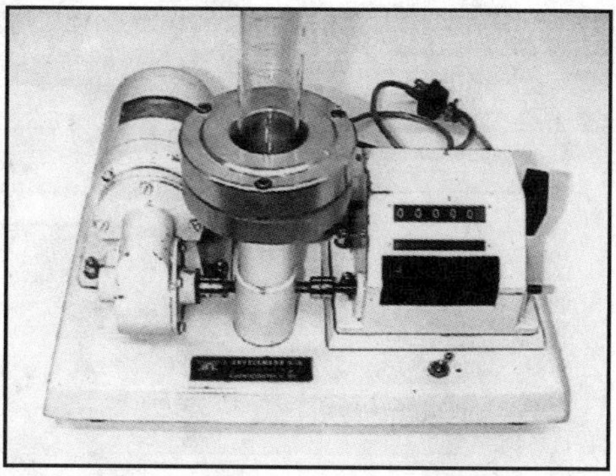

Figure 3.24b. Photograph of tapping apparatus.

3000 taps for you, kindly stopping when the tapping is completed. We say "kindly", because it is a rather noisy device.

GREEN STRENGTH

Green strength describes the strength of the P/M part as it comes out of the press. Why does it need to be strong? Green strength helps keep the part from cracking as it is ejected from its tooling. Then, when the feed shoe pushes it aside, it ought not to crack! Someone stacks the parts on a tray and then unloads them onto a sintering furnace belt. Green strength helps get the part into the furnace in one piece. How is green strength measured? ASTM B 312 and MPIF 15 teaches how to break a small bar in three point bending. You support a 1 ¼ x ½ x ¼ in. thick bar on 1 in. centers. Then press on the top of the bar, in the very center. At a certain load the bar breaks. We know the pounds to break it, as well as the width and the thickness. Green strength is the stress in the under side surface when the bar broke.

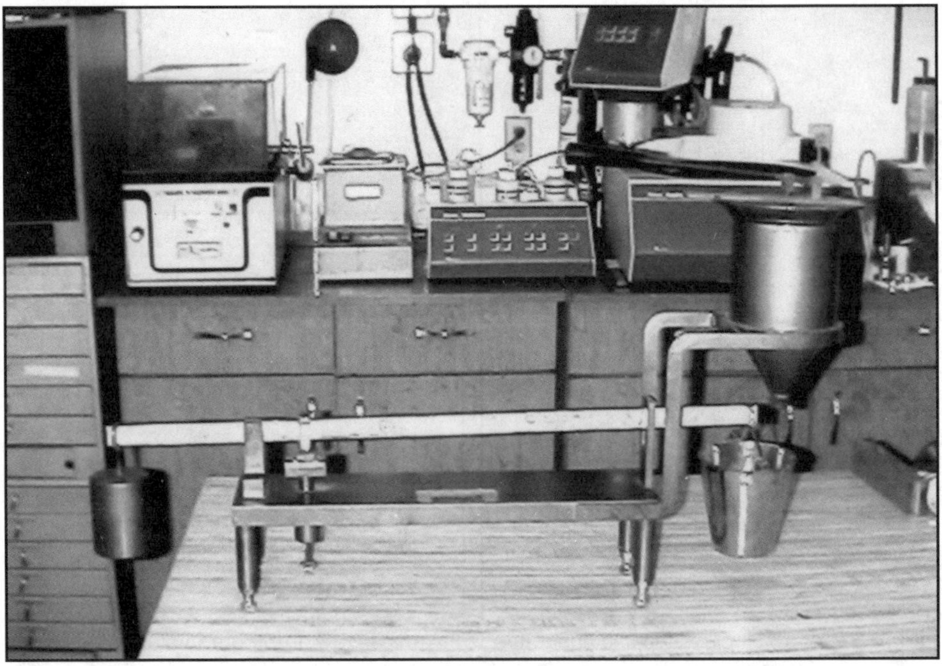

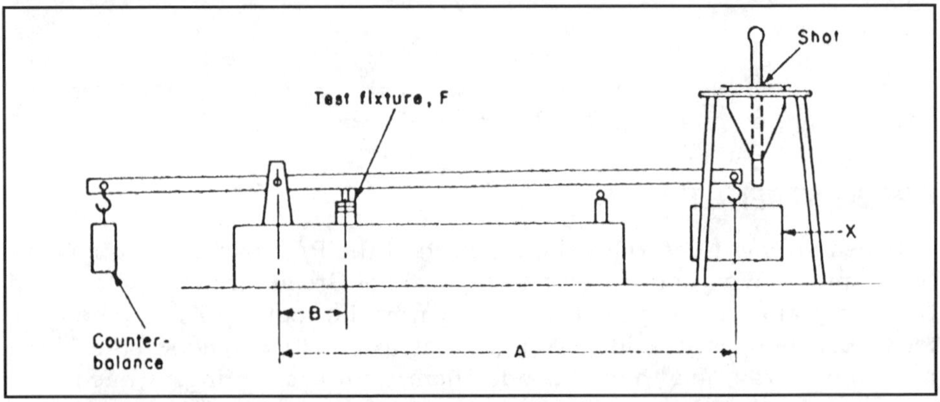

Figure 3.25. Constant loading beam device to apply force to an unsintered test bar, so as to measure its green strength, test method ASTM B312.

Green Strength = (1.5 x load, lb)/(thickness x thickness x width, inches)

Typical irons are 1200 to 2400 psi. Below 800 or 900 psi, you cannot handle the part. ASTM B 312 shows a simple beam device and Figure 3.25 shows the one in our lab. The beam that breaks the part is loaded at one end by gradually pouring shot into a bucket. When the sample breaks, just weigh the bucket to get at the pounds of force. Our device uses a 10 to 1 beam so we multiply the bucket loads by 10, to get at the force. The force is then used in the formula above to get the green strength.

COMPRESSIBILITY

Compressibility describes the ease of pressing metal powder to high density. For a given powder, the higher the density the stronger the sintered part. If we actually were able to somehow squeeze out all the pore space during powder compaction, we surely would! Once you have a compacting press and a tool set, measuring compressibility is pretty easy. The test methods are ASTM B 331 and MPIF 45. A typical die is the transverse rupture bar tool set: 1¼ in. x ½ in., Figure 3.26 You must first decide if you are testing lubricated or unlubricated powder. If a powder contains no lubricant, you have to put some lubricant on the die wall. You can mix up some stearic acid and alcohol and paint it on the die wall. The alcohol evaporates and leaves the soapy stearic acid in

Figure 3.26. Tooling to make transverse rupture bars 1 ¼ x ½" die size. Die floats on springs to give equal compaction on top and bottom punches.

place. Then carefully dump in about 15g of iron powder and level it. Insert the upper punch in the die. If you are running a test to see the density when pressed at 30 tsi, the force to be applied to the upper punch will be 60,000 lbs. times the die area (0.5 x 1.25 in.) = 37,500 lbs. If you are using a universal testing machine to press the bar, just dial in the force to 37,500 lbs., reading it to the nearest 100 lbs. Eject the bar, weigh it to .01g, measure it to .001 in. (length, width and thickness). The density is then:

$$\text{Weight (g)}/(l \times w \times t \times 16.38)$$

The 16.38 is the number of cubic centimeters in a cubic inch and converts our inch measurements into centimeters so that the density comes out in g/cm^3.

If the density were 6.9 g/cm^3 we would say that the compressibility at 30 tsi is 6.9 g/cm^3. This allows you to sort out iron powders to find the one that is the easiest to press. Interestingly, the density measured geometrically, above, is always lower by about .03 g/cm^3, than if you measure density by the water immersion method (ASTM B328, MPIF 42). You can think about why there should be this difference, and maybe we will consider it again later.

DIMENSIONAL CHANGE DURING SINTERING

In Chapter 5 you will learn why and how compacted metal powders change size during sintering. You can imagine the chagrin of the P/M parts producer when he carefully designs and builds a tool set, only to find that the parts shrink or grow too much during sintering. That is why he needs a way to test his powders and keep them under control. He may use ASTM 610 or MPIF 44 to perform the test. It is quite easy: make up some ½ x 1¼ x ¼ in. bars using the transverse rupture tools. Note the die dimension to .0001 in. on the 1¼ in. length. Sinter the bars in a production or lab furnace. Measure the now sintered bar's 1¼ in. length, and calculate the difference from die size. Divide this difference by the 1¼ in. die size dimension and express it as a percent of growth or shrinkage. Clearly the dimensional change will depend on the sintering conditions. To be sure sintering conditions have not changed since the last time you did this test, run a standard powder and its bars at the same time as your test. If the standard has its usual size change you can depend on the data for the new powder lot. Some P/M houses keep statistical records on how the standard bar behaves. They can establish an upper and a lower limit for the standard's size change. If the standard lot falls outside this range, something funny happened at sintering and you need to rerun the trial or investigate the sintering.

Some P/M houses run standard bars in each furnace, once a day, to keep the sintering under control. These bars would be tested for size change, transverse rupture strength, and hardness, (top and bottom). The iron - 2% copper - .9% graphite mix works well for this purpose. At the outset, we noted the plight of the parts producer whose size change during sintering wandered astray. Where did he get the size change data that were used to design the tooling? The starting point could be the data from the transverse rupture bars, made at the same density as the parts. Other designers prefer a larger disc or a ring to learn about size change. Even better would be data from a similar sized part with the same composition. He learns from his experience and past tooling errors/successes. Where parts are repressed or heat treated as part of the production process, one will usually carry the test specimen through all the same steps, and get the size change from die to the final state, e.g. heat treated.

MIXING METAL POWDERS

First, what is the difference between mixing and blending? Most P/M parts producers mix powders: they mix a single lot of iron with 2% copper .9% graphite and 3/4% Acrawax. Mixes have different chemical species just like cake mixes. People who sell powders often blend different lots of pure iron together. The lots may differ a little in apparent density or flow rate and they need to reach a closely controlled target apparent density. Blending consists of thoroughly intermingling different lots of the same basic chemistry. Once the powder producer has a large amount of this homogeneously blended powder, he can take a few tons and put it in a mixer to create the iron copper lubricant mix needed by the P/M house.

Are the machines for mixing and blending much different? Not that we can see. Figure 3.27 shows a "V" blender, Figure 3.28 shows a double cone mixer. We wouldn't pretend to tell powder producers how to blend but we might offer parts producers some suggestions on mixing. First, each mixer has a rated capacity in pounds of powder of volumetric capacity for powder of a certain apparent density. They think big when it comes to apparent density, like 187 lb/ft^3. We think small: ordinary iron powder is $3g/cm^3$. If a mixer is rated for 1,000 lbs., it is a poor idea to make a lot of 75 lbs. because of segregation as the powder slides long distances against the wall of the mixer. Also, when it slides, it does not tumble over itself and mix or diffuse together. The powder should occupy 2/3 of the volume of the mixer, which is about even with the axis of rotation. Then, the order of adding the ingredients matters. What does not

Figure 3.27. Cross-Flow™ "V" blender for mixing metal powders. The uneven legs produce a strong axial exchange of material between the two legs. Starting with equal amounts of powder in each leg, in one revolution, 25% of the material is moved from one leg to the other. Courtesy of Patterson-Kelley, East Stroudsburg, PA.

work well is to put in the iron powder and then dump in your pounds of graphite lubricant copper and nickel. It will take a long time to mix in a large pile of lubricant, and it could get pounded by the metal powder mass, leaving lubricant balls (politely: agglomerates). The preferred way is to have a feeding system above the mixer opening and simultaneously pour in the iron and the additives as separate streams. This gives the mixing a head start. You could also make a mastermix with a concentration of the additives in the iron powder. Once it looks well mixed, it could be diluted with the rest of the iron powder by further mixing.

Once the mix is done, you may empty the powder into one or more barrels. Some care should be paid to the last powder exiting the mixer. If you rap the mixer to coax out the laggards, you may find that the last powder is mainly graphite and was sticking to the walls or floating on the iron. The last bit should be discarded or recycled. Powder producers sometimes deal with this problem by adding a small amount of liquid

Figure 3.28. Double cone mixer. Capable of mixing several thousand pounds of metal powder at one time. The powder to be mixed can be brought to the mixer in 500lb. fiber drums, attached to the mixer, the mixer is then inverted, and a valve opens to empty the powder into the chamber. This controls dusting of the powder. After mixing, the procedure is reversed and the special valve can empty out the mixer, one barrel at a time. Courtesy of Gemco, Middlesex, NJ.

polymer and solvent to the powder mix. The polymer glues the graphite, nickel and copper to the iron. This tends to prevent segregation or separation of the ingredients in the bonded mix. Segregation could be a problem in a non-bonded mix during shipping, handling, and transfer up to the press hopper, as well as during the descent into the tooling. You could measure the tendency to segregate by flowing powder down an inclined ramp. The ramp has some small holes or slots that lead to small containers underneath it. As the powder descends, some falls down into these containers at various points in the descent path. In a perfect world, the large particles would not go racing ahead down the ramp and be captured near the bottom. Instead, each container would have the same chemistry and particle size distribution. So, if you check the chemistry and particle size distribution in each small container, you will learn if the powder segregates. (How do you check chemical composition? There are ASTM test methods and you can DIY, or you can send it out to a laboratory)

PRODUCTION OF METAL POWDERS

In Chapter 2 you learned how to make sponge iron powder by two different methods: hydrogen reduction of mill scale (at Pyron Corporation) and carbon reduction of iron ore (Hoeganaes and Höganäs). (If you are new to P/M, a little pronunciation guide is in order for the latter two producers: Hay'-gin-us and Hur' guh ness', and it helps to be of Swedish descent). In North America, iron powder is water atomized by Hoeganaes, Québec Metal Powders, Kobelco, and Pyron (now part of North American Hoganas). In 2,000 they produced about 400,000 tons. The three U.S. producers used selected scrap steel for their raw material while QMP uses hot metal from their sister company Québec Iron and Titanium. Figure 3.29 shows a diagram of a device to water atomize iron. The iron is melted at 3,000°F in air or protected by nitrogen. It is then poured into a ceramic lined tundish with one to three 5/8 in. openings in the bottom. The metal falls through one nozzle at the rate of 500 lb/min and is struck by a stream of water. The chamber may have some nitrogen in it, to partially protect the iron. To disintegrate the 30,000 lb/hr. of iron, requires 5 times the iron weight or 18,000 gallons/hr. of water. The

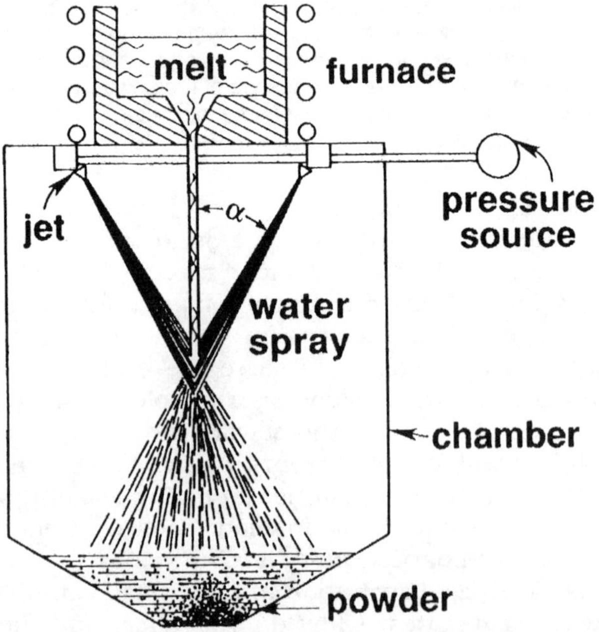

Figure 3.29. The water atomization process, where a molten metal stream is disintegrated by multiple water jets. The angle determines the atomization efficiency.

water is pressurized to about 14 megapascals which equals 2,000 psi, to produce a powder with an average particle size of 42 micrometers. At 150 megapascals (22,000 psi), the powder would be about 5 micrometers average particle size.

The collected water and powder must be filtered to remove the water and then the iron is dried. During atomization, the iron powder does pick up some Fe_3O_4 iron oxide on the surface. Therefore the powder is passed along a stainless steel belt through a furnace containing some hydrogen at 1400 - 1800°F. This hydrogen carries off the oxide as water vapor, leaving a clean iron surface. The powder is also softer (annealed) because the fine, strained metal grains have relaxed and grown in size. Any fine particles have become bonded to the larger ones. The material exiting the furnace is stuck together, like a cake. It must be carefully broken up so as not to rework or reharden the iron.

We have been speaking of plain iron powder but alloying elements like molybdenum could be added to the melted metal, prior to atomizing. In that way each metal particle would be fully alloyed (pre-alloyed) with the added metals. If we added 17% chromium, 12% nickel, 2.5% molybdenum, we would have made 316 stainless steel. Adding the following elements to pure iron: .83% carbon, 4% chromium, 5% molybdenum, 6% tungsten and 2% vanadium, would have given us M2 tool steel. In the low alloy steel the powders are made with low carbon (less than .03%) so that they will be soft enough to press. Interestingly the molding grade of M2 is water atomized with its full complement of carbon, .83%. Dr. P. Nurthen of Powdrex, explains that the atomized powder is annealed in a carbon containing atmosphere that removes the oxides that came from the water atomization. That atmosphere is used to adjust the carbon level to .83% ±0.05%. By slow cooling, the powder is somewhat softened by this annealing process, and made more compressible. If the powder were made with 0% carbon, significant amounts of graphite would have to be mixed in to give the right carbon content after sintering. As the carbon diffused into the low carbon tool steel powder, it would find carbide forming elements (Cr, Mo, W, and V) and immediately form their carbides near the surface of the particle. This has the potential to produce a non-uniform distribution of carbides. To produce very clean, as-atomized, low oxygen metal powders, one must atomize with something other than water, such as inert gas. (But gas atomized particles cause poor green strength in compacted parts.)

In gas atomization the molten stream of metal is struck by high pressure gas. This breaks up the metal stream in the way shown in Figure 3.30, page 103 P/M Science by R. German. Figure 3.31 shows a photo of the atomizing process. There is considerable technology, mostly proprietary, in designing the gas inlet direction and the atomization nozzle so

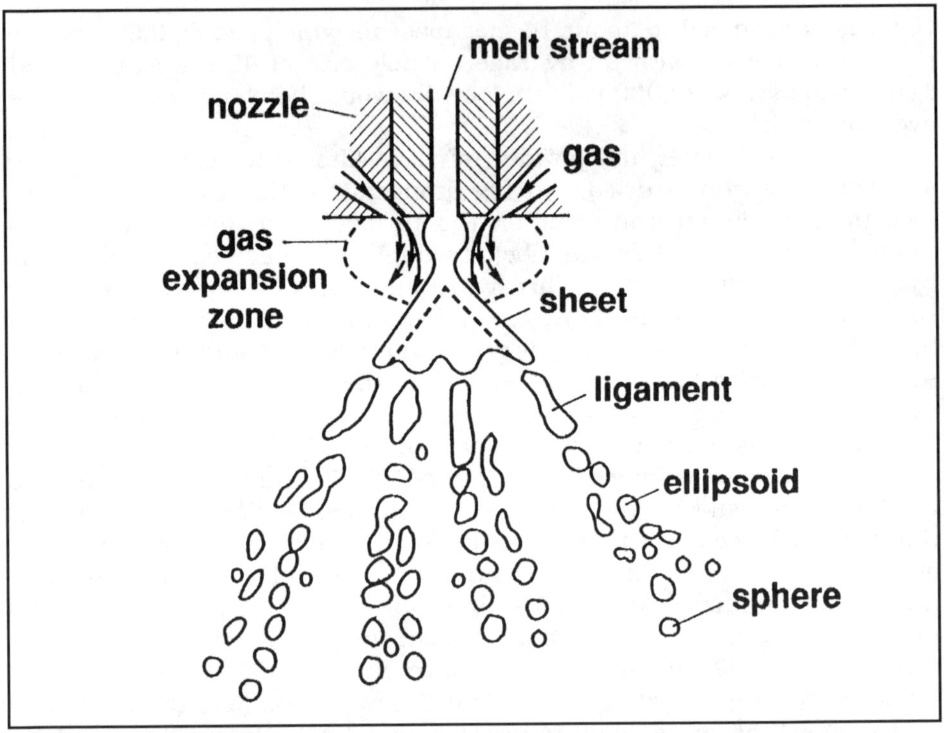

Figure 3.30. The formation of a metal powder by gas atomization involves the break-up of the liquid stream by the rapidly expanding gas. Because of a suction pressure in the gas expansion zone, the stream first forms into a thin hollow sheet, and subsequently forms ligaments, ellipsoids, and spheres.

as to maximize the effect of the gas in breaking up the liquid metal. The superalloy materials that go into jet engine parts in Figure 1.24 are made by gas atomizing.

COPPER POWDER

High purity copper scrap can be melted above 1981°F. The liquid is atomized or converted into coarse copper shot, about .020 in. in diameter. The liquid stream is broken up by low pressure air or water. The process is called shotting when coarse spherical powder is produced. You will immediately recognize that the coarse spherical powder will not be very useful for making P/M parts! Well, the shot is just an intermediate stage in the process of making an irregular powder with good green strength. How to break up the shot? You can see the problems of pounding on copper BB's and creating pennies rather than powder. The copper

Figure 3.31. Gas atomization of a stream of liquid metal, courtesy of Crucible Research Center.

is too ductile to break up. Therefore the copper shot is loaded on a stainless belt and roasted in a furnace in air at 1200°F, converting most or all of the copper to a nice brittle copper oxide. Remembering how Pyron and Hoeganaes make sponge iron, you will not be surprised to learn that after the copper oxide cake has been broken up, it is a fairly easy process to reduce it back to pure copper. The furnace atmosphere has some hydrogen to react with the oxide and carry it away as water vapor. This oxide reduction occurs at 800 to 1200°F. The pure copper cake exiting this second belt furnace is then broken up and sieved to -100 mesh particle size. The copper can be quite irregular in shape, with good green strength. Spongy copper is the usual powder that is combined with 10% tin, for compaction into bronze bearings. In such bearings, the oil impregnated into the bearings can be stored both between the particles as well as inside the particles. On heat up, the oil between the particles will weep out first and lubricate the shaft which runs against the bearing. Some oil is held in reserve, in the small pores inside the particles, and does not leave the bearing unless things get very hot.

Copper particles are also made by water atomization. Although somewhat rounded as atomized, subsequent annealing sort of cracks them and increases the surface area. That provides a compacting grade of copper. Copper powder is also made by electroplating copper onto a substrate. The substrate is periodically scraped, to remove and save the copper. This plating is done under conditions that give a very rough plating, compared to the nice smooth electroplated surfaces that you are used to seeing. This rough copper plating contains copper particles that have a high surface area because they look like little trees (dendrites) or ferns. The electrolytic powders have terrific green strength and are very low in apparent density. They are often mixed with graphite, and compacted to form the brushes used for carrying current in electric motors. Remember Chapter 1, page 50.

There are many other metal powders that are extensively used in P/M: nickel, tin aluminum, cobalt, tungsten, and tungsten carbide to name a few. Some of these have interesting and exotic methods of manufacturing. If we list any more you may give up and never get to the next chapter which teaches you how to do something useful with metal powders, so press on. Or if you need more information on how to make powders we refer you to Volume 7 of the *ASM Handbook,* 1998.

CHAPTER 4

Compaction

Now that you have learned how metal powders are produced, we should explore the next step - forming. Bill's wife, Jane always tells people "Making P/M parts is just like making aspirin or Alka-Seltzer". In fact, the folks at Stokes Machine were making presses for aspirin tablets before the presses were ever used to press metal powders.

This is a good comparison, especially for very simple parts. When a P/M part is in the "as molded" or green state, its strength is comparable to an aspirin. Making discs or washers like these tablets from powder metal is simple. It's when you attempt to make complex parts with multiple levels in the direction of pressing [molding] that life gets interesting.

Powder metal does not flow well around corners under pressure; it's not a fluid. When you mold a part with multiple levels, complex presses and multiple punches are required. These are necessary to provide proper powder fill, compaction and ejection for the various part levels. [See Figure 4.1 for diagram of part categories from the Basic P/M Short Course and Figure 4.2 "Tool Motions During Powder Compaction Cycle" from Figure 4-1, page 87 of the *Powder Metallurgy Design Manual.* Notice the difference between the powder fill height and the compacted part height. The ratio between these heights is called the "fill ratio" and is usually about 2.3 to 1. Also notice in figure 4.2 that the powder is always fed into the die and leveled off at the top of the die by a feed shoe.

Mike Jaffe, a well known consultant who spent many years in the P/M parts business, discovered a unique way to demonstrate molding action. He was seeking a material that he could easily use in seminars to show how P/M parts are molded. What did he come up with,? Popcorn!

Popcorn has some great properties that make it useful for this purpose. It is irregular in shape, sort of sponge like, highly visible, flows reasonably well into a die, [unless you use the caramel kind] and you can eat the green scrap!

Seriously, Mike built a small press that was operated by a hand crank and could change easily from single motion to dual motion. He made a

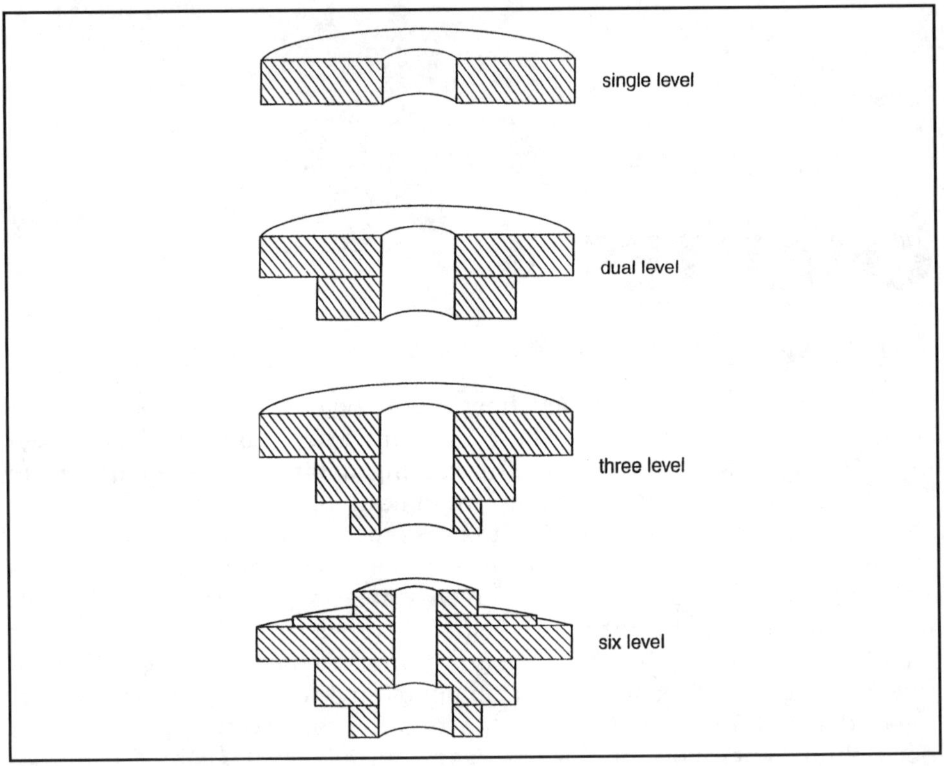

Figure 4.1. Part Categories.

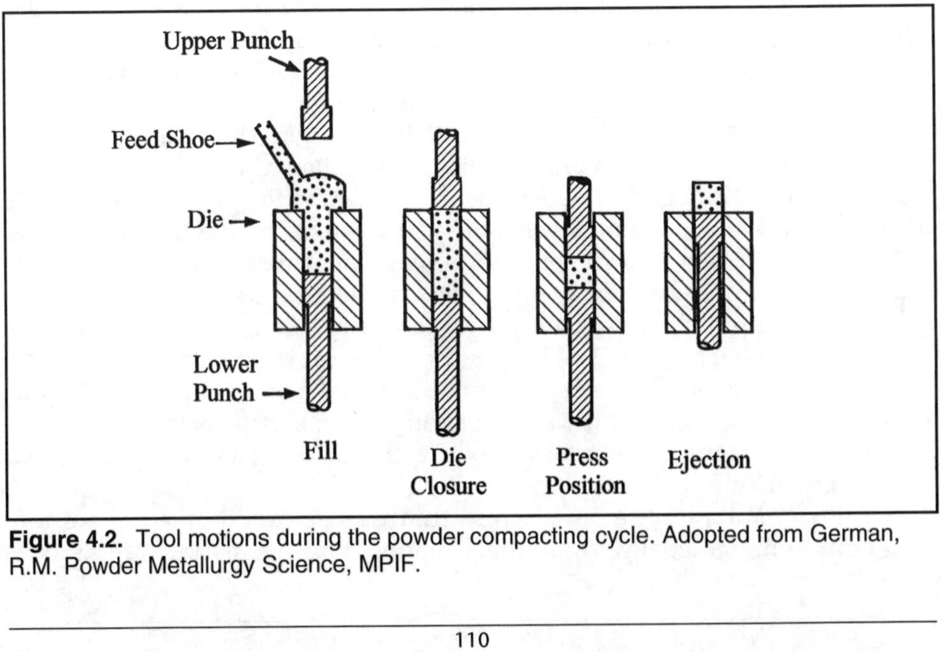

Figure 4.2. Tool motions during the powder compacting cycle. Adopted from German, R.M. Powder Metallurgy Science, MPIF.

Figure 4.3. Popcorn Press.

die out of clear plastic and proceeded to press popcorn into square blocks. A core rod was added to provide blocks with holes and by changing to dual motion, stepped blocks could be compacted. Figure 4.3 shows a photo of the popcorn press. Many people have seen his demonstration and it's a wonderful way to show how simple parts are molded.

Before we can mold parts, we need to make the tooling that will produce the part. We mentioned earlier that discs or washers are simple. You need a die that forms the outside diameter of the part, an upper punch which forms the top of the part and a lower punch which forms the bottom of the part. If you are making a part with a hole in it, like a washer, you need a core rod to form the hole. The punches will have holes through them, to accept the core rod. Figure 4.4 shows a bushing and its molding tools.

We have been discussing molding or compacting parts but as we move

Figure 4.4. A bushing and its molding tools.

towards more complex shapes, ejection from the die becomes a very important consideration in part design. Most P/M parts are molded at pressures of 20 to 50 tons per in^2 of top surface area. A lot of this pressure is transmitted sideways in the die. When the part is compressed in the die, the die will expand slightly due to this sideways force. When the compaction pressure is released, the die tries to move back to its original shape, compressing the part during ejection. The part, which is being compressed sideways, has to be pushed up out of the die. When clear of the die, the part expands, thus relieving the sideways stress. Because this expansion is caused by residual compressive stress, the amount of expansion is related to compacting pressure. In any event, this sideways force must be overcome in order to eject the part from the die. For multi-level parts, we need multiple punches to form and eject the various levels.

Some parts with one step or a flange on the outside can be molded by utilizing a step or "choke" in the die. The issue of apparent density and fixed fill enters here and careful calculation is necessary to establish the depth of the step in the die. There can't be an extreme overhang of the step on the part because too much surface against the die wall will cause cracking in the overhanging section when the part is ejected, This is because the frictional resistance of the overhanging section against the die walls exceeds the green strength of the part in that location. One

Figure 4.5. A flanged part and its tooling which includes a stepped die.

to two degrees of draft are required around the outside vertical surface of the overhanging section and that helps to reduce the ejection force. A radius around the edge of the step to be formed in the die is also required. All of the above are helpful in ejecting the part from the die and are more critical when working with low green strength materials. Figure 4.5 shows a flanged part and its tooling which includes a stepped die.

Figure 4.6 shows why multiple punches are used rather than a step on the punch in order to achieve even density. Multilevel parts that have large overhangs or steps with a lot of surface area in contact with the die, like gears, really need multiple punches. Figure 4.7 shows a diagram of cluster gear tooling using multiple punches. Sometimes springs are used under a lower outer punch to help with part ejection. The lower outer punch would be under the overhang of the part and would move downward during compaction, compressing springs located underneath it. When the part is ejected, the spring pressure on the punch helps to eject the part. Spring selection is critical because if there is too much pressure, the part will crack when the upper punch moves away . If the springs are too weak, the overhang will crack and stay in the die. The spring technique is used with single or dual action presses that do not have the ability to positively control the lower outer punch. Multi-action

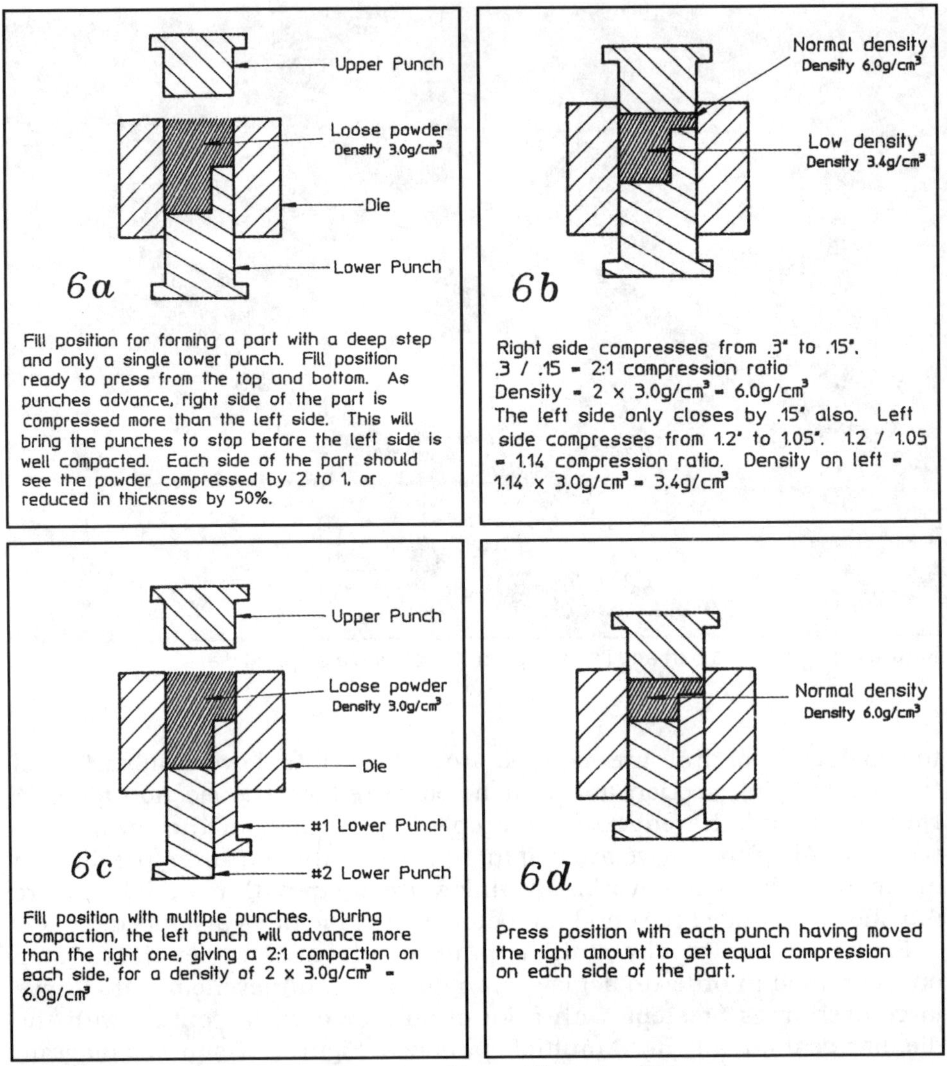

Figure 4.6. Use of multiple punches to achieve uniform density.

presses are a much better solution because there is positive control on all punches during powder fill, compaction and ejection. This means that each punch can be moved independently by a cam, hydraulic or pneumatic cylinder. The fill position of each punch can be set so that fill ratios in each section of the part are the same. This same flexibility of adjustment and control is available during compaction and ejection.

Figure 4.8 shows a multilevel part that requires three lower punches

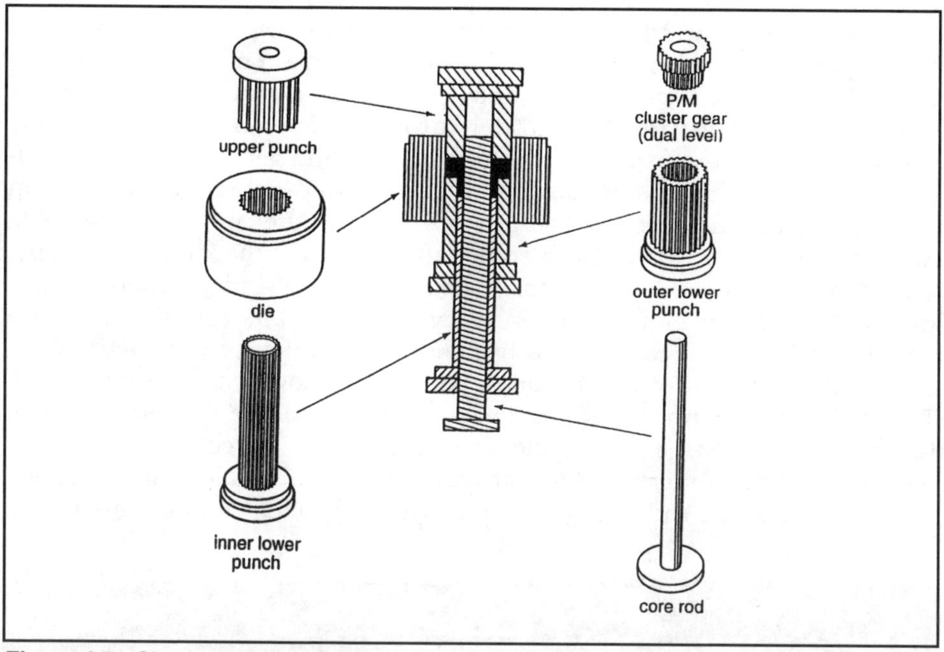

Figure 4.7. Cluster gear tool diagram.

Figure 4.8. Multilevel part requiring 3 lower punches and 2 upper punches.

and two uppers. This complex tooling allows good control of density throughout the part as well as proper ejection support. If these punches get out of adjustment, cracks can be formed during molding or ejection and these can cause part failure. We will discuss green cracks shortly.

Figure 4.9 is a multilevel part that requires two lower punches and two upper punches. It also requires a step on the core rod. During molding, after the upper punches enter the die, the core rod and lower inner punch move up to transfer powder into the upper outer punch. This transfer allows us to put powder where needed and gets around the problem of the die always being filled level. The powder is then molded to form the upper hub on the part. At the same time as the upper hub is being molded, the rest of the part is also being molded by other punch motions. These motions are computer controlled for timing and position so that uniform parts are molded on every cycle. The relative timing of these motions is critical. If compaction begins in one level of the part and not in the adjacent level, powder particles will try to move into the next level. This "pre-compaction" results in shear cracks which can cause failure of the part. Figure 4.10 shows a micrograph illustrating green shear cracks.

Figure 4.11 is a spiral bevel gear with a hub and a hole that is stepped from square to round. This part requires 2 lower punches, an upper

Figure 4.9. Multilevel part requiring 2 lower punches and 2 upper punches.

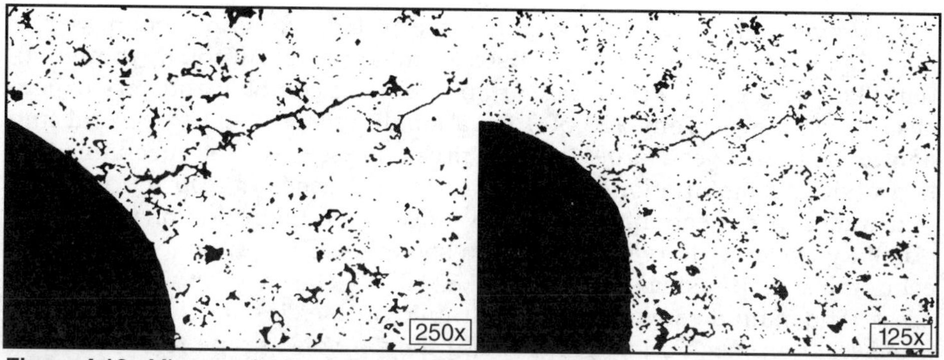

Figure 4.10. Micrograph showing shear cracks in green parts.

Figure 4.11 . Spiral bevel gear.

punch which rotates during molding and ejection and a stepped core rod. The lower outer punch has protrusions on the face which we call densifiers. These help to equalize part section thickness and density in the gear teeth.

Now that we have discussed some examples of parts and tooling, let us investigate some of the presses used to mold parts. Although they function like the Popcorn Press they are much more complex and very expensive. Priccs range from $20,000 for a small simple press to several million dollars for a very large complex press. Presses can be mechanical or hydraulic or combinations of the two. Most presses of 200 ton capacity and under, are mechanical and may have some hydraulic or pneumatic auxiliary systems. Larger presses are sometimes hydraulic for reasons of economics or physical size. Figure 4.12 is a large tonnage hydraulic press. The main disadvantage of hydraulic presses is that their cycle time is slower than mechanical presses and they are also prone to oil leaks.

Simple thin parts can be made on single action presses. Thin parts can be done this way because there is not enough die wall friction to substantially lower the transmitted pressure from the upper punch. Compaction in these presses is achieved by the upper punch only. The die is stationary and the part is ejected by a single lower punch. The lower punch is stationary during compaction and since all compaction is done by the upper punch, we call these presses single action. Examples of single action presses are Stokes T and R press models. Parts with one step on the outside need two molding motions. These can be achieved by using a

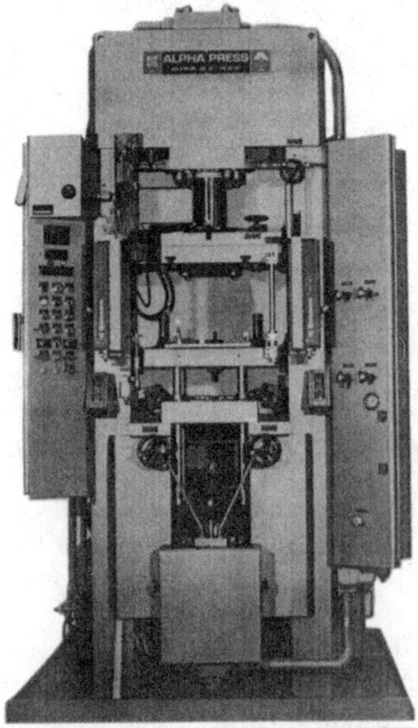

Figure 4.12. 200-ton hydraulic press.

double action press in which the die moves downward or the lower punch moves upward during compaction. [See Figure 4.13 for schematic of press actions and Figure 4.14 for photo of floating die press.]

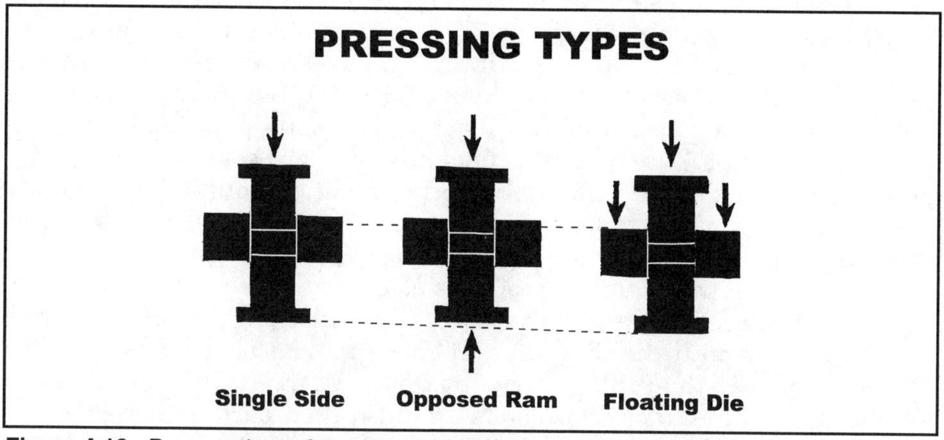

Figure 4.13. Press actions. Courtesy of Cincinnati Inc.

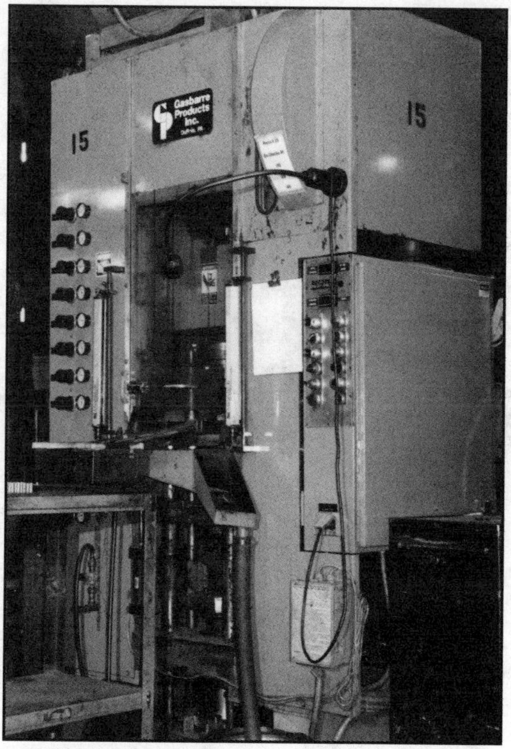

Figure 4.14. Floating die table press.

As we mentioned earlier, multilevel parts require multiple punches that can be independently controlled for fill, compaction and ejection. Multi-motion presses are much more expensive than single motion presses but are necessary when compacting complex parts. For example, a 100 ton floating die press which can make dual level parts very effectively costs about $170,000. A multi-motion 100 ton press with two die sets for quick tool set-up costs about $550,000 (see Figure 4.15). It is not possible to achieve uniform density in various part levels without positive control of fill and compaction, so if you are going to make complex parts, you need a substantial investment in equipment. Density that is not uniform from section to section can cause cracks, poor strength and size variation during sintering.

Ejection of the part from the tooling is as critical as the compaction step. Long thin walled parts and parts like gears which have a lot of surface area in contact with the die wall and core rod, can have ejection forces which are greater than the compacting force. We have experienced several tool failures due to ejection loads and also seen parts that were compacted by the lower punch during ejection because the die wall friction load was so high.

Figure 4.15a. 110-ton multi-motion press.

Figure 4.15b. Removable die set for multi-motion press.

Ejection from conventional presses is accomplished by the lower punch [or punches] moving the part up until it is out of the die. There are also withdrawal presses which move the die downward towards a stationary lower punch during compaction and continue that motion until the part is out of the die. In general these presses have additional lower punches and produce multi-level parts.

We mentioned in the beginning of this chapter that green parts have strength comparable to aspirin. This may be a bit simplistic but it gives you something to relate to. Green strength varies with material, particle shape and density. [See Figure 4.16 for some typical values.]

Handling of green parts is a very important part of the compaction process. We don't want to open the aspirin bottle and have a lot of chunks of tablet fall out because the bottle was mishandled. We certainly don't want P/M parts going into the sintering furnace with cracks or corners and edges chipped or missing. Therefore, parts have to be carefully loaded on trays or carrier plates for storage or conveyance through the furnace.

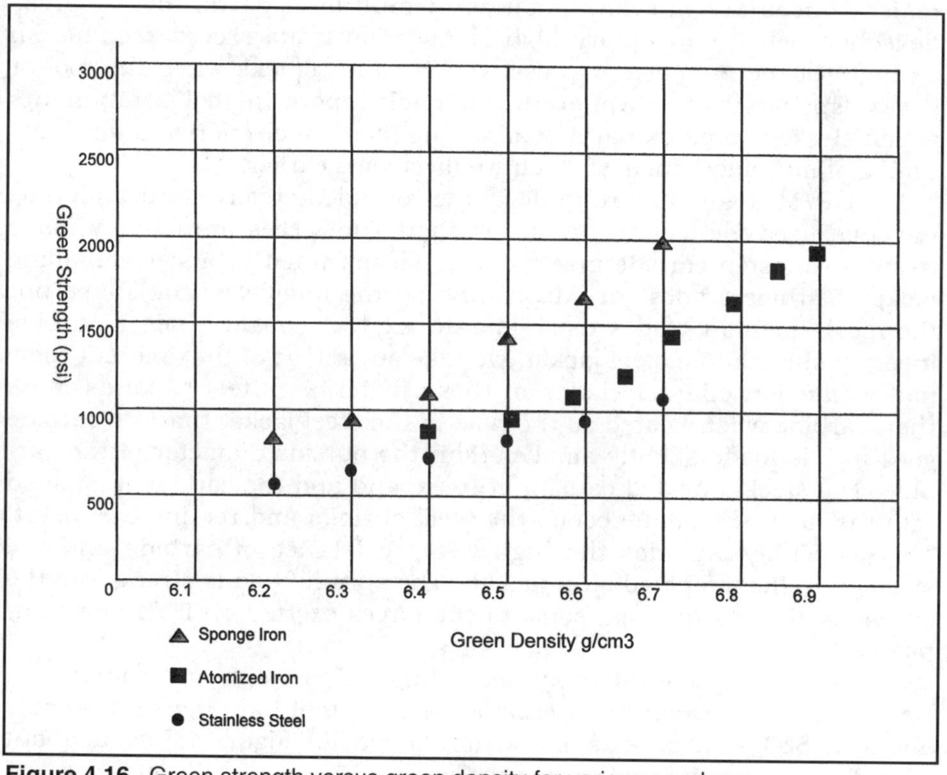

Figure 4.16. Green strength versus green density for various powders.

Once the green parts are clear of the die cavity, they are usually pushed off the lower punch by the filler shoe. They then gently slide down a chute where they are removed by an operator and are stored in preparation for sintering. Sometimes, very complex parts are compacted single cycle and the operator, human or otherwise, removes each part from the lower punch when the part is clear of the die cavity. This is usually necessary when the lower punch has steps or protrusions that prevent the part from sliding easily off the punch. Many parts have enough green strength to allow them to slide off the chute into a container without being handled by an operator. In any case, green parts can be stored for long periods of time without deterioration, but most are sintered as soon as possible to minimize work in process and production lead time.

Many opportunities exist for parts to develop cracks during the compacting cycle. Extensive research has been done on this subject at Worcester Polytechnic Institute and a text entitled "Handbook of the Common Cracks in Green P/M Compacts" has been published by MPIF that describes green crack locations and probable causes. Some of the major causes are uneven densities in multilevel parts; tool or press deflection [when momentary high compaction loads exceed the column strength of the punches or press] which cause cracks when the tool or press "springs back"; undercuts or back tapers in tool components which stress and crack parts by pinching them on ejection and improper timing of multiple punches which we discussed earlier.

Most P/M materials are quite abrasive and therefore hard materials must be used for tooling - even for short runs. Dies are usually made using a tungsten carbide insert that is shrink fitted in a steel retaining jacket. Carbide grades vary depending on the fragility of the shape but the most common grades used contain 12-13% cobalt. When a carbide insert is shrunk in a steel jacket, we take advantage of the different thermal expansion characteristics of these materials. Steel expands more than carbide when heated so the hole in the steel jacket that the carbide goes into is made slightly smaller than the outside diameter of the carbide. The steel is heated causing it to expand and the carbide is pushed into the hole. As the die cools, the steel shrinks and retains the insert. This assembly provides the high wear resistance of carbide and the strength of the tempered steel jacket . The shrink fit adds strength to the assembly helping to resist some of the forces exerted on it during compacting.

Tool costs vary considerably depending on part size and complexity. Here are some examples. A simple bushing tool like Figure 4.4 costs about $1,500. A larger bushing with a flange like Figure 4.5 costs about $2,600. Tools for the part shown in Figure 4.8 which has 5 punches and a splined core rod cost about $7,500. The cluster gear tool for the part in Figure 4.7 [gear O.D. 3in. and overall thickness 1½ in.] costs about $9,500.

When thin protrusions in the die are necessary, tool steel inserts in the form of keys or blades are usually fitted in the carbide die because carbide is too brittle to be used for very thin sections. Sometimes steel such as CPM-9V is used instead of carbide for the die insert when part quantities are low or the die shape has very thin sections.

Core rods can be tool steel or carbide depending on their size and shape as well as the part quantity. When steel is used for any tool component, periodic checking for wear is a necessity. Worn cores and dies will produce parts that are out of tolerance and can lead to scrap loss. Unlike the popcorn scrap, the P/M scrap is pretty hard on your teeth! Tool wear can also cause undercuts where the part is compacted and these result in green cracks because the parts are squeezed during ejection. See Figure 4.17.

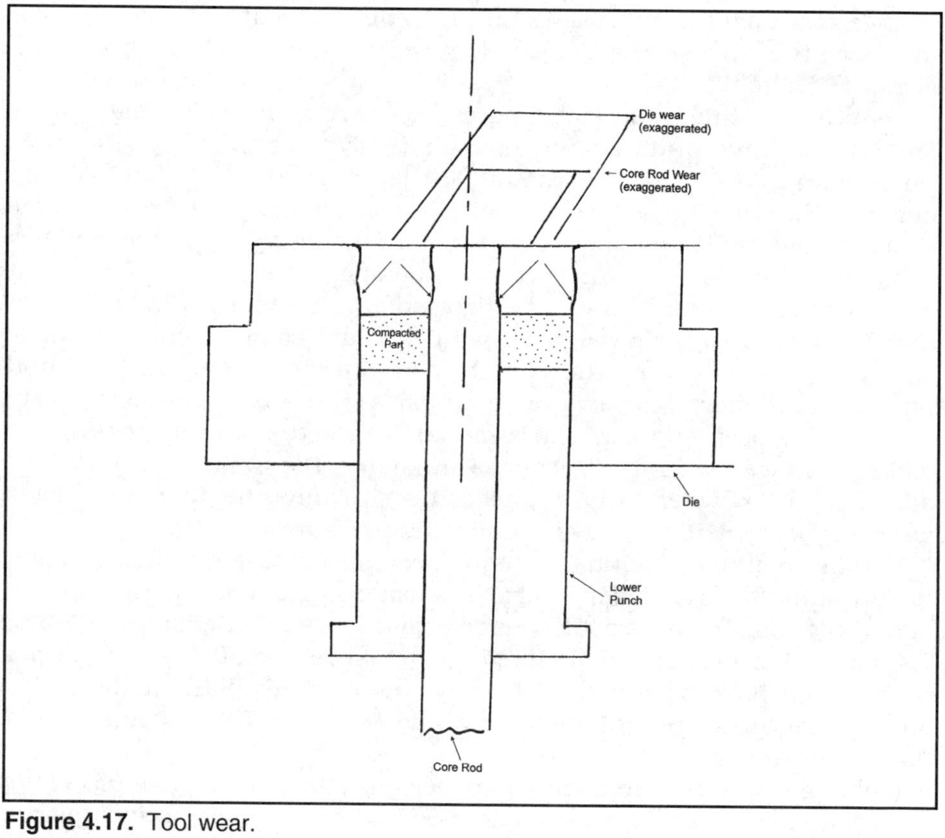

Figure 4.17. Tool wear.

Punches that have substantial cross section and no fragile projections are usually made from high-carbon, high-chrome tool steel. Lower alloy types are used if the punches are thin or have fragile shapes. Tool steel punches can usually produce at least 200,000 parts but they must be periodically sharpened or refaced because of wear on the faces and edges. Steel dies can also produce between 100,000 and 200,000 pieces especially when some of the newer alloys are used like CPM 10V. [Interestingly this material is a P/M product - sort of like pulling yourself up by your own bootstraps] Carbide dies and core rods can make millions of pieces and they usually break before they wear out. As we mentioned earlier, carbide is a brittle material and overloading or misalignment can easily cause breakage. Recent developments in coatings have improved the life of steel tool surfaces. These coatings are relatively inexpensive and can provide up to ten times the wear life of untreated tool steels. Coatings are vapor deposited in vacuum chambers and compositions like tungsten carbide with carbon and titanium carbonitride produce excellent results. The thickness of the coating is minute so size change of the tool member is not a problem. The coatings not only offer increased wear resistance but also lower ejection force due to lower coefficient of friction.

Tool components, both steel and carbide, are subjected to high stress for thousands or hundreds of thousands of cycles. Molding pressures range from 20,000 psi to over 100,000 psi depending on material and density. Figure 4.18 shows pressure versus density curves for some common materials. You should take note of the flattening of the curve as densities approach solid or pore free density. We should mention that one of the contributing factors to high molding pressure is the lubricant that is mixed with the powder. Without this lubricant, usually a metallic stearate or organic wax, the powder would score or gall the tools and production of more than two or three parts would be impossible. There has recently been success with a method of electrostatically coating the tooling surfaces with a lubricant. This allows the amount of lubricant mixed in the powder to be reduced dramatically and therefore yields much higher densities for a given molding pressure.

Another method which has recently been successful in reducing molding pressure for high density parts is warm compaction. The raw material and the tooling are heated to approximately 295°F. Densities of 7.3 to 7.4 gm/cubic centimeter at molding pressures of 50 to 55 tons per square inch have been achieved using this method. Without the 295°F and special lubricants, the density would be about 7.25 gm/cubic centimeter maximum.

Molding pressure also affects part design. The obvious goal of P/M is to provide near net shape as molded. Sometimes we have to incorporate

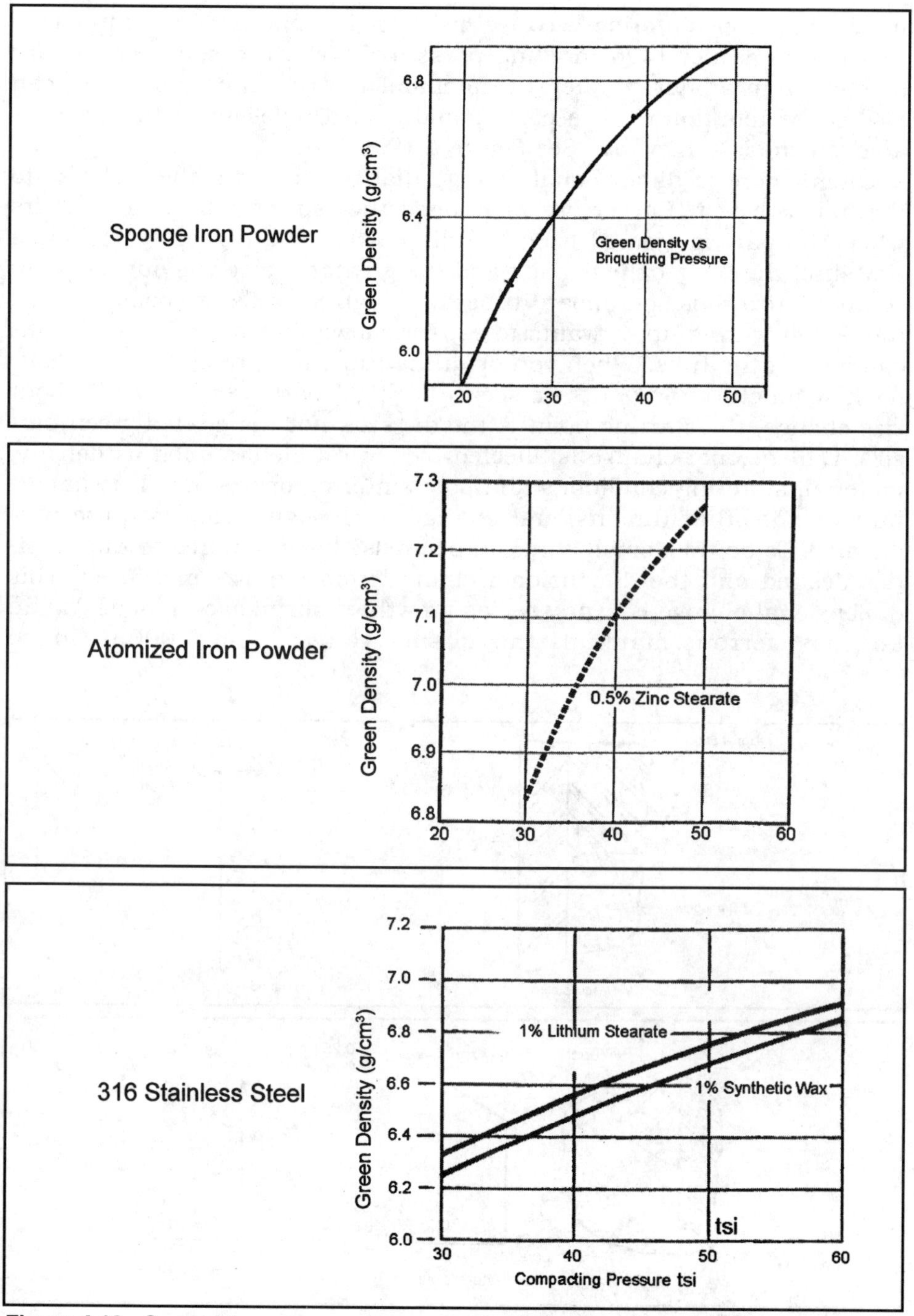

Figure 4.18. Compacting pressure versus green density.

machining after sintering because molding thin part sections can cause tool failure under high molding pressure. We will discuss secondary machining in a later chapter. Steep chamfers or radii on part edges can also cause tool failure. The parts usually need redesign so that reasonable tool life is achievable. See Figure 4.19.

Often, there is dimensional change that occurs from the tool size to the finished part. Earlier, we mentioned the "spring out" which occurs when the part is ejected from the die cavity. The higher the molding pressure, the higher the density and the greater the spring out. You can compare it to those cardboard tubes of dough for rolls or cookies. When the dough warms up, it wants to expand slowly but it can't because it's confined in the tube. When you open the tube, it "springs out". No matter how much "spring out" we see in the P/M part, when we talk about size change, the starting point is the die size, not the ejected green part size. Dimensional change is affected by several factors such as density, material, sintering conditions and post sintering operations. It is important to note that dimensional change during sintering is quite predictable. To do this, small bars can be pressed and sintered to the densities desired and the dimensional change from die size measured. This dimensional change is expressed as growth or shrinkage in inches\inch. For most ferrous materials this change ranges from 0.000in./in. to

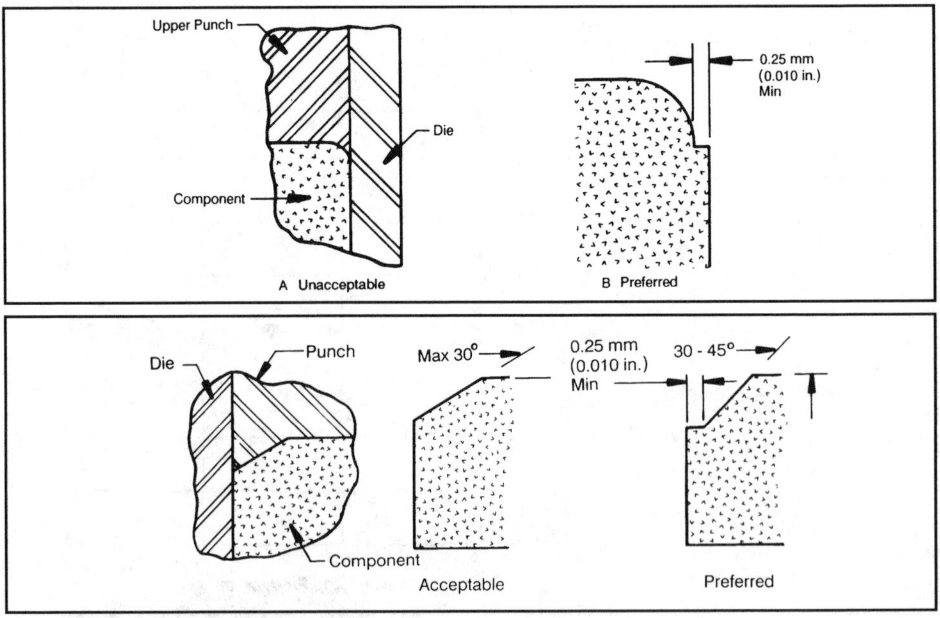

Figure 4.19. Part edge re-design for improved tool life.

0.003 in/in of growth. Some nonferrous materials such as 80/20 brass shrink about .016 in/in. When the tools are designed, these dimensional changes are factored in so that the production parts meet the print size specifications. Well, we have talked a bit about sintering in these last few paragraphs so you should go to the next chapter to learn about the "sins" of sintering!

CHAPTER 5

Sintering, as Seen Through the Eyes of a Metallurgist!

How is sintering defined? How does it work? Why does it work?

The American Society for Testing and Materials, ASTM, has a subcommittee, B09.01, that defines terms related to powder metallurgy. In their glossary, ASTM B 243-00, we learn: "3101 Sinter v. the bonding of adjacent surfaces of particles in a mass of powder, or in a compact, by heating"

Well, if that is perfectly clear, you can now go on to re-pressing in Chapter 6. However, a few more words of explanation may be in order.

It is a lot easier to tell you how to sinter, or to show you the results of sintering, than it is to answer the question of why the process of sintering bonds metal particles together. Were you really pleased with the explanation in Chapter 2 that the driving force for sintering is minimizing the surface area and surface tension of a heated heap of powder? At room temperature, a pile of iron powder just sits there (contentedly). Particles don't become bonded during your lifetime. Why should the particles bond together when we heat them up but don't actually melt them?

Let's start with, "Why do the atoms of iron in a bar of solid steel stick together? Why don't they just roll around like so many small marbles?" Well, the *metallic bond* holds them together! "Small marbles" is not a really good way to think about atoms. A better model would be a positively charged core, +, and a cloud of very small negative, -, electrons flying around the core (nucleus). Bring two atoms of iron near each other and the outer electrons can move about both cores. In fact, the two iron atoms will fairly fly at each other, consumed with the idea of sharing electrons, Figure 5.1. Technically, one says that the two atoms have moved into a lower energy condition (state). To separate them, we have to raise their energy level back up to what it was before they became joined. To raise their energy we have to do work on them, and, as every

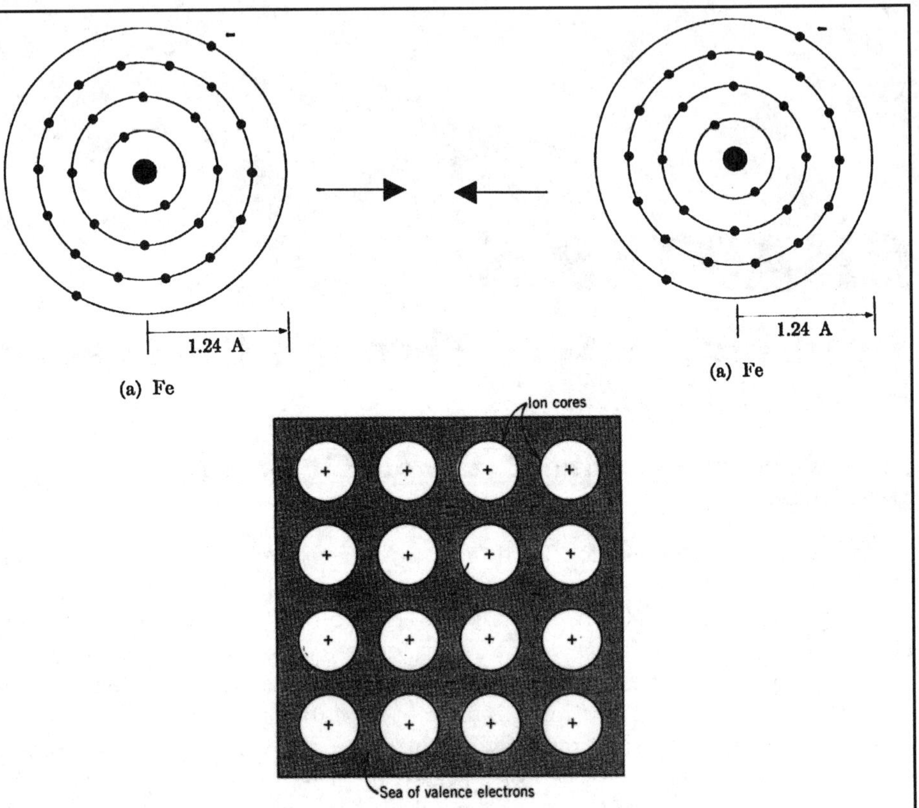

Schematic illustration of metallic bonding.

Figure 5.1. Attraction between the iron atoms so that they can share the outer electrons and form a metallic bond as shown.

good physicist knows, the only way to do work is to apply a force and move an object. Well this work is hard, and bars of iron are strong! Try pulling a bar of iron apart, and you will soon be a believer in the idea that it takes work to break metallic bonds.

When we tell you that the metallic bond is created by two atoms of iron sharing their electrons, we are not telling you why this happens. Physicists have models that let them calculate the strength of the bonding forces. The models describe *what happens,* but *not why it happens.* The question is like asking why the universe was created in such a way that positive electrical charges attract negative charges. We are now reduced to answering, "That's the way it is!"

Now if you really believe that two atoms of iron in very close proximity can form a strong bond, lets consider the loose pile of iron powder. We think the particles are only lightly touching each other, but, consider the plight of an atom trapped at the point of contact between two particles, Figure 5.2A. It is under a huge stress, (small force/very small area). Furthermore, as an atom sitting on the surface of an iron particle, its electrons don't have as much opportunity to circulate around nearby atoms as one in the middle of the particle. So, it is not as well bonded as those secure interior atoms. It wouldn't mind finding a more secure perch. At room temperature, our atom does not have much energy and it does not vibrate back and forth, any great distance. But, entering the sintering furnace, it heats up, and can vibrate greater distances. Spurred on by the force of the encroaching particle and now vibrating greater distances, our atom spies a place on the encroaching particle where he can better surround himself with other atoms. He extends his electron cloud, (a very short distance), happily lowering his energy level while sharing electrons with the atoms in both his new and old home. *Voila!* We now have a very small metallic bond between two particles, Figure 5.2b.

One atomic bond does not a very strong P/M part make. To continue on, atoms on the surfaces of the touching particles jump up and down and move around on the surface. If they are lucky enough to move toward the first bond and join it, they are gaining bonding on more sides, while lowering their own energy and lowering the combined surface area of the two particles, Figure 5.2c. Gradually, a full scale bond develops between the two particles, with about 40 billion atoms of Fe in one particle bonded to 40 billion on the other. While this bond is building, the empty pore space between the particles is changing from very sharp and angular (high surface area) to well rounded (low surface area), as shown in Figure 5.3. We take a shortcut in our thinking and say that sintering is driven by the high surface area of particles and the tendency at elevated temperature to minimize surface area. The longer way to remember what happens in sintering is to think about the surface atoms with their weak connections to underlying atoms, hunting around for lower energy sites, where they are more securely surrounded by other atoms. The obvious place to find such sites is where one particle is contacting another. If you go back to Chapter 2 and look at Figure 2.8, you will see how the atoms build the bonds between the particles.

Although it may not seem obvious, when particles form bonds by atomic movement over their surfaces, the center distances between the particles do not change. To prove it to yourself, make a little sketch with three or four particles in contact. Then move a few atoms over the surface toward the points of contact and develop some bonds. Keep moving the atoms and you go from sharp pores to rounded pores like Swiss cheese, Figure 5.3, (Swiss Cheese with only two holes!). But the distance

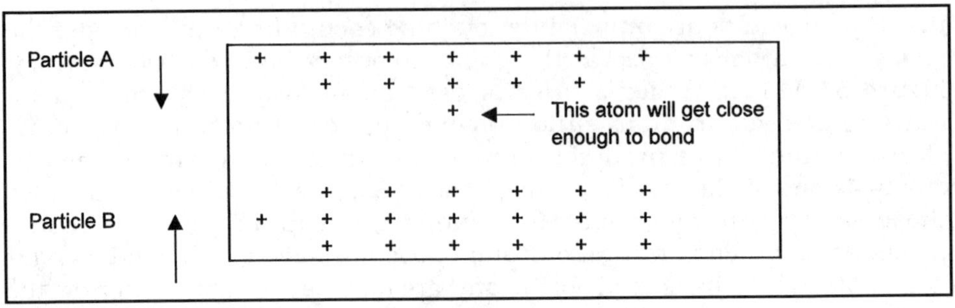

Figure 5.2a. One particle encroaches on another.

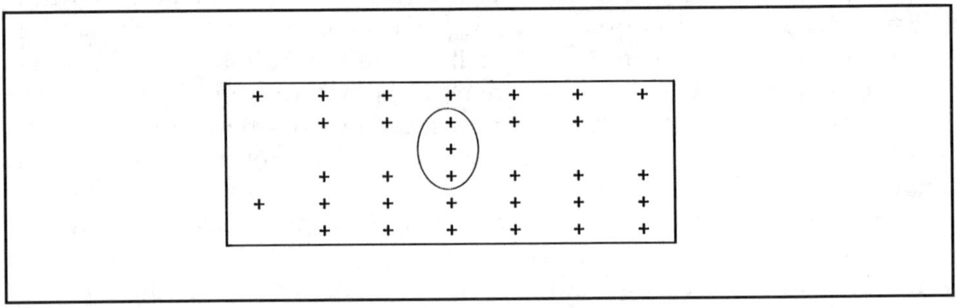

Figure 5.2b. Forming first metallic bond.

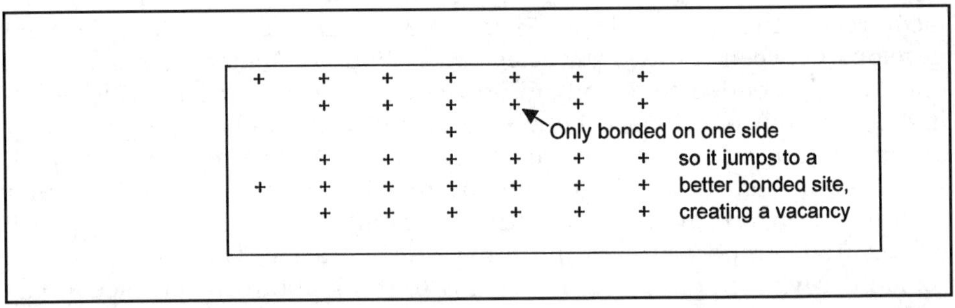

Figure 5.2c. Second atom read to join band.

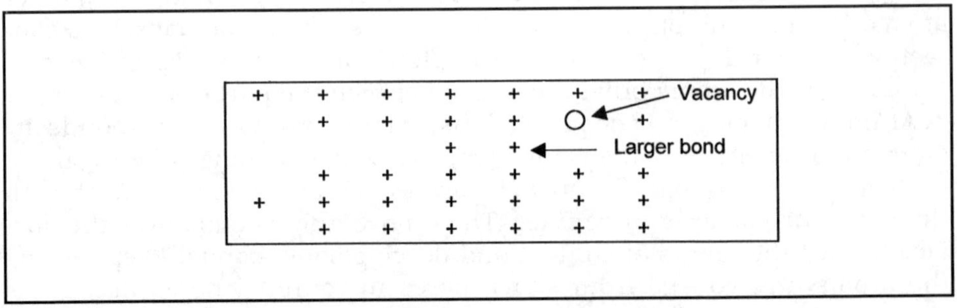

Figure 5.2d. Larger bond, vacancy created.

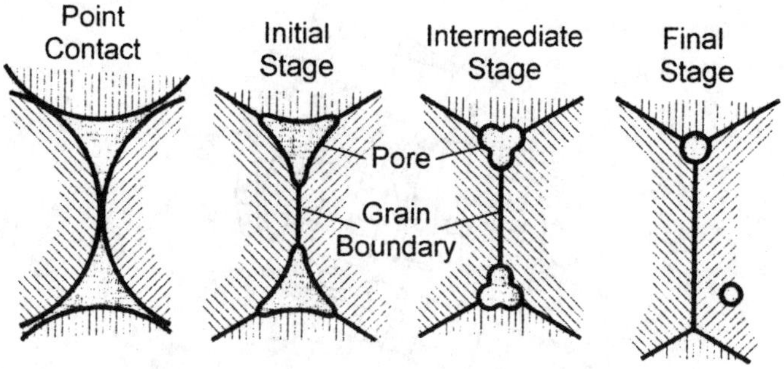

Figure 5.3. Particles in contact create very sharp angular pores. As atoms migrate over the surface to the points of contact in initial stage sintering, the pores become rounder and the boundary between the particles enlarges.

between the particles stays the same. Well, if this is true, why do our tool designers spend so much time making the tools larger or smaller than the final part? Parts must change size during sintering! The answer to this conundrum is that there is more than one way to build bonds between particles and to round out the porosity.

First, you have to acknowledge the existence of metal grains. You saw grain boundaries on the particle surfaces in Chapter 2, Figures 2.8 and 2.9. They aren't just painted on the surface by the electron microscope. The grain boundaries run into and throughout the particle. We suggest that you think of a metal grain as an orderly stacking of atoms in layers, like a deck of cards, Figure 5.4a and 5.4b. Picture the disorder and discomfort among those poor atoms where one grain intersects another, as in Figure 5.4b. The atoms are not as securely bonded as they would be in the middle of the grain. For iron, above 1600°F, the atoms are arranged in a cubic structure where each atom has six abutting neighbors. That's true for the inside of the grain, but not out at the edge where the grains push into each other. (For those with a geologic bent, grain boundaries or intersections are like a very small version of plate tectonics!) One can imagine that under the influence of the furnace heat, atoms might leave one grain and jump over to the other. Some grains would grow, others would shrink. But, some of the atoms might work their way along the grain boundary until they find the security of ... a small pore. What's to like about a small pore? See Figure 5.5 a and b. The bonding on the small pore's surface could be more secure than in the grain boundary. Put another way, the atom could lower its energy by coming to rest on the inside of the small pore.

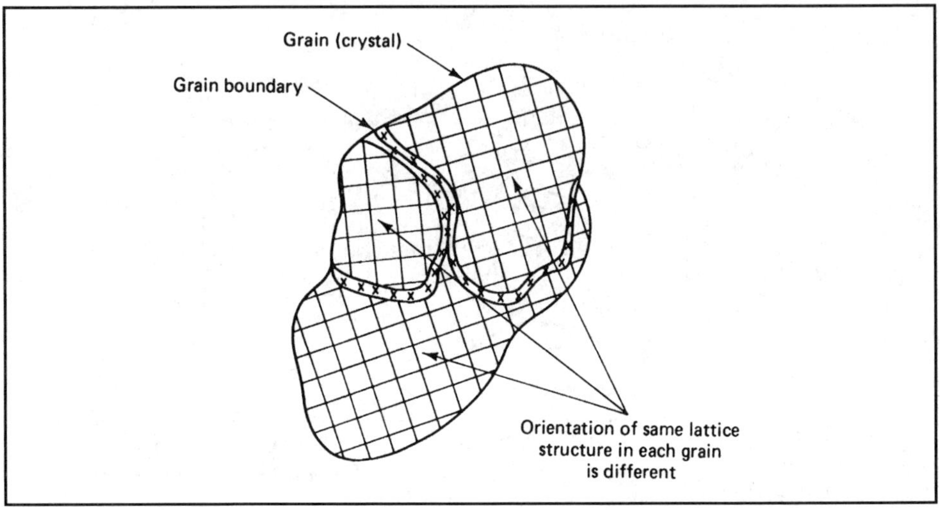

Figure 5.4a. Grain boundaries.

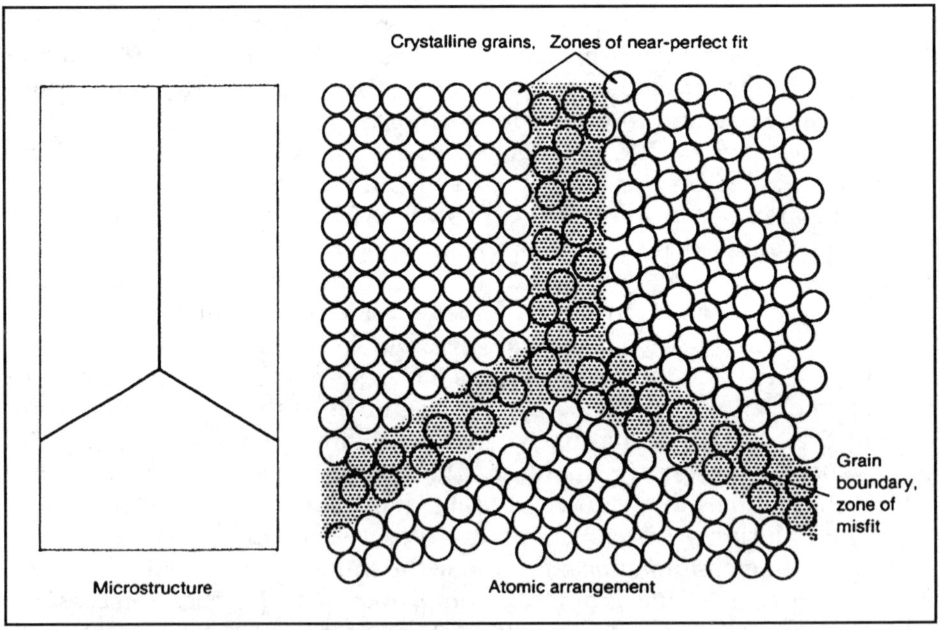

Figure 5.4b. Appearance and nature of grain boundaries.

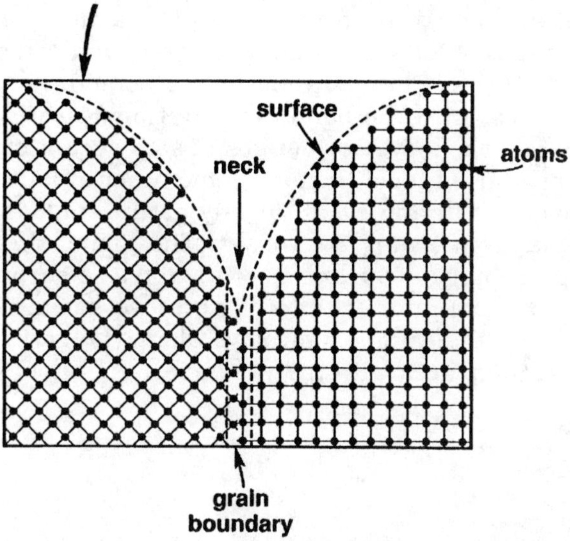

Figure 5.5a. Atoms in the irregularly bonded grain boundary can migrate to the pore surface and have more secure bonding, thus increasing the neck bond between particles and reducing the size of the pore. Courtesy of Metallurgy for the Non-Metallurgist, ASM International.

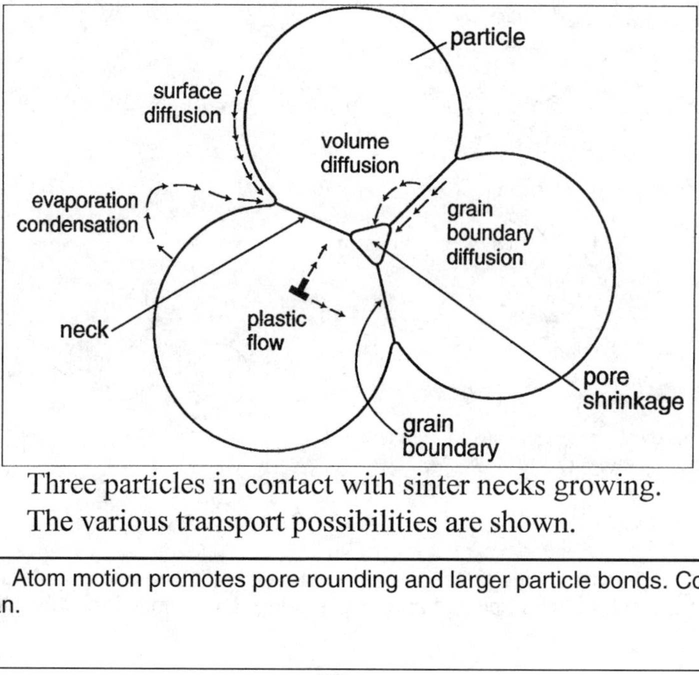

Three particles in contact with sinter necks growing.
The various transport possibilities are shown.

Figure 5.5b. Atom motion promotes pore rounding and larger particle bonds. Courtesy of Rand German.

To complete this story and to bring closure to the idea that a porous object could shrink, we need the concept of a "vacancy", Figure 5.6. In reality, our bar of solid steel is not perfectly solid when viewed from the atom's point of view. The vibrating atoms sometimes jump out of their secure little sites, with the bonds to the 8 or 6 nearest neighbors. This empty space is a vacancy. But, it doesn't stay vacant much longer than a parking place in downtown Boston. Some other atom, or the same atom, will jump back in. In any event, at any given temperature, a small percentage of the atomic sites are vacant. The higher the temperature, the higher the percentage of vacancies. You won't be surprised to learn that along the grain boundaries, there are more vacancies than average, Figure 5.5a, and that atoms can move along the grain boundaries, continually changing places with the vacancies. The atom goes one way, the

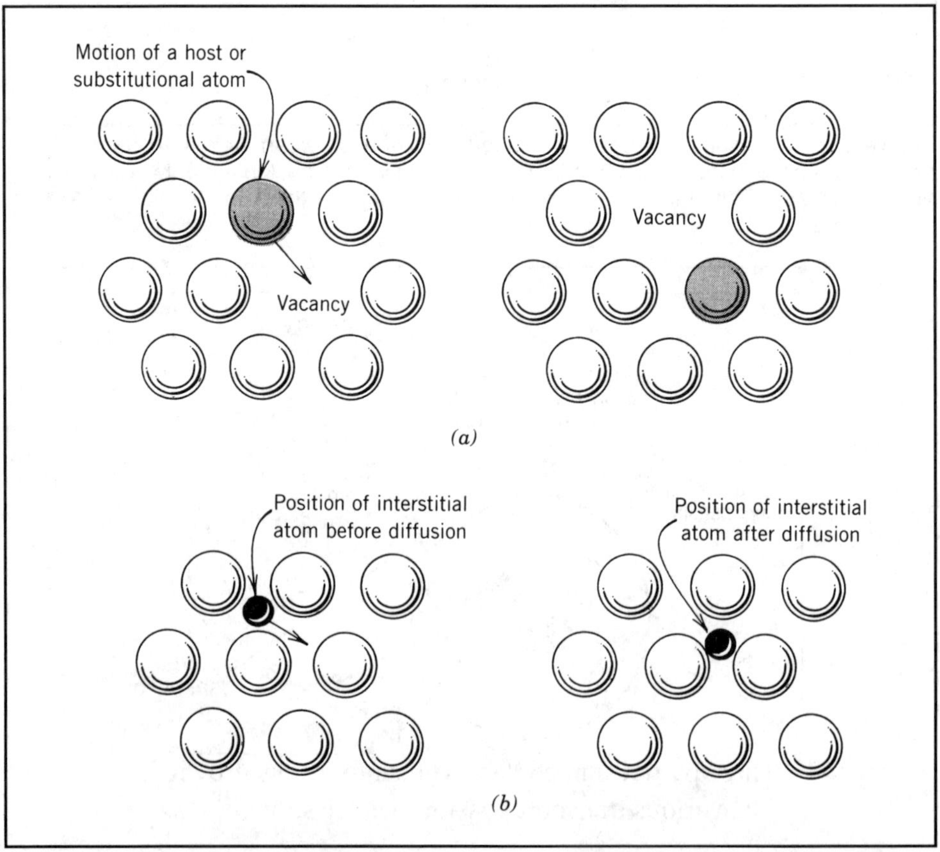

Figure 5.6. Schematic representations of (a) vacancy diffusion and (b) interstitial diffusion. Courtesy of Materials Science and Engineering, an introduction, John Wiley & Sons.

vacancy goes the other way. Remembering that a perch on the surface of a small pore provides better bonding than on a large pore, you can now picture what happens when a grain boundary connects a small pore to a large pore: atoms move to fill in the small pore and make it even smaller: it disappears. The vacancies move toward the larger pore. When you add a little extra space from a vacancy to the space in the pore, the pore grows a little bigger. What happens if the other end of the grain boundary is attached to the outside surface of the part? The vacancy reaches the world outside the part, and adds a little space to it, or, viewed the other way, the part gets a little smaller!

Well this shrinkage technique works OK as long as there are grain boundaries to act as channels. But the atoms know that they are not in their most secure state as long as there are those disorganized grain boundaries around. The atoms move across grain boundaries from one grain to the other, in such a way that some grains shrink and some grow, but the total area of grain boundaries goes down. Sinter long enough, and you have no grain boundaries, just one perfect single grain (crystal). (But nobody ever got rich sintering long enough to turn a P/M part into a fully dense single crystal!) Once the grain boundaries are gone, the process of pore elimination and densification slows down. How does densification go on at all? Vacancies. There are still a few vacancies around. Although it seems kind of improbable, atoms and vacancies interchange places in such a way that vacancies move out to the surface of the part while atoms move in. Ultimately, every last pore is filled in, the part has shrunk to the pore-free condition, and the surface area of the pores has been reduced to a minimum (zero).

When the pores are all gone, the atoms on the surface will eventually rearrange themselves to minimize the part's surface area and turn it into a sphere! Although that seems unlikely when you are sintering iron at the usual 2050°F, go back and look at the rounded edges on the part in Chapter 2, Figure 2.9, sintered at 2050°F. Then think about what happens when the metal melts, Figure 2.1, Chapter 2, ... little spheres! So, it should seem at least plausible that if you were pretty close to the melting point of the metal, that the atoms would be very mobile and the part could turn into a sphere in less than a person's lifetime! The practical problem with very high temperature sintering is that the steel is not very strong and the force of gravity acts on the part to cause it to slump or distort. (An effect noted by humans at the end of a long day!)

Up to now, we have been thinking about sintering as if the metal particles were all solid, and all of one kind, like pure iron. But only the soft magnetic folks like pure iron! We need to think about two things that happen in the real world. First, we need to strengthen our pure iron by adding in some other ingredients. Second, we need to consider what happens if one of these added ingredients causes some partial melting during the sintering process. Well, second things first: let's add some copper to the iron powder, say 2%, as in Figure 5.7a, and heat up the part to the normal sintering temperature of 2050°F. On the way up, copper melts at 1981°F. What do you think happens to the melted copper particles? You suspect that they form little spheres like Figure 2.1, Chapter 2 and just sit there? Wrong! They take note of all the iron atoms on the pore surfaces and observe that these surface atoms are not very well bonded and not in a low energy state. The copper atoms decide to share electrons with these superficial iron atoms, lowering everyone's energy level and making for greater stability. The copper atoms do this by spreading out over the surface of the iron particles, Figure 5.7b. When the liquid copper layer gets near the narrow region between two neighboring iron particles, it sees that it can do double duty, wetting the iron particle surfaces on both sides. This lowers everyone's energy level even faster. To the casual observer, it looks as though the copper is being sucked into the tight space between the iron particles. The tighter the space, the faster the copper moves into it. Tight spaces are called capillaries and the rushing in effect is called capillary action, which is effectively, attraction. At home, when you dip one end of a sponge in water, the water wicks its way along through the sponge, even going upward, in defiance of the law of gravity. Capillary forces are not trivial. Now every time you look at a wet sponge you won't be able to avoid thinking about how the water is rushing about lowering the energy of the atoms on the surface of whatever sponges are made out of.

Meanwhile, back at the iron- 2% copper part, the action continues. The 2% copper basically ends up in all the tight spaces between the iron particles. If the iron is atomized, with fairly solid particles, this in-rushing of copper forces some of the iron particles apart, Figure 5.7b, causing the whole part to grow, maybe even 1%. (Well, 1% is a lot when you are trying to control size change to .1% so you can sell the part.) The in-rushing of copper happens quickly, perhaps 1 to 3 minutes after the copper melts. We did not add the copper just so we could have problems with growth. The copper has some positive benefits. In those first 2 or 3 minutes, the copper sort of brazes or bonds the iron together, which strengthens it. If someone really cheats on the sintering time, you can see thin little ribbons of melted copper bonding the iron particles together, producing much better strength than if pure iron were sintered for such a short sintering time. (If someone really cheats on the sintering temperature and does not melt the copper, the copper particles just sit there in the pores, maintaining their original stoic, blocky or irregular shape.)

After a few minutes, the iron atoms on the surface of the particles notice that they are no longer breathing the furnace atmosphere, but are sitting on the bottom of a copper lake. They note that the copper atoms swim freely about in the liquid. Further, they see that when they jump off into the copper lake, a copper atom can take their place on the iron particle. Everyone has lowered his energy. The iron atom can move freely about, and find the sharp corner where its particle is gradually forming a bond to the adjacent particle, Figure 5.7c. Remember that before the copper came, this bonding process was going on sort of slowly, because the iron had to hop over its own surface. In the copper lake, it can easily move to the point of bonding and lower its energy by exiting the lake and increasing the bond area. Put simply, the liquid copper dramatically speeds up the bonding of the iron particles.

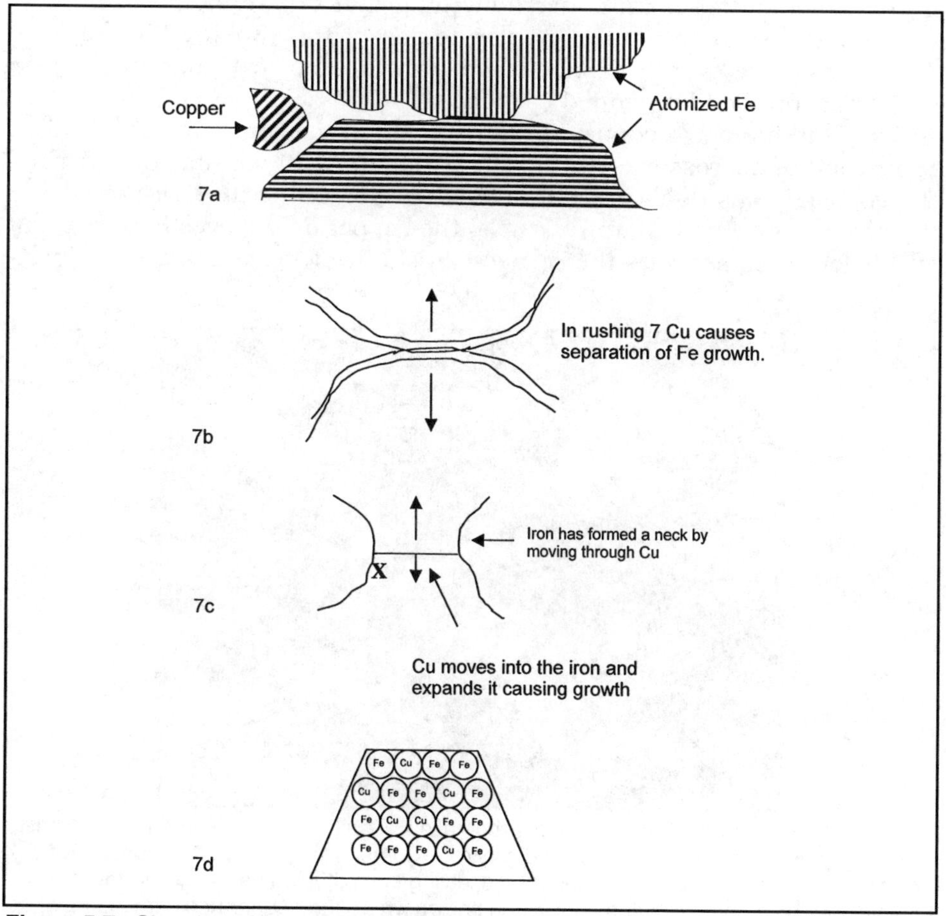

Figure 5.7. Sintering of iron and copper.

Now while all this bonding is going on, the copper lake begins to disappear or dry up. The copper atoms move off into the iron particles, substituting themselves for the occasional iron atom. (Substitutional solid solution! see Figure 5.7d) After about 5 or 10 minutes, the copper lake is gone and the 2% copper is securely residing inside the iron particles. Figure 5.7c. (Transient liquid phase sintering!) However, the assembly of iron particles (the part) is now somewhat larger, because the copper atoms that are inside the iron are still taking up space. You might imagine that the iron particles accommodate the copper visitors simply by taking a little more space from the pores. This is surely true to some extent. But those liquid copper atoms that rushed in and initially separated the iron particles ended up in a particularly good vantage point to cause the two particles to further separate. Copper atoms that enter the iron at the bonding point cause a swelling that separates the iron particles because there is no nearby pore to accept the swelling, Figure 5.7c. This causes the overall expansion or growth seen in the atomized iron with 2% copper and no added carbon.

Once the atoms of copper dissolve in the iron, you can't see reddish copper in the microscope. But, in the usual 20 to 30 minutes of sintering, the copper atoms don't move from the surface of the iron all the way to the center. For a typical iron particle, the copper only moves in about half way before the part exits the furnace. If you are fond of doing metallurgi-

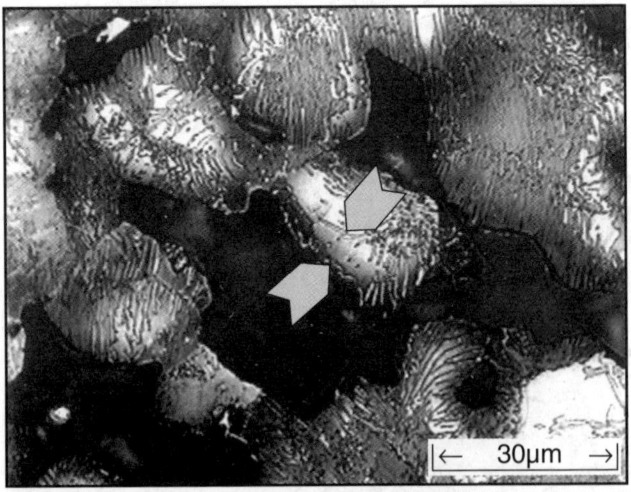

Figure 5.8. An iron 2% Copper, .75% carbon alloy. The white area, lower, right is ferrite, pure iron, in which very little copper is dissolved. Some ferrite, adjacent to pores, see arrows, is stained gray or tan, by copper dissolving and diffusing part way into the particle. The gray color is called coring and shows where the copper has diffused in. For reference, the 30 µm scale is the thickness of a human hair. (≈800x)

cal things (and who isn't) you can polish a cross section of a piece of sintered iron copper, etch it heavily with a mixture of alcohol and 2% Nitric acid (Nital), look at it at 200 to 1000X in the microscope, and see that there is a definite layer on the surface of the iron particles that marks how far the copper went in. (coring), Figure 5.8. Well you now know that the copper helps the bonding go faster and that copper dissolves to some degree in iron. So what? Well, if you are looking at sponge irons at low density, say 6.1 g/cm^3, that 2% copper increased the tensile strength by 33%. "A pretty good bang for the buck", as they say. (To measure tensile strength, we make a small bar out of the iron copper mix, sinter it, measure its cross sectional area, (.23 in. x .25 in.) and pull on it until it breaks apart, 1,460 lbs. To calculate the tensile strength, divide the breaking force by the area, (1,460 lbs.)/(.23 in. x .25 in.) and get 26,000 lbs./in^2. Think of the 26,000 lbs./in.^{2} this way: if you had a bar of this iron - 2% copper mix with a cross section of 1 square inch, you (or something) could lift a 26,000 lb weight (like 10 small cars.)

While you were making the 1 square inch bar of sponge iron you cleverly noticed that it did not grow with the copper addition, the way it did with atomized iron. Why was that? It's like Thomas' English Muffins. The sponge iron particles themselves are filled with many small internal pores. The copper lakes go rushing into these pores impelled by the same capillary forces that urge them into the tight spaces between the iron particles. The copper that rushes into a pore, gives up the chance to separate two particles by in-rushing between them. Also, when the iron starts to expand because of the copper dissolving in it, it simply expands into the pore, rather than causing the overall expansion that occurs when the local growth is in the bond or joint between two particles.

Why did we find the 33% strength increase in the iron when we dissolved the 2% copper in the iron. Two reasons: One: the copper helped build bigger bonds between the iron particles. And secondly, because copper dissolved in iron strengthens the iron itself: (solid solution hardening). To explain this effect, we must understand how the plain iron itself deforms and breaks. A familiar small bar of iron is a paper clip. If you just pull on it with your fingers, you don't see much happening. But in fact, you are separating the atoms by a small amount, (a *very* small amount!). Release the clip, and the atoms spring back to their original position. This is elasticity on a very small scale. Not as easy to see, as doing the same thing to a rubber band, but the effect is still elastic. Next, you give into the urge to bend the paper clip until it is opened into a more or less straight piece of wire. You carefully note that it no longer bounces back to its original shape, but it is still not broken. Why didn't it bounce back elastically? When you did the original stressing, you moved the atoms a little bit away from their normal position, and when you released, they sank back to their normal position. When you unwound the paper clip, you pushed part of a plane of atoms so far that it jumped a whole atom position to the next stable point and rebonded to a new group of atoms. You were able to do this with just your fingers, because it was a very small area where this jump occurred, as shown in Figure 5.9. (Screw Dislocation, Barrett Structure of Metals, p396.) In fact, unwinding the clip created millions of these local slips of one atom plane over another. Add up enough small slips and it equals significant deformation or movement of the surface of the metal. Deformation that you can see. So, in real metals, we don't slide one whole plane of atoms over another, all at once (holus-bolus), but we do it in the little steps or dislocations shown in Figure 5.9. Now if you believe this idea of how metals stretch and deform, then you can easily picture the effect of a few copper atoms occupying sites that were formerly occupied by iron. The copper atom is a little different size than iron, and sets up local stresses around it, as it bonds to the iron. These stresses interfere with the little slip motions of Figure 5.9 and so we perceive that the Fe with 2% copper is stronger. That's all there is to it.

So if iron with 2% copper is good, why not just add more copper and make it even better? Well, all things in moderation. If we keep adding more and more copper, we find out that at 2050°F, 5 or 6% is all the copper that will dissolve. The rest just hangs around as copper rich lakes with a little iron dissolved in them. The other interesting thing is that as you cool down from the sintering temperature, not all that

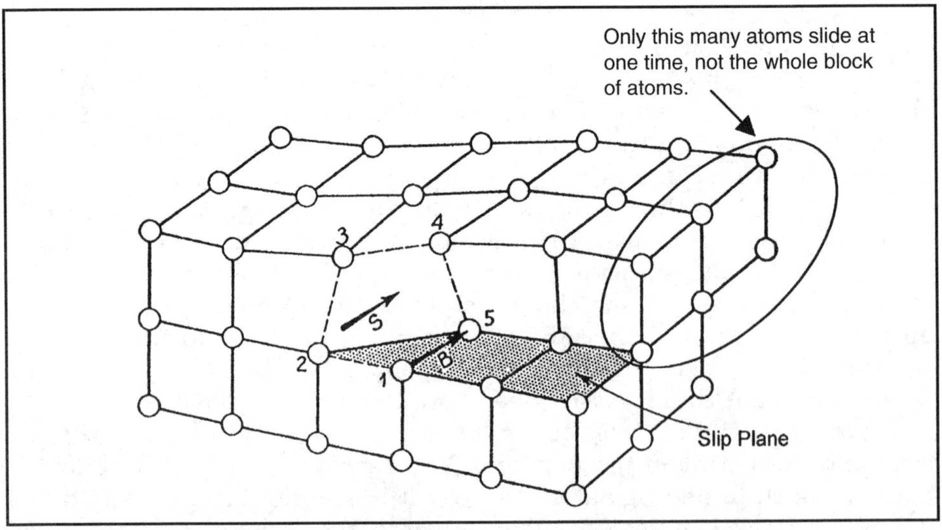

Figure 5.9. Screw dislocation at atoms 1, 2, 3, 4, 5, the bonds are stretched and under high tension.

copper wants to stay dissolved in iron. Some of it comes out as little copper-rich particles which are usually too small to see with the average microscope. Metallurgists like to summarize or collect data on things like how much copper can you dissolve in iron at a given temperature, if you wait long enough for it to happen (i.e. reach equilibrium). The information is stored in phase diagrams like the one in Figure 5.10 which shows the iron-rich side of the iron-copper system. From it, you can see what to expect if you heat up any given iron copper mixture to a particular temperature and hold it for a long time at temperature, i.e., reach equilibrium and stability.

If you were continuing to bend the paper clip in your fingers while reading this long dissertation you may be wondering why it finally broke in two. (In my younger days I used to think it was because of the heat generated.) You have already seen that one iron particle can have many grains. The paper clip has millions of grains, too. As a grain deforms and the dislocated area moves as in Figure 5.9, it can run into another dislocation or into another grain that can only deform in a different direction. The dislocations pile up on each other and don't just move cleanly out onto the surface as a small step. It gets harder and harder to move the dislocated areas through the other dislocated areas in the metal (work

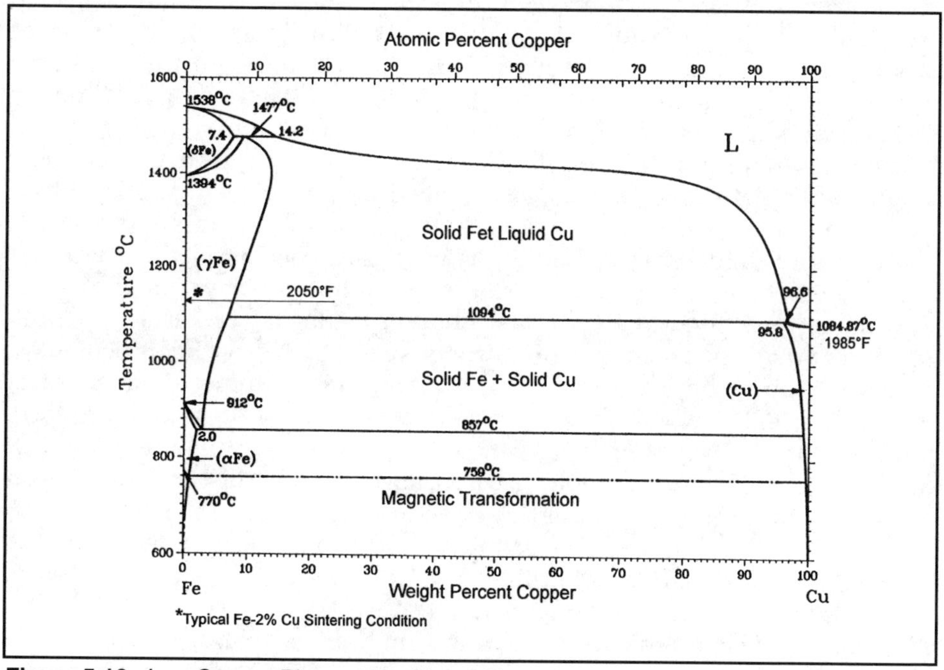

Figure 5.10. Iron-Copper Binary Alloy Phase Diagrams.

hardening). The paper clip seems to be getting stronger. Sooner or later there will be no easy places for the grains to slip and deform, and they will respond to your bending by splitting apart; opening up a crack that then moves through the metal to fracture it. Having produced this agonizing outcome for the clip, we trust you will never do it again!

Well, now you have a pretty good fundamental idea of what happens when you heat up two metal particles and sinter them together. Also you have been forced to wade through a little more basic metallurgy than you really wanted; don't worry, it is rarely fatal! You probably thought we would never get to how real people make sintering occur. Read on!

Sintering: How to Do It

Atmospheres

For folks who aren't involved with the heat treatment of metals, the most familiar source of high heat may be an oven for firing ceramics or the interior of a charcoal fire. The ceramics oven or kiln is certainly hot enough to sinter metals, and it indeed does do a good job of sintering or bonding the ceramic particles together. The problem for metals is the presence of air. The oxygen in the air does not affect the ceramic particles, because they are usually unreactive oxides or nitrides anyway, and will not pick up any more oxygen on their surfaces. They happily bond together, pretty much following our earlier description of how surface atoms move around. But putting iron in the kiln in air at 2050°F has a different result. Starting at about 400°F, the air reacts with the iron to form a thin shell of iron oxide. The iron atoms don't move easily through this shell. In principle though, we could sinter the iron oxide shells together and get something that would be iron particles surrounded by a nice brittle network of iron oxide. It would be about as strong as the nicely glazed fired ceramic that your aunt gave you last year. (Ceramics are brittle because they lack the neat deformation mechanism of dislocation motion we showed in the last chapter.) Just so you will not think that this business of sintering in air is just another airy theoretical exercise, remember the Metal Injection Molding folks. Some of them inject mixtures of metal powders and thermoplastic and wax into a die to form a useful shape. They then burn out some of these organic binders in air, at 400°F, and temporarily hold the iron particles together with some iron oxide. Of course they get rid of the oxide in a later sintering step.

Well if you can't sinter iron in air in a ceramic kiln, what about the nice hot fire on the charcoal grill? We've never actually tried it, but there is some ancient history to suggest it might work. Between 1300 B.C. and 200 B.C. the Egyptians were heating lumps of iron ore (iron oxide) in charcoal fires and then beating the hot metal into shapes like knives, axes and hoes. The space around the glowing charcoal was not filled with air, but rather it contained carbon monoxide (CO), carbon dioxide (CO_2), some water vapor, some nitrogen and some oxygen. This mix of gases was an early and primitive form of *atmosphere.* The oxygen in the air was reacting with (burning) the charcoal and creating the carbon monoxide and carbon dioxide and water vapor. With sufficient CO present, the CO reacts with the iron oxide in the lump of iron ore. That is, the oxygen in the ore reacts with the CO to form carbon dioxide (CO_2), leaving fairly pure iron in the form of a spongy lump. Those lumps are sort of like the ones that Hoeganaes/Hoganas make from the Swedish iron ore and then grind up to form sponge iron powder. That is why many authors point to this early Egyptian work as an early form of powder metallurgy. But, the Egyptians never actually ground up the lumps to make iron powder. What they did was take two hot lumps out of the fire and hammer them together, squeezing out the porosity, and bonding the two lumps together. It was like an early form of blacksmithing. When the lumps were hot and soft, they could be more easily beaten into the form of a knife or an axe. (This historical excursion reminds me of asking my sons what happened, only to get the response "First the Earth cooled, then the dinosaurs came ...")

We hope that by now you are convinced that if we are going to sinter iron or bronze at a high temperature, we are going to have to protect the metal from reacting with the air in the room. To this end, we create and maintain specialized atmospheres (gases) in the furnaces. Nowadays, the common furnace atmosphere for sintering iron is nitrogen + 10% hydrogen. Nitrogen, as N_2, as molecules, does not react with the iron. It is basically inert. In principle you could use 100% N_2 as the furnace atmosphere. But as nice as the modern iron powders are, they still have about .1% oxygen on and in them. And that oxygen can get in the way of the iron atoms that are trying to migrate over the particle surfaces and form bonds between particles. The 10% (by volume) of hydrogen combines with these oxides and carries them away as water vapor, H_2O. The squeaky clean particle surfaces are then an easy path for the migration of the surface atoms of Fe. It would be nice if the water vapor just up and left the furnace. But, if there is enough water vapor around, it can react with an iron and carbon (graphite) mixture to form CO or CO_2, reducing the amount of carbon in the iron. The iron needs about .8% carbon for good hardness. If the water vapor removes the carbon from the surface of the iron part, the part will be too soft. (Metallurgists like to call this a decarburized surface) To every problem, there is a solution (we hope). If one adds about .5% of natural gas (methane, CH_4) to the atmosphere, the water vapor will combine with the carbon in the methane to form CO or CO_2, leaving the carbon in the iron unattacked.

You should be getting the idea that furnace atmospheres are crucial to sintering and must be carefully chosen and controlled. There is one more component in iron parts that must be considered: the lubricant. Most P/M parts have about .75% of organic waxes in them. The lubricant allows the part to be compacted without sticking to the die. The lubricant is on all the particle surfaces and must be cleanly removed if the iron atoms are ever to have a chance to migrate and form good bonds. As the part heats up, lubricants like Acrawax melt, then evaporate and break down into smaller molecules between 475° and 1100°F. Waxes are made up of long molecules with lots of carbon and hydrogen atoms. When all goes well, they evaporate from the part, enter the furnace atmosphere, breakdown into the smaller molecules and are carried along by the flowing atmosphere to the front entrance door of the furnace. There, they are burned in air to form water and carbon dioxide. When all does not go well, the lubricant gases break down in the furnace between 900° and 1300°F to form carbon soot and hydrogen. The carbon soot can form on the part surface (a nuisance) or inside the part (a disaster). When soot forms inside the part, there is physically no room for it, and it separates the particles, leaving a weak blistered surface. It seems that there can be a local build-up of organic vapor around the part. This can physically restrain subsequent lubricant vapor from evaporating, or it can intrude into the part surface, building up the organic vapor content between the particles of iron. At a critical concentration, the organic vapor cracks to form the soot between the particles. To combat this, the atmosphere in the area of the furnace where the lubricant comes out of the parts is enriched with gases that oxidize (burn) the lubricant vapors and reduce their concentration. What material is used to oxidize? Air, carbon dioxide and water vapor are what are commonly added to the atmosphere, but only in the furnace region where the lubricant is evaporating. If the lubricant molecules are converted to carbon dioxide and water vapor by this "burning" process, they can not break down to form soot.

There are a few other aspects of the sooting/blistering problem that are worth mentioning. The formation of the soot only occurs between 900° and 1300°F and it takes some time to occur. An elegant solution to the problem is to use an accelerated lubricant burnoff zone. If this zone is heated by burning gases, there will be lots of oxidants like water vapor and carbon dioxide to combine with (burn) the evaporating lubricant. The rapid heating minimizes the time available for the sooting reaction to occur. Some factors that cause sooting are the presence of admixed nickel, or zinc from zinc stearate, or a high surface area powder or an end-of-the week furnace with carbon deposits on its innards. All of these materials can cause or catalyze the sooting reaction.

The nitrogen base atmospheres have been increasing in usage since about 1975. The nitrogen gets delivered as refrigerated liquid and stored in special tanks. The nitrogen boils off and the associated gas pressure drives it through pipes to the furnaces. Hydrogen has lately been delivered in liquid form to large scale users and is handled like the nitrogen, being mixed in at about 10% by volume. Smaller users make their own hydrogen by breaking up (dissociating) liquid ammonia (NH_3). This is done by heating the NH_3 to 1850°F in a chamber with a nickel catalyst present. The ammonia breaks down to form 75% H_2 and 25% N_2. This dissociated ammonia (DA) is then added to the nitrogen going into the furnace.

Prior to 1975, folks generated their own furnace atmospheres. They would burn natural gas or propane with some air. We are used to the idea that burning natural gas with air gives heat. It only does that if you give it enough air. If you hold back some of the air, the burning or reaction does not occur unless you supply extra heat to it. This kind of reaction that requires heat to be added is called endothermic and the atmosphere is familiarly called "endogas". It contains 40% hydrogen, 40% nitrogen and 20% carbon monoxide. There are small but very important amounts of water vapor and carbon dioxide present. By now you probably realize that if we don't keep the water vapor and carbon dioxide at low levels, they will either oxidize the iron surfaces or steal the carbon. What is complicated is that the maximum amount of water vapor that is permitted in the furnace depends on the temperature. For instance, at 2050°F, .7% water vapor would gradually reduce the carbon in the iron until it reached only about .2% C. You would say, oops, too much water vapor to make a .8%C steel! But wait! As the parts cool to around 1500°F, the .7% water vapor is in equilibrium with .8% C so the carbon leaves the furnace atmosphere and moves back into the iron. With a little luck, what you lost at 2050°F gets restored at 1500°F! This is what we meant when we said that the correct amount of water vapor depended on the furnace temperature. Were you wondering where the restoring carbon came from, when we were mainly talking about water vapor? It came from the 20 volume % carbon monoxide, CO. The reaction is such that the CO breaks down on the surface of the iron part and forms carbon, which then moves into the part. It also forms carbon dioxide, CO_2 that stays in the furnace atmosphere. There is a nice balance or equilibrium in the furnace atmosphere among the gases: H_2, H_2O, CO, CO_2, and carbon in the iron. That is the stuff of thermodynamics. It is used to predict how much water vapor will be in equilibrium with the C in the iron, and you will be pleased to learn that you will have to look elsewhere to find all the equations! (References like Volume 7 of the ASM Handbook and Prof. German's *Powder Metallurgy of Iron and Steel*)

If you go out in the shop and ask the sintering foreman if he runs his endogas with .7% by weight of water vapor, he may not know, because he measures the water vapor content in another way. He uses the term "dewpoint." You know this term because your local weatherman, at least in New England, always tells you the dewpoint. You remember well from last summer, a dew point of 68°F is what they understatedly called "oppressive humidity." You probably realize that in the evening, when the air cools down to 68°F, dew will condense out of the air . Many P/M houses use a "dewpointer" to measure the water vapor content of the furnace gas. Basically, they cool some furnace gas and look for the temperature at which dew forms. Your basic summer dewpoint of 68°F corresponds to 2.3% by weight of water in the air. When that clear Canadian air moves in and drops the dewpoint to 50°F you feel better because the air now contains only 1.2% by weight of water vapor, and you can more easily evaporate your own excess moisture! Water vapor content falls off very fast with dewpoint: the .7 weight % water vapor in endogas corresponds to a dewpoint of about +40°F. Furnace atmospheres based on liquid nitrogen often have dewpoints of -30 to -40°F with only .02 to .01 weight % water vapor. With so little water vapor in the nitrogen based systems, there is not much floating around to react with the carbon in the parts, and very little carbon is lost. The parts do not go through the carbon loss and restoration that has to occur with the endogas.

All this talk about furnace atmospheres may seem like a lot of trouble, just to avoid contacting air during sintering. What happens if we put the parts in a tight box and pump out all the air? We will have invented a vacuum furnace! But there are still a few details. What happens when the Acrawax lubricant evaporates into the vacuum furnace? Historically, terrible things. It coats the inside of the furnace and gums up the pump which was creating the vacuum. However, the folks who sinter tungsten carbide, whose green parts contain copious amounts of wax, have learned to install cold traps to catch the waxes before they get to the pump. The same logic has been employed in MIM parts to catch all their funny binders, or, their binders may be reacted with low pressure hydrogen gas. Without these specialized vacuum furnaces, it is necessary to burn out the binder in a separate step before putting the parts in the vacuum furnace. If there is no atmosphere in the vacuum furnace, how will it get rid of that small amount of oxygen which coats the surfaces of even the best powders? Sometimes the oxides just break down to form metal and oxygen, and the oxygen gets pumped away. Also, many parts contain a little or a lot of carbon. The carbon can react with the surface oxides to form CO or CO_2 which then gets pumped away. The result is a clean surface over which the atoms of Fe can easily migrate and form strong particle bonds.

Do the metals ever evaporate in vacuum? Yes. Try Fe with 2% Cu at 2050°F and look for the nice red copper color on the inside of the vacuum chamber. We try to prevent this by not pumping out all the air. We remove about 99.94% of the air and then replace it with an inert gas like nitrogen or argon. The modest pressure of this backfilling gas suppresses the tendency for the metal to evaporate. Does this tendency for things to boil and evaporate at low pressure have any common kitchen analogs? It does if you live in Colorado. At high altitudes, there is less air pressure on heated water, and it boils (i.e., water vapor escapes) at a lower temperature than good old 212°F. So, either you put up with this low temperature, slower-cooking, boiling water, or you carry the pot down off Pike's Peak to a point where there is decent air pressure and water boils at 212°F.

This issue is especially important with stainless steels that have a distressing tendency to evaporate chromium into a vacuum at 2400°F. Low chromium = low corrosion resistance. The answer is two fold: first backfill with 1/1520 th of the normal atmospheric pressure with argon. (That pressure is familiarly known as 500 microns or less familiarly as 66 Pascals.) Since that back pressure does not completely suppress the loss of chromium, follow OMG Americas' Professor Klar (now retired) and Dr. Samal, who offers advice later in this chapter, and hold at 2100°F for a few minutes, on the way down in temperature. This will allow some chromium from inside the part to migrate out to the surface and restore what little chromium was lost.

Sintering Furnaces

You are probably wondering how anyone could write ⅔ of a chapter about sintering and give you no better idea of what a sintering furnace looks like than the kiln at the local ceramic studio. (It probably comes from one us (LP) having the distressing tendency to view all of P/M through a microscope.) From what you have learned so far, you can imagine that a furnace must have a controlled atmosphere to keep out air, and it must be divided up so as to first take out the lubricant from the part, then remove the surface oxides on the particles, continue by heating the part enough for the particles to bond together, not do violence to the carbon content, and then cool to room temperature so that the part comes out nice and shiny. In your dreams, you could see the parts coming off the compacting press onto a conveyor that leads them into the furnace. They exit the furnace and go into a box for shipment. For some folks, that is reality.

Finally, Figure 5.11 shows you sketches of the six usual types of sintering furnaces. The mesh belt conveyor has a coarsely woven stainless steel belt, often 314 stainless steel. The parts can sit on the belt or on wire trays or baskets. The furnace is loaded on the left, the work passes through, gets heated, and exits on the right. At the exit end, the parts can be carefully removed into tote trays or allowed to dump into a barrel, or conveyed on to the next operation. Such furnaces typically sinter 100 to 1,000 lbs. per hr. at a cost of $.30 to .60/lb., for the smaller furnaces. The nitrogen, hydrogen, natural gas and electricity for a small furnace can run about $25./hr., but it depends a lot on your local costs.

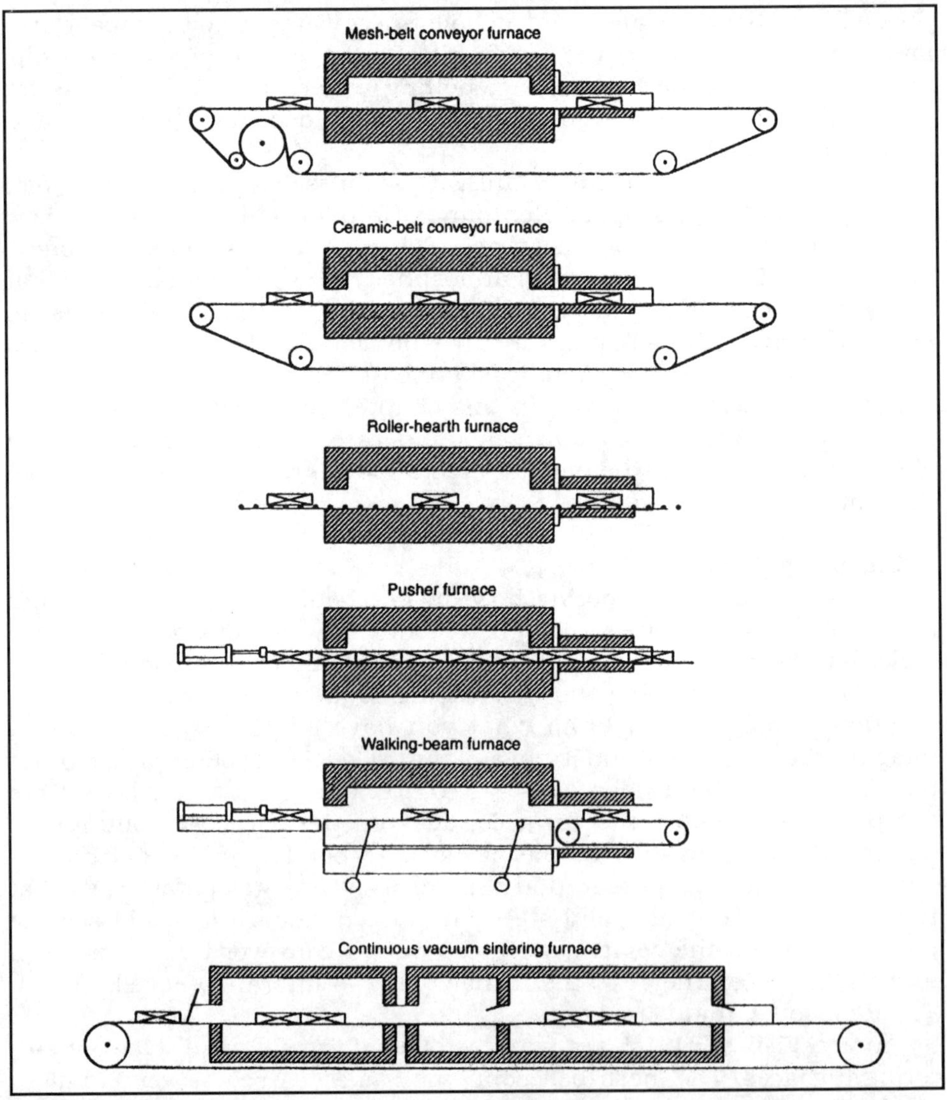

Figure 5.11. Schematic of various continuous furnaces. Courtesy of ASM, Vol. 7.

An 18 in. wide, 300-400 lb./hr. furnace running with nitrogen, might cost $35./hr. + labor + overhead. With 100% hydrogen, it would be about $50./hr. Such a furnace might sinter about $2 million/yr. worth of steel parts. Stainless steel mesh belts begin to stretch and weaken when the temperature exceeds about 2100°F which explains why we have learned to accept the mechanical properties of steel parts sintered for 20 minutes at temperature at 2020 to 2050°F. The other 5 furnaces represent the clever ways the P/M industry has discovered to economically sinter at temperatures above 2100°F. The most recent innovation is the ceramic belt furnace where the belt is made up of small links of silicon carbide, SiC. This relatively strong ceramic cheerfully runs at up to 2400°F. If a ceramic link cracks, you just replace it. The steel parts sit on alumina ceramic trays, so that they will not react with the silicon carbide belt. You may be wondering why all steel P/M parts are not sintered at 2400°F to take advantage of the better pore rounding, toughness and densification that occurs. This high temperature sintering costs about 15% more per lb. of parts produced, and the parts are more subject to dimensional change and distortion, so the tolerances are not as closely controlled. The high temperature sintering is only used when the usual 2050°F sintering does not produce a strong or tough enough part.

The third furnace is a roller hearth. The parts sit on trays that rest on a belt that is moved through the furnace by power driven rolls. The rolls can be made of a strong stainless steel alloy and can be big enough to support a high load. Roller hearth furnaces run up to 2200°F comfortably and with some sacrifice of the belt, 2300°F. They can support 28 lbs. on each square foot of hearth space. A very large version, 60 in. wide by 200 ft. long, is used by the iron powder manufacturers to anneal powder. The pusher furnace is based on the idea of having the parts sit on heavy ceramic plates or baskets. An air or hydraulic cylinder at the entrance end pushes the trays through the furnace. There are clever designs that allow the trays to be pushed slowly and continuously for one tray length. Then the pusher retracts and another tray is automatically loaded in front of the pushing mechanism. And so on. The ceramic trays can stand a lot of heat and pushers can run at 3000°F, hot enough to turn iron into a puddle. Pusher furnaces can convey parts at the rate of 10 lbs./ft^2 on 5 lb trays or 250 net lbs./hr.

The walking beam furnace can convey heavy loads at temperatures well above the usual 2050°F sintering temperature for steel. The parts are carried through the furnace on a thick, strong ceramic beam. The beam fits up through the floor of the furnace. The pusher on the left pushes the ceramic or metal tray onto the beam and the floor of the furnace. The beam rises up a fraction of an inch, lifts the tray off the surrounding furnace floor, and carries it forward a fraction of an inch.

The beam then drops gently down and sets the tray on the furnace floor which abuts each side of the beam. As the beam continues downward a fraction of an inch, the tray of parts sits on the furnace floor and spans the gap above the beam. The beam then backs up a little ways, and then continues on with its next cycle of rising up and moving trays forward. Such furnaces can sinter 2,000 lbs. of steel parts per/hr. at temperatures approaching 2400°F.

The last furnace in Figure 5.11 is a continuous vacuum furnace. It has to have one section that removes the lubricant, and separate sections with sealable doors to heat, cool, and exit the furnace. Such furnaces can run at 2400°F and can produce 350 lbs./hr.

Figure 5.12 shows a schematic drawing of a typical mesh-belt sintering furnace. The entrance side and loading table on the left lead into a pre-heat or burnoff zone where the lubricant is driven out of the green parts. The decomposed lubricant burns in a flame as it flows out the loading door. For iron parts, the pre-heat zone will bring the parts up to about 1400 - 1600°F. The lubricant-free parts then enter the high heat zone where the sintering occurs. The belt speed is adjusted so that the parts spend 15 to 30 minutes at 2020 to 2050°F in the high heat zone. The high temperature zone is usually heated by electrical heating elements located outside the sintering chamber. (It should also be noted

TABLE I. OPERATING CHARACTERISTICS AND CAPITAL COST OF SINTERING FURNACES

Type	Maximum Operating Temperature °F	Capital Cost $1,000's	Pounds Parts Produced Per Hour*
Mesh Belt	2100	$275	450
Ceramic Belt	2400	$350	250
Roller Hearth, 36"	2200	**	2,000
Pusher	>2400	$375	250
Walking Beam	>2800	$750	500
Continuous Vacuum	>2400	$850	350

* Production rate depends on the ability of parts to be packed closely enough to yield this rate.
** Capital cost is about 50% more than a well equipped belt furnace of the same size.

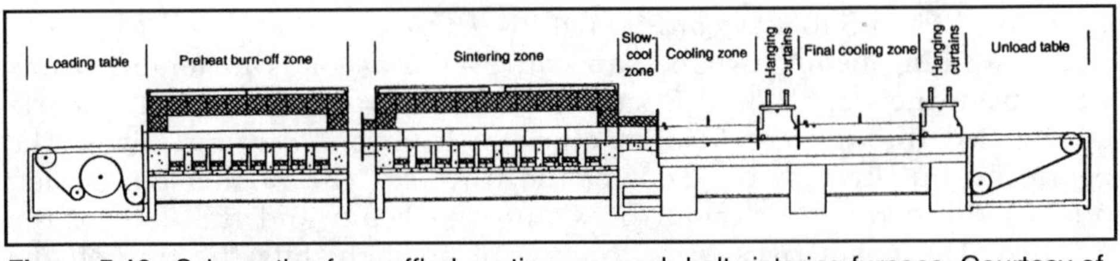

Figure 5.12. Schematic of a muffled continuous mesh-belt sintering furnace. Courtesy of ASM, Vol. 7.

that many sintering furnaces are heated by burning natural gas outside the sintering chamber, which can lower maintenance and operating costs.) The sintering chamber itself (the muffle) is shaped like a tunnel and is made from a special stainless steel. This muffle serves to keep the furnace atmosphere inside the furnace and near the parts. To be sure that the parts are exposed to the right time and temperature, a temperature probe (thermocouple) can be inserted inside a part and can then ride part way through the furnace, continuously indicating the actual temperature inside the part.

The right side of Figure 5.12 shows chambers that are used to cool the parts back to near room temperature. These chambers are surrounded by a water jacket that serves to carry away the heat. In some cooling zones, the hot atmosphere gas is rapidly pumped out, cooled off in a separate water cooled device (heat exchanger) and then pumped back into the zone. The rush of cold gas returning to the cooling chamber serves to cool the parts very quickly. This rapid cooling serves to further strengthen or harden steel parts, compared to the usual, slower cooling rates.

Figure 5.13 shows one of the methods by which parts can be rapidly delubricated in the burn-off zone. A mixture of air and gas are burned on the left. The hot gases flow to the right, to a baffle, and are then redirected back over the green parts, toward the door where the parts are entering. The hot gases serve to rapidly heat the parts. Some extra

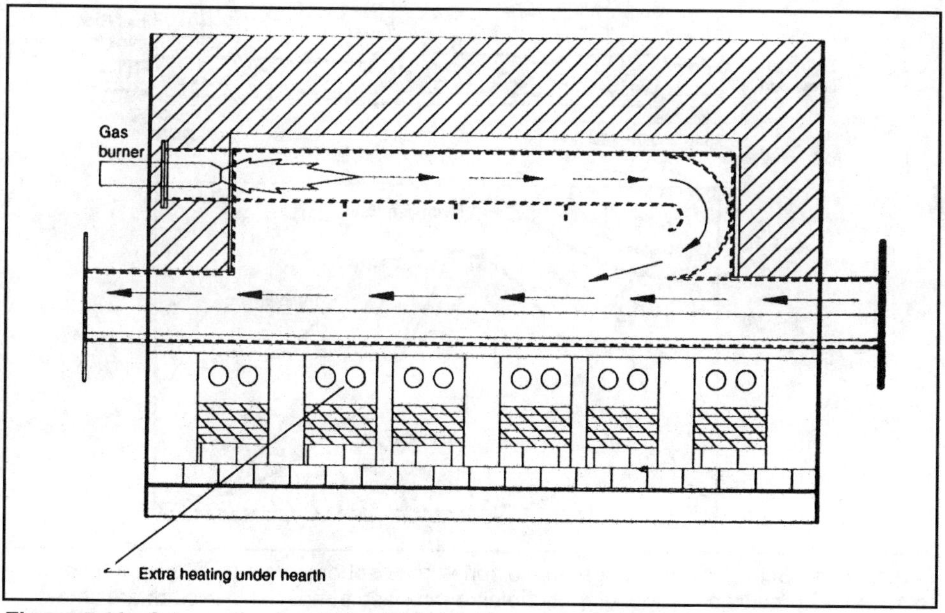

Figure 5.13. Schematic of an enhanced delubing system. Courtesy of ASM, Vol. 7.

heat can also come from the electrical heating elements located below the furnace floor. The hot gases from the burning natural gas contain substantial amounts of water vapor, H_2O and carbon dioxide, CO_2. These are the oxidants or oxidizing agents that can react with (burn) the lubricant vapors. The lubricant vapors have molecules with many carbon and hydrogen atoms. These carbon atoms combine with the oxygen in the water vapor H_2O to form CO, carbon monoxide and carbon dioxide CO_2 gas. That gas safely moves to the entrance of the furnace where CO gets burned to form CO_2. By quickly getting the lubricant vapors converted to these other gases, the lubricants can not then break down to form soot inside or on the parts.

Figure 5.14 shows a sketch of a sintering furnace where different atmospheres are injected into specific sections of the furnace, a zoned atmosphere system. In the preheat zone on the left, water vapor is added to the nitrogen gas to act as an oxidant to help burn the lubricant

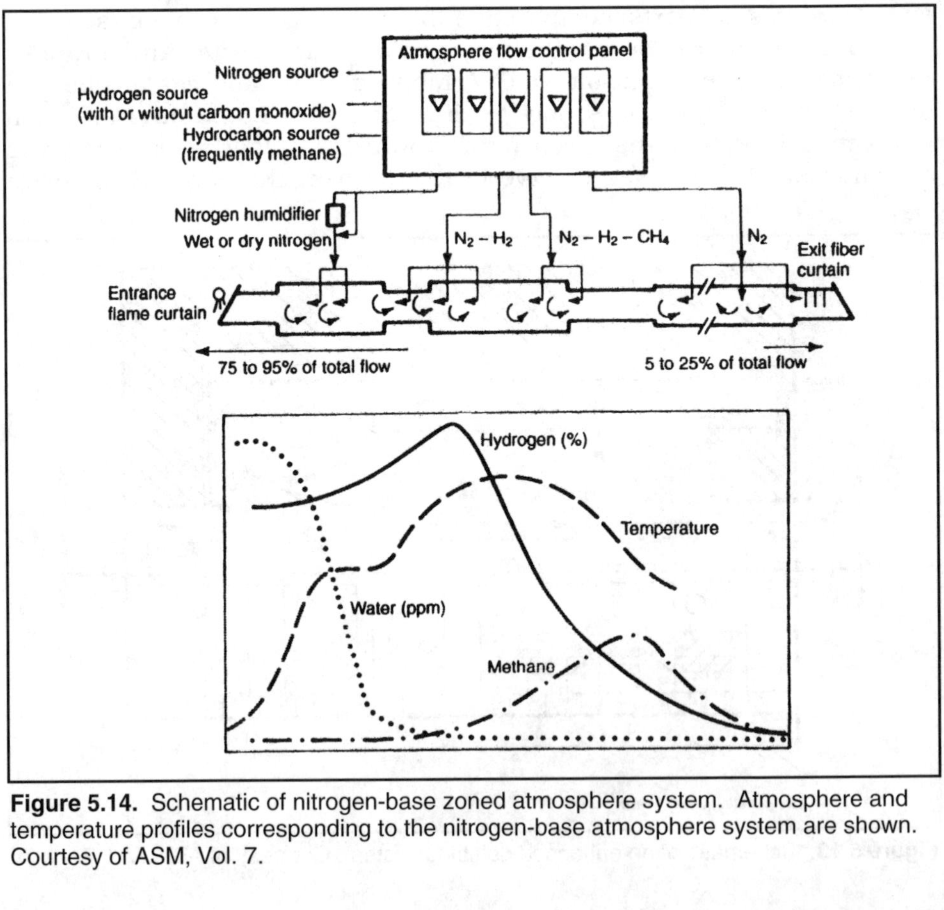

Figure 5.14. Schematic of nitrogen-base zoned atmosphere system. Atmosphere and temperature profiles corresponding to the nitrogen-base atmosphere system are shown. Courtesy of ASM, Vol. 7.

vapors. That is an alternative to the gas flames described above. The high heat zone receives nitrogen and hydrogen with very little water vapor in it. The hydrogen cleans up (deoxidizes) the iron particle surfaces and permits atom migration over the surfaces and the resultant formation of metallic bonds between the particles. A small amount of methane, perhaps .5%, makes sure that no carbon is lost from the surface of the parts. Finally, the cooling zone has mainly nitrogen with very little water vapor. This allows the parts to cool without forming any colored oxides on the surface. The graph below the furnace sketch shows how the different atmosphere gases vary in amount from the entrance delube zone, all the way to the exit door. This system is somewhat removed from the ancient Egyptian's charcoal fire 1300BC!

Sintering Bronze

Interestingly, in 1923, 3,223 years later, we were still using charcoal to sinter, as taught in Harry M. Williams' patent 1,642, 349, for producing bronze bearings, assigned to General Motors: "After the compression step, the compressed article is packed in a carbonaceous material and heated to a temperature of about 1350°F for a period of about 5 hours. After the article has become cool enough to handle, it is removed from the carbonaceous packing, and, if a bushing, is sized, for example by forcing through a die upon a mandrel. The latter operation simultaneously sizes the bushing, both inside and outside." Bronze was packed in charcoal, and perhaps put in a can, which was then put into a furnace operating in air. The air leaked in around the charcoal and formed CO which was a reducing gas that could protect the bronze that was being sintered.

What is bronze? Its an alloy of copper (Cu) and tin (Sn), usually being 90% Cu and 10% Sn. Do you just mix the copper and tin powders together, add a little lubricant and press them? That works, and is still frequently practiced today. When you heat this mix, the tin melts at 449°F or in the neighborhood of 231.9681°C. The melted tin flows around and coats the copper particles. The copper is still pretty solid, because it doesn't melt until 1981°F. You have already read how liquid copper can diffuse into iron. Well, the liquid tin starts to move or diffuse into the copper particles. And then things get complicated: η ζ ε δ γ β α! If this looks like Greek to you, it is, and it represents the conditions or compounds that the tin atom goes through as it dissolves in the copper. (Just as you learned the rare earth elements in Chapter 1, all good students will want to learn the names of the lower case Greek letters (phases): eta η, zeta ζ, epsilon ε, delta δ, γ gamma, beta β, and alpha α, which are the first six letters of the Greek alphabet, in reverse order!) So, what are these compounds, (intermetallic compounds) that the tin suffers through on its way to becoming dissolved as a nice orderly solid solution in the copper?

Compounds are stable substances with a fixed relation between the number of atoms of copper and tin. For example epsilon, ε, has 3 copper atoms for each tin atom. Compounds also have distinct crystal structures. The copper and tin atoms are always arranged the same way, in a repeating pattern. By contrast, in the α, (alpha) solid solution, all the tin atoms are still there, substituting for the copper atoms in copper's crystal structure, but they are placed at random points.

As the tin atoms start to move into the copper, instead of forming a random solid solution, they form the compound closest in composition to tin: namely eta, η which consists of about 61% tin. If we cut off the sintering at a very early state, and looked at the structure, we would see copper particles surrounded by a layer of eta phase and then surrounded by tin. But this is not a strong structure! As the temperature goes up and time passes, the tin migrates further into the copper and the initial Cu-Sn compounds dissolve giving rise to other compounds which are richer in copper. After about 10 minutes at 1550°F, the tin has reached the center of the copper with 10% tin dissolved in the copper, and all the intermediate compounds that were formed along the way have now dissolved. (We are tempted to give the copper tin phase diagram as a student assignment!) All that is left is alpha bronze, a single phase with the color you associate with bronze. If you really short circuit the sintering, you leave some of the intermetallic compounds in the bronze. These are hard and brittle, and thought to degrade bearing performance and to hurt machinability. If you under sinter a little, you leave some copper-rich grains that look red under the microscope, not bronze colored, because the tin did not diffuse in all the way. Also, as the tin moves into the copper, the initially small cold worked copper grains begin to grow into bigger alpha bronze grains. The better the sinter, the larger the alpha bronze grains.

Sintering bronze and controlling the product for both strength and close dimensions is no picnic. Sintering for about 10 minutes at 1550°F in nitrogen base, endothermic, or exothermic atmosphere will give the required minimum strength. That's easy. The real problem is controlling dimensions. Almost everything affects growth and shrinkage of the bronze bearings. Figure 5.15 shows how a 6.5 g/cm^3 bronze premix can grow 4%. During sintering, bearings grow and then eventually shrink back below die size. The trick is to finish a little above original die size. Coarse tin powder makes bearings grow, fine tin allows them to shrink. High green density and rapid heating accentuate growth. High sintering temperature produces rapid growth followed by rapid shrinkage. Sometimes the bronze powder is pre-alloyed copper and tin. This makes for rapid shrinkage. In Figure 5.16 we see that the bearings would shrink 2.5 to 3.5% from die size! To preserve some open porosity, bearings are not usually made from well sintered prealloyed bronze. Such bronzes are best reserved for structural parts that need high density with good strength and ductility. Sometimes the bronze powder is a mixture of copper, tin and copper-tin compounds. We see this in the lower

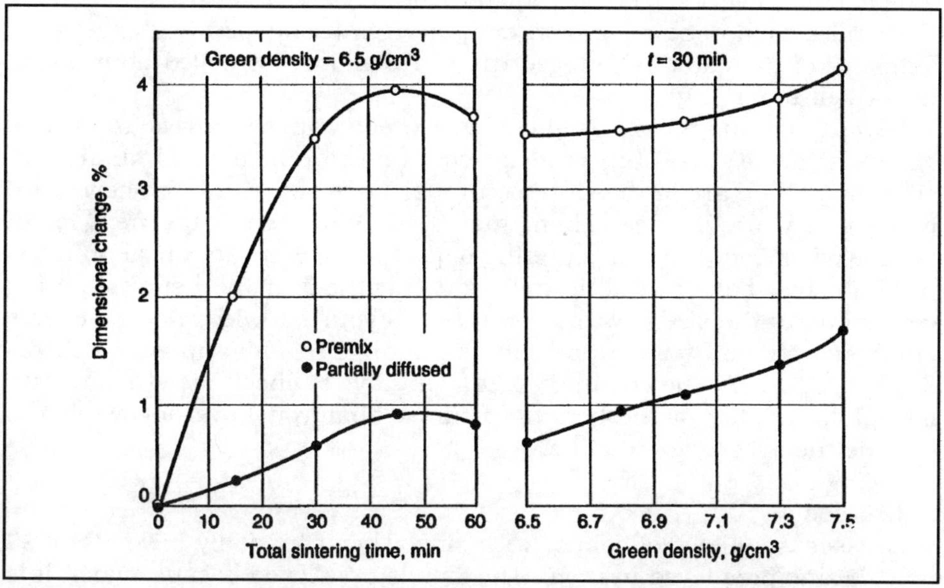

Figure 5.15. Effect of properties on dimensional change of bronze. Sintered at 820˚C (1500˚F), under dissociated ammonia atmosphere. Courtesy of ASM, Vol. 7.

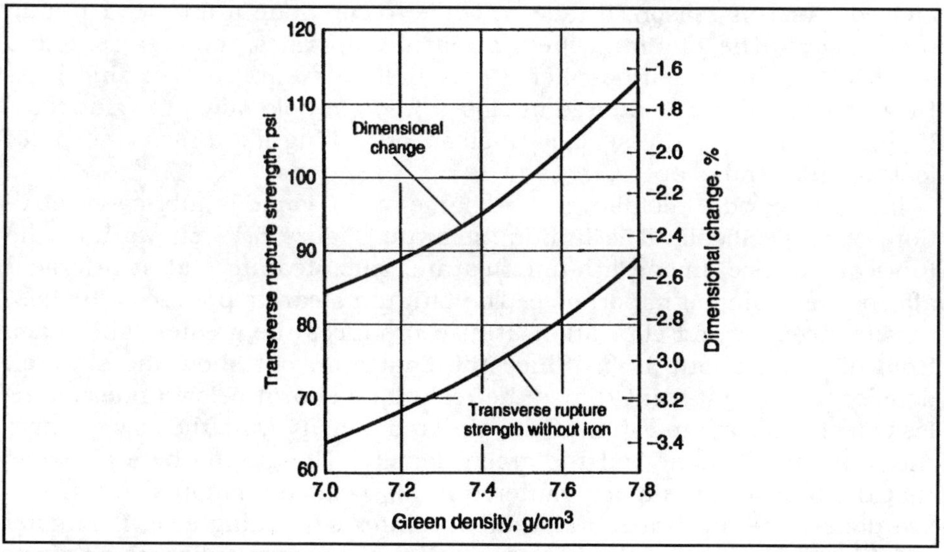

Figure 5.16. Properties of Prealloyed bronzes. Sintered 30 min at 840˚ C (1550˚ F), under DA Courtesy of ASM, Vol. 7.

part of Figure 5.15. That mixture produces just a little growth above die size, .2% to .9%. The right side of Figure 5.15 shows how green density affects sinter growth. You can see that the green density can cause the dimensions to vary by 1%. If your bearing has a low density middle portion and a high density at each end, after sintering, you will expect the diameter of the ends to be bigger than the middle, so-called hour-glassing or bell-mouthing.

Powder manufacturers use all of these techniques to arrive at a series of mixes, all 90Cu/10Sn, graded according to dimensional change on sintering. If you are in the sintered bronze business, you learn how these behave in your furnace. Then, just to be sure, when the next lot of bronze parts comes to the sintering department, you run a trial on a few parts ("a test bake"). If all is well, you run them at the usual sintering conditions. If the size is wrong, you use all your knowledge of the effects of sinter time and temperature on size change to come up with a modified procedure. At the end of the run, you have to check the strength and oil holding ability of the bearing to prove that you have not wandered outside the specification limits.

Brasses

Brasses are alloys of copper with zinc. They commonly have 10 or 20 or 30% zinc prealloyed in them. The zinc is prealloyed because when it is present as the pure elemental powder, it tends to evaporate at the sintering temperature of 1600°F. In fact, it even tends to evaporate when prealloyed. For this reason, it takes a while to condition a furnace running brass alloys. The zinc from the first parts evaporates and creates a zinc gas pressure in the atmosphere. This then prevents further zinc loss. Brass parts are often sintered in closed trays to hold down the zinc loss. If zinc is lost from the surface, instead of looking like brass, the parts look reddish and coppery.

In brass powder metallurgy, the name of the game is lubricant selection, and specifically it is lithium stearate. Figure 5.17 shows how the lubricant choice among lithium stearate, zinc stearate and stearic acid affects everything of importance. The Lithium stearate produces the best tensile strength and elongation. It also produces the greatest shrinkage from die size, about 1.5%. Since the charts do not show the sintered density, you might want to use the technique shown below under stainless steel sintering to estimate the sintered density from the size change, the lubricant content and the green density. The graphs here all show that the brasses are usually sintered in dissociated ammonia. When you are done sintering brass and want to sinter something else, it is good practice to run junk in the furnace until the zinc atmosphere is all dissipated. This avoids contaminating the next material with unwanted zinc.

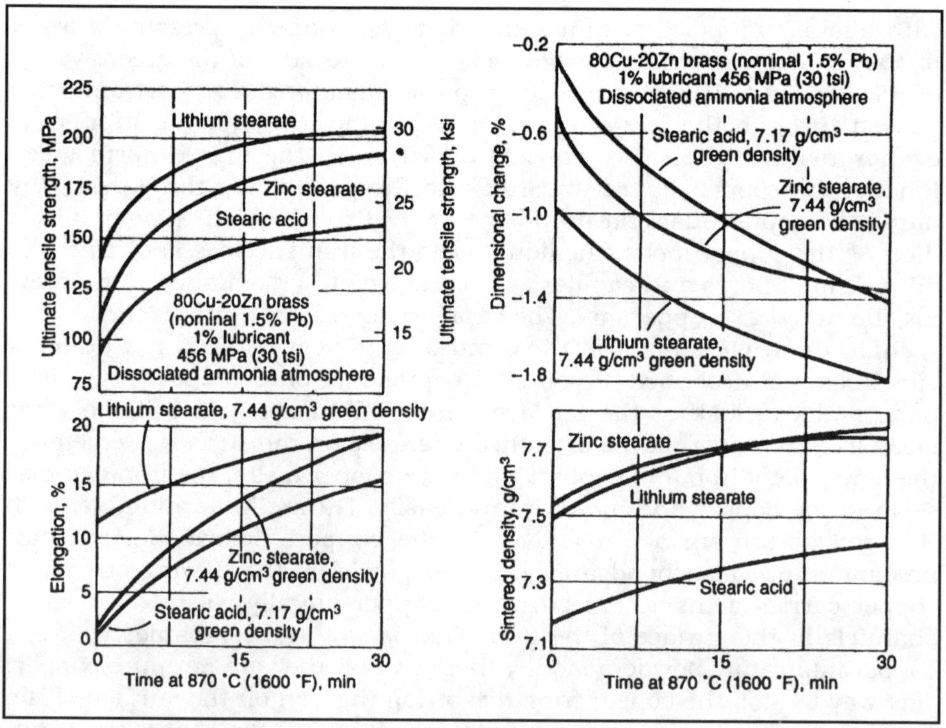

Figure 5.17. Effects of lubricants and sintering time at temperature on tensile properties, sintered density and dimensional change of brass compacts. Courtesy of ASM, Vol. 7.

Copper Infiltration of Steel

Copper infiltration refers to the filling of the pores in a heated iron or steel P/M part with liquid copper. Why did powder metallurgists set out to do this? Why not just mix the iron and copper together? One possibility is that they thought that the copper would braze or join the steel particles together and increase the strength. Well, that proved to be true. Another thought has to do with the sponge iron powders that were in use prior to 1968. They were hard to press, and so it was difficult to make a dense, high strength part. Since they were hard to press, it took a big press to make a relatively small part. If you could use less pressure, a given press tonnage can make a bigger part, but it is sort of low in density. Then you strengthen this fairly porous, low density part by filling the pores with copper.

As originally practiced with sponge iron powder, the iron was pressed to a fairly low density, about 6.1 to 6.2 g/cm^3. This left 22% of the space inside the part available to be filled with copper. Typically, they infiltrated enough copper to fill pores and bring the density up to 7.3 g/cm^3, or with about 10% porosity remaining. This was done by pressing a wafer of copper or copper alloy powder weighing 18 to 20% of the green weight of the iron part. This was placed atop the unsintered iron part and then passed through the sintering furnace. During heat up, the lubricants evaporated and then the graphite dissolved in the iron to form steel. Finally, the copper melted and ran into the part, filling the pores. Why didn't the copper just end up on the floor of the furnace? (Sometimes it did!) As the copper melts, the liquid wets the iron surfaces and runs into all the small pores and capillaries, just as we described above for sintering the iron - 2% copper mix. The capillary forces are quite strong.

Infiltration has quite a few ins and outs, and it really is not quite as simple as just described. If you sit a compact of pure copper on the iron part, and you look at the surface, after infiltration, you will see what looks like little river marks or eroded channels on the surface. Remember the sintering of iron and copper: when the copper melts, the iron can dissolve in the copper and move around easily. During infiltration, some of the iron at the surface dissolves in the copper. Before the iron can precipitate out at a bond area between particles, the flowing stream of copper carries it inside the part. The copper literally washes out small channels in the surface of the iron. That is why the companies that sell copper infiltrating powders modify them so that they are not pure copper. One way to stop the copper from dissolving the iron on the surface of the part is to add some fine iron powder (carbonyl Fe) to the copper infiltrant. When the copper melts, the entrapped iron dissolves first and saturates the liquid copper with iron so that it can not absorb any more iron from the part. Other additions to the copper include manganese, cobalt, zinc, and graphite. Some of these additives do not dissolve and move into the iron part. Instead, they maintain a kind of rough form or skeleton on the surface of the part. This residue is then blown away or tumbled off, after infiltration. If you are unlucky, it sticks firmly to the part. In the old days, it was found that a small piece of toilet tissue placed under the copper alloy wafer kept the residue from sticking. Now-a-days, many of the copper alloys for infiltrating do not leave a residue.

Typically, there is 9 or 10 volume % (v/o) porosity left in an infiltrated part, if the final density is say 7.3 g/cm^3 and the pore-free density of the part is 8.1 to 8.2 g/cm^3. How would you go about filling most of the pores, to make it really dense? Put a really big copper alloy wafer on top of the part? That is not the best way. You can fill pores pretty well infiltrating from the top, but the most thorough way is to fill the part from the bottom. That way, as the copper is carried up into the part by the capillary forces, the air in the pores is pushed out ahead. If the copper wafer under the part melts so fast as to form a liquid puddle, it can just sit there until it gets sucked into the part. If the big lump of copper alloy is sitting on top, and melts too fast, it runs down over the part like frosting on a cake. It traps gas inside the parts and many pores do not get filled. Further, you end up with excess mounds of copper sticking to the outside of the part. Working from the bottom, it is possible to infiltrate up to about 7.9 g/cm^3, leaving only 3 or 4 v/o pores.

Remembering that the strongest attractive forces for the liquid copper are found in the smallest spaces will help you avoid a trap. Suppose you wanted to infiltrate a complex shape that had high and low density regions. The intent would be to add some copper in those weaker low density regions and build up their strength. Unfortunately, Mother Nature sends the copper to the high density areas first, where all the small capillaries and tight spaces reside. This can leave the low density regions with less copper than you had hoped for. The cure is not much fun: you have to keep adding excess copper until even the largish pores are filled. Another practice that helps infiltration is to use at least some sponge iron which has internal pores that attract and hold liquid copper. In the MPIF Standard for infiltrated material, the 20% copper alloy is based on a 50/50 mixture of sponge and atomized iron. If you want some real fun, try to make the 20% Cu infiltrated iron with no added graphite (carbon). With atomized iron as a base this combination tends to cause the parts to grow far too much during sintering. The answer is to use a very spongy iron, e.g. the hydrogen reduced mill scale. The copper can run into all those small pores inside the sponge iron and not force the iron particles apart. Harold Harrison, famous former President of Pyron, and an unabashed promoter of their hydrogen reduced sponge iron, once told us how good this low carbon, 20% Cu infiltrated material really was. If you take a piece of it, case carburize it, quench it, and temper it, you can hit it with a sledge hammer!

With the advent of nice compressible atomized iron in 1968, P/M houses had the idea to save on the copper alloy and compact the iron to 6.6 to 6.7 g/cm^3 before infiltration. This worked well and produced stronger and harder parts. To give an example of the benefit of copper infiltration note that a P/M steel part with .8% carbon at 6.6 to 6.7 g/cm^3 would have a tensile strength of 42,000 psi. If we fill the pores with copper and drive the density up to 7.3 g/cm^3 the tensile strength goes up to 87,000 psi. This also increases the impact strength or impact energy by 75%.

If we drive the infiltrated density up to about 7.5 g/cm^3, the pores are no longer connected together and the material is pressure tight. This makes it useful for pumps and devices where there is air or liquid pressure. Further, the part can be electroplated with no problem from subsequent bleed out of the electroplating solutions. (In a typical porous P/M part, this bleed out problem is prevented by filling the pores with a plastic resin. Of course, the copper infiltrant can withstand much higher temperatures than the plastic resins). We promised economics: our friends who make copper infiltrated parts say it adds about 40% to the cost of a part.

Stainless Steel Sintering

Stainless steel powder has been available for at least 40 years, but has not grown in usage as has iron powder. Usage during the 1980's and early 1990's was about 3,000 tons per year while plain iron powders were 100 times that. The recent savior in terms of tons of usage has been the automotive exhaust flange. The auto industry can not make an accurate stainless steel stamping of the thickness and size required and so has been forced to fall back on P/M stainless. This flange must be welded into the exhaust system. Looking ahead, this flange could consume many tons of stainless steel per year, mainly in the grade 409 Cb which contains 11% chromium, less than .03% carbon, and at least 8 times whatever the carbon level is, as columbium (niobium). After welding, the columbium will react with the carbon on slow cooling to form columbium carbides, and so the chromium will not form any carbides, and will not lose its corrosion resisting ability.

Considering that there are hundreds of thousands of tons of wrought stainless products sold each year in the US, why was P/M limited to only 3,000 tons of this huge market? The P/M material was often not very corrosion resistant and the 316L powder costs $2.09/lb vs. solid bar stock at $1.50 to 1.75/lb. Further, under conditions of high temperature sintering for good corrosion resistance, there is 2% linear shrinkage and difficulty in holding dimensions. Finally, there is no way to harden the corrosion resistant grades by cold working, as is commonly done in bar stock. Well sintered corrosion resistant P/M stainless steel is soft and ductile, with a low yield point.

Are there any ways around this problem? We really don't know how to solve all of them at the same time, but, fortunately, for many applications you don't need to. The biggest problem that P/M stainless steel faces is appearance. If stainless steel is going to be up front in a consumer item, it better not stain or rust. If the stainless steel part is inside a piece of machinery, the issue becomes one of durability, not surface appearance. Durability has to do with how fast the corrosion eats away the stainless steel. A stain may not bother your application, but corroding away .01 in. per year may make the parts fit together badly and then make them thin enough to break. The current test methods being developed for P/M stainless focus on staining and appearance. You soak the part in 5% salt water solution and look for the first rust stain. The usual 300 series grades of stainless steel, e.g. 18%Cr and 8%Ni, sintered at

2050°F in dissociated ammonia will show rust stains in less than a day. Some of the newer non-standard grades with added copper and tin last longer before they stain.

What causes this lower staining and corrosion resistance? Why is stainless steel stain resisting in the first place? These alloys resist the formation of rust because the chromium atoms react with the oxygen in air to form a thin, invisible, protective oxide layer on the surface. Fortunately this layer is not red, like rust. The layer does not continue to thicken with time, nor does it fall off. So there is no real loss of material. Unfortunately that same tendency for chromium to react with oxygen also applies to chromium reacting with nitrogen and carbon. If all the admixed lubricant is not thoroughly removed from the green part during the delubrication cycle, some of the carbon in the lubricant can react with the chromium to form chromium carbide. Further, the 25% nitrogen in the dissociated ammonia furnace atmosphere can react with the chromium to form chromium nitrides. To the extent that the chromium on the surface of the part is tied up forming the nitrides and carbides, it can not form its protective passive oxide layer at room temperature. With a less protective layer on the surface, the iron in the part can react with water to form red rust spots.

The role of carbon is more complicated than it first appears. The stainless steel powders arrive at our door with some residual oxides on the surface. These are "thick" oxide patches (about .5 micrometers thick) and should not be confused with the very thin passive oxide layer that forms at room temperature. The oxides on the particles both tie up chromium and interfere with sintering. The hydrogen in the furnace atmosphere can react with the oxides on the particles and carry the oxygen away as water vapor. Also, *small* amounts of residual carbon from the lubricant can react with the oxides to form carbon monoxide which gets carried away. The problem occurs when there is too much carbon, and the carbon ends up as chromium carbide. As a rule, when sintering in atmosphere, delubricate so as to get rid of as much carbon as possible, and let the hydrogen clean up the surface oxides. In a vacuum, there is no hydrogen, and so a very small amount of added carbon can help reduce the surface oxides and carry them away as carbon monoxide.

In the year 2,000, there are three grades of the 300 series stainless steels listed in MPIF Standard 35. The grades are N1, N2, and L. They apply to all of the various 300 series chemistries where 316=17% Cr, 12% Ni, 2% Mo; 304=19%Cr, 10% Ni ; and 303=18% Cr, 10% Ni, and .22%S (for machinability). The N1 and N2 grades allow .2 to .6% nitrogen. The L grade keeps nitrogen and carbon under .03%. N1 is typically sintered at 2050 to 2100°F for 30 minutes. N2 is sintered 30 to 60 minutes at 2350°F. Both grades are sintered either in dissociated ammonia or nitrogen base. They pick up the .2 to .6% nitrogen from the sintering atmosphere. On cooling to room temperature, some of the nitrogen combines with chromium to form chromium nitride which reduces corrosion resistance. The remaining nitrogen, which is dissolved in the stainless alloy, is useful and serves to strengthen it. If there were a convenient way to cool the parts very quickly from 2100°F to room temperature, the chromium nitrides would not have time to form and the parts would have good strength and corrosion resistance. This effect is actually practiced by some parts producers who sinter in vacuum with a 500 micrometer back pressure of nitrogen. Some vacuum furnaces are equipped with high pressure gas quenching systems. When the sintering time is up, high pressure nitrogen is admitted to the hot zone and rapidly quenches the parts. To date, the method is not widely practiced and has not had the related mechanical properties standardized.

The L grades of 300 stainless steel are made by sintering in a nitrogen free atmosphere for one hour at 2350 to 2400°F. This is usually done in a vacuum furnace with 1000-1500 micrometers of argon back fill. Prasan Samal and Erhard Klar of OMG Americas explain that lately, parts fabricators have been adding a 10 minute cooling "hold" at 2100°F on the way down, to restore the chromium content that has been lost at the surface of the parts. The pressure is 10,000 to 20,000 micrometers during the hold, and the pressure is read on a separate gauge from that used for the lower pressure sintering. During the hold, the chromium inside the part moves back out to the surface. This sintering gives linear shrinkages of 1.5 to 2.0%, and densification to 7.0 g/cm^3, while the pores become well rounded and the particles well bonded. For the 316L, this gives a yield strength of only 22,000 psi but an elongation of 21%. The same time and temperature in a nitrogen containing atmosphere gives a yield strength of 38,000 psi and elongation of 13%. This data is for parts at a sintered density of 6.9 g/cm^3. The L grades give the best corrosion resistance. An increasing number of P/M houses are acquiring the ceramic belt furnaces which will cheerfully process the stainless steels at 2400°F in 100% hydrogen. This should prove to be more economical and to provide closer control of tolerances than batch sintering in a vacuum furnace.

Late in 1998, the PMPA Standards Committee did a very small test on hydrogen sintering of 316 stainless test bars at 2100°F in 100% hydrogen. One group of bars survived more than 1000 hrs. in 5% salt solution with no evident rusting. That participant attributed part of the success to very good delubrication of the parts. Some P/M houses sinter stainless steel in hydrogen at 2200°F in mesh belt furnaces. The belts don't last very long, but over all, the economics pay off. There are some P/M parts fabricators who insist that the only way to press and sinter stainless steel is in a dedicated plant. In that way, no iron powder gets mixed into the parts or lands on top of them, later to cause rust spots.

Some of the end users of stainless steel parts are not really pleased with the fact that they end up actually having to try the stainless steel in their application, rather than knowing from data or a handbook that P/M stainless will really succeed. Why aren't we more like the big boys that make the wrought stainless steel bars and pipe? Probably because we are not "big boys" and because there are thousands of different environments to which stainless steel might be exposed. As noted above, our first "baby steps" have been to test out the stainless steels in 5% salt solution. In 1999, the 7,000 tons of stainless steel powder sold for P/M use had a powder value of only $20 million. Those sales are divided up among three different suppliers. Nobody has the necessary financial incentive to try stainless in all the possible environments. Also, having accumulated all the data, you would then remember that the real corrosion resistance depends on how the stainless steel powder is pressed and sintered at the P/M parts fabricator. There are many different levels of competency in processing the stainless steel powder into P/M parts. And, the 2000 edition of MPIF Standard 35 only lists typical corrosion results, not mandatory properties.

The 410 stainless steels with 12% Cr and .2% carbon have good hardness and strength. Designated SS 410-90HT, its properties are obtained by adding about .2% -.25% graphite to the 410 powder, compacting to 6.4 to 6.5 g/cm^3 and sintering for one half hour at 2050°F in dissociated ammonia. The material hardens during cooling from sintering and must be tempered to relieve stresses, at least one hour at 350°F. This gives the 90,000 psi tensile strength shown in Standard 35. For good magnetic properties, the 410 is processed without carbon additions and is called 410L, (for low carbon). That is sintered at 2350°F in vacuum or hydrogen. The high temperature sintering densifies it to 6.9 g/cm^3. The high density and low carbon give the product 10% elongation, but it yields at 20,000 psi. Its amazing what a difference a little carbon makes! The 409Cb grade of stainless steel has 11% Cr and a %Cb = to 8 × % carbon, where carbon is 0 to .03%. The Cb, columbium, is added to tie up any unwanted carbon as columbium carbide. This creates a weldable grade of stainless steel. On welding, there is no available carbon to react with chromium and form chromium carbide, which would deplete local areas of chromium and allow corrosion. It is sintered at elevated temperatures, like 2350°F, to densify it to over 7.0 g/cm^3. All of the stainless steel powders are prealloyed, with the elements like Cr, Ni, and molybdenum dissolved in solution. This makes them hard to press. Fortunately, at 2350°F, they rapidly densify, so they can start off at a fairly low green density.

There are some developments in stainless steel processing that are beyond what has yet been standardized in MPIF Standard 35. In ASTM B 853, there is a stainless alloy consisting of iron plus 23% Cr + 18% Ni + 3.5% Mo + 0.35% B (boron) with less than .05% C and less than 0.1%N. This was originally developed by Dr. O. W. Reen and has the remarkable property of sintering to near full density and having excellent corrosion resistance. In another area, at the less than fully dense end of the spectrum, companies that make stainless powder have been experimenting with grades that contain additives to help deal with the lack of corrosion resistance usually found in the 2050°F sintered stainless steels. The Metal Injection Molded folks have standardized 17-4PH stainless steel in the MIM section of Standard 35. It contains iron + 4% Ni + 16% Cr + 4% copper + .3% (Nb+Ta). It is a hardenable grade in the sense that you can heat it up again after sintering (to re-dissolve some of the elements), quench it and reheat or age at 900°F. The aging precipitates something out of the structure which considerably hardens and strengthens it. (If you have been wondering why some stainless steels contain columbium to prevent formation of chromium carbides, and others use tantalum, Ta, you should know that our faithful 1963 Funk and Wagnalls dictionary defines columbium as the *former* name for tantalum. The PMPA and ASTM standards committees are, in 2000, considering an upgrade to tantalum!)

Sintering, and the Resultant Properties of Parts

Lets start with something we all need to do: sinter iron and graphite to make steel. Figure 5.18 shows early work by Athan Stosuy, (retired from Hoeganaes) mixing sponge iron with enough graphite to end up with .8% carbon in solution, a high carbon steel. The green density of 6.1 g/cm^3 is low by today's standards, but was typical for a sponge iron powder in the 1960's. The vertical scale shows the growth from the original die size, in %. For reference, .1% growth meant that a one inch part would grow .001 inch. Sintering time was from 5 min. to 120 min. and probably meant the time that the parts were in the lab furnace hot zone and not the actual time at temperature. The first thing to note is that it looks like the parts grew like crazy in the first five minutes of sintering. We need to remember that the parts expand as they come out of the die after being pressed. A good guess is that the parts were .2% greater than the die size, as they emerged from the die. In that first five minutes in the hot zone, the graphite dissolved in the iron and caused some growth from green size. Then all the shrinkage and sintering mechanisms we described above take over and the parts begin to shrink with time. The longer the time, the more the shrinkage. Note also that the higher the temperature, the faster the shrinkage back toward zero and the size of the die. Athan ran the sintering time out to 120 min. to show you the trend. But if you are sintering at 2050°F, you will not be in the hot zone for much more than 30 min. So, your parts will come out about +.28% above die size. This will be a happy occurrence if you thoughtfully made the die undersize by about.28%! [If, Δ, delta, is the size change from die, on sintering, expressed as a percent, and P is the dimension of the finished part after sintering, to get at the die dimension, D, that compensates for this size change, multiply P by 100÷ (100+Δ) When Δ is near zero, e.g. +.28% as above, you can get away with just deducting the .28% from the finished part size. But do not try that shortcut when Δ is large, as in stainless steel, -2% or MIM, -14%! Get out your calculator and try the MIM example to convince yourself.]

The neat thing about powder metallurgy is that each and every imaginable mix of metal powders has its own set of dimensional change curves like these. And if you make parts, you have to know these curves, and you have to work with your powder supplier to be sure that today's iron shrinks or grows the same as it did last month. People who sell powder or make parts spend a lot of time trying to work out mixes that will produce a part near die size, so that you don't have to compensate the size of the die with the long formula! Of course, also, mixes with large dimensional changes are, overall, more variable in the final dimensions of parts.

Prof. German has an example in Figure 5.19 of a modern material, atomized iron with 2% nickel and .5% carbon, pressed to 7.2 g/cm^3.

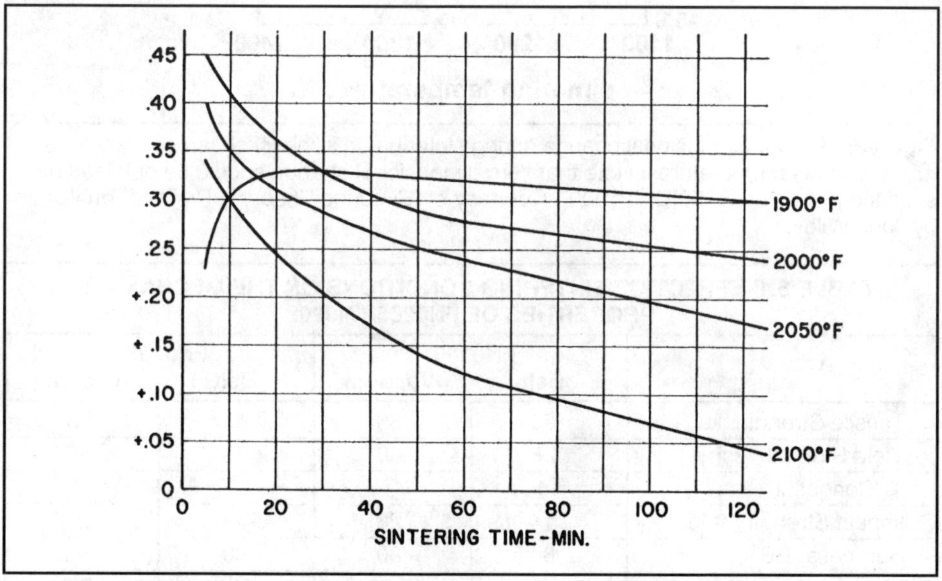

Figure 5.18. Growth of iron-1.5% graphite compacts, green density 6.1 gm/cc, at various times and temperatures. (from A. Stosuy, Metallography of Sintered Steel, Hoeganaes Corp., Riverton, N.J.).

That density is more or less the upper limit for a green part, requiring 50 tons per in.^{2} to compact it. Sintering was for 30 minutes at temperature with a range of temperatures shown from 2000 to 2400°F. This gave size changes from near 0 at 2000°F to a .3% shrinkage at 2400°F.

Howard Sanderow, presently Chairman of the PMPA Standards Committee, reported the mechanical properties in Table 5.I for iron with 2% nickel and .5% carbon steel, pressed to 7.2 g/cm^3. He compared belt furnace sintering at 2060°F for 30 minutes with vacuum sintering at 2300°F for 2 hours at temperature. The high temperature sinter increased the tensile strength by 64% and more than doubled the yield strength. But that was not the important change! (Its easy to get that 80,000 psi tensile strength with a heat treated iron copper carbon mix

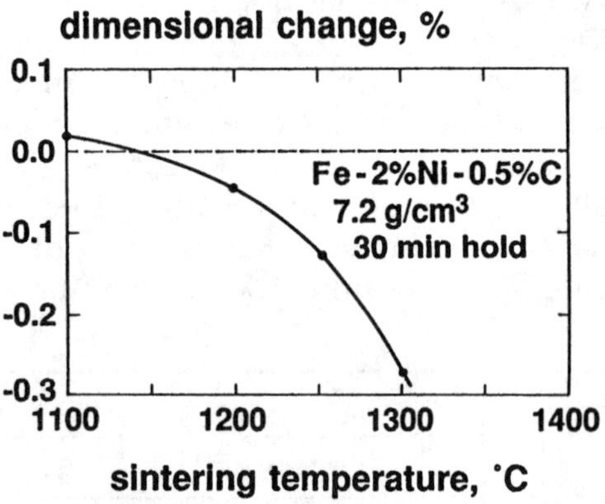

Figure 5.19. Net dimensional change during sintering an iron-nickel-carbon alloy for 30 min at various temperatures. Note that zero dimensional change should be obtained by sintering just below 1150°C.=2100°F. Courtesy of "Sintering Theory & Practice" published by John Wiley.

TABLE 5.1 EFFECT OF SINTERING CONDITIONS ON THE MECHANICAL PROPERTIES OF FN0205/FN0208

	FN0205		FN0208	
	Belt	Vacuum	Belt	Vacuum
Tensile Strength, ksi	55	80	65	110
Yield Strength, ksi	28	60	48	85
% Elongation	4	7	2	4
Impact Strength, ft-lb	14	28	8	24
Hardness, RB	64	80	80	90
Sintered Density, g/cc	NR	7.32	NR	7.3

Notes: All samples pressed to a green density of 7.2 g/cc; Belt - 30 minutes @ 2060°F in nitrogen/endo atmosphere; Vacuum - 2 hours @ 2300°F with nitrogen backfill; NR - Not reported; Data from Ref. 4

that has been sintered in a belt furnace.) The real trick was getting 7% elongation, a significant amount of ductility and an indicator of excellent toughness. Notice that the density went from 7.2 g/cm^3 green to 7.32 g/cm^3 sintered. Did the part shrink, in order to reach this higher density? Yes. How much? He doesn't say, but we could make a guess. If we assume that the part with a green density of 7.2 g/cm^3 had .5% lubricant in it, then without the lube, it was only 7.17 g/cm^3 To go from 7.17 to 7.32 g/cm^3, the density increases by 2.05%, so, if the weight stays the same, the volume must decrease by about 2.05%, or more precisely 2.09%. There are two ways to figure out how much the diameter changes, when the volume shrinks 2.09%. One is to solve a cubic equation. No thanks. The other is to approximate it by dividing the volume % change by 3: 2.09% /3 = -.7%, shrinkage.

To picture what happens in sintering 316 stainless steel at a green density of 6.85 g/cm^3, it is hard to improve on some data from Harry Ambs, now retired from OMG America, and Don Dyke, retired from SSI Technologies, in Figures 5.20 – 5.24. These examples are intended to

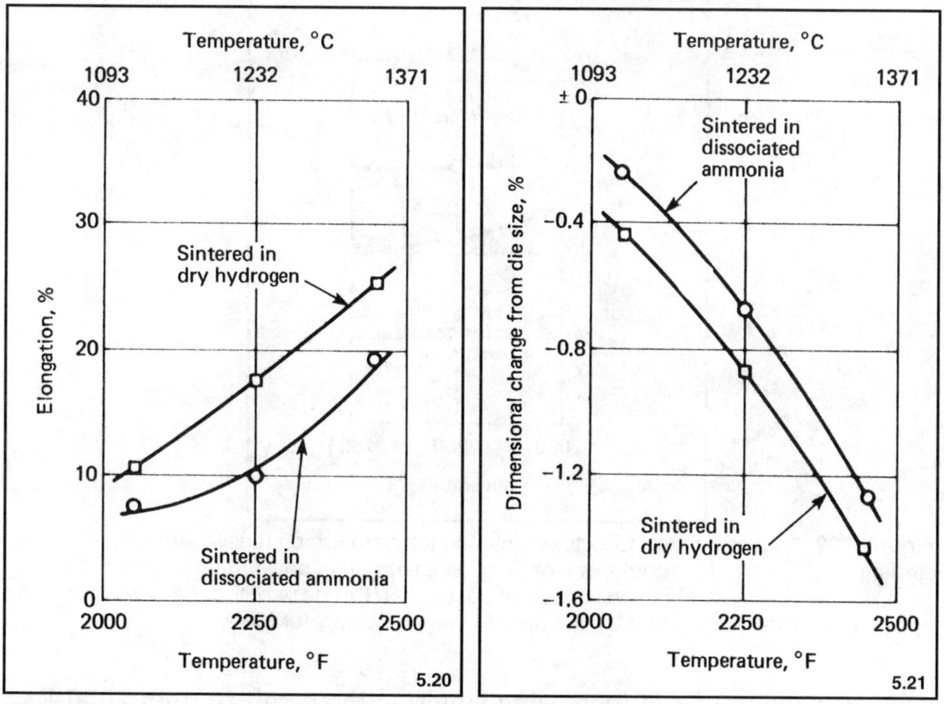

Figure 5.20 and 5.21. Effects of sintering temperature on elongation and dimensional change during sintering of type 316L stainless steel powder pressed to 6.85 g/cm^3 and sintered for 30 minutes in various atmospheres. Courtesy of Powder Metallurgy, Applications, Advantages and Limitations, American Society for Metals.

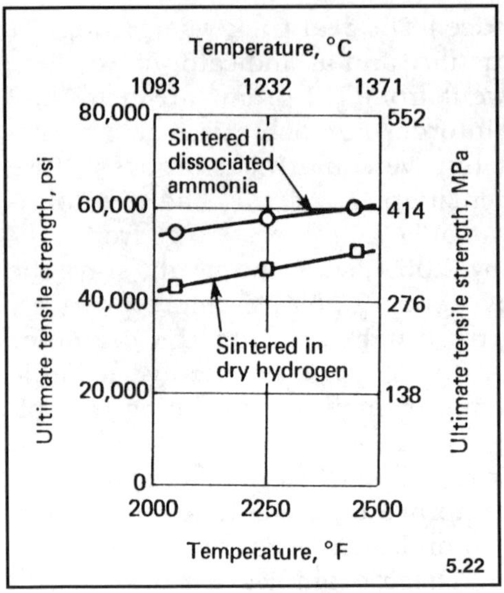

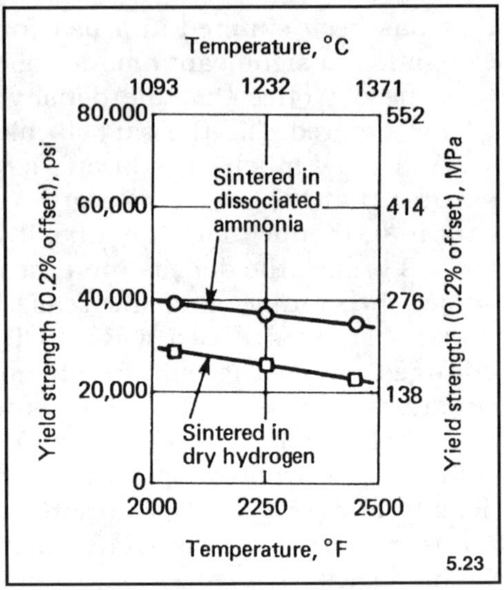

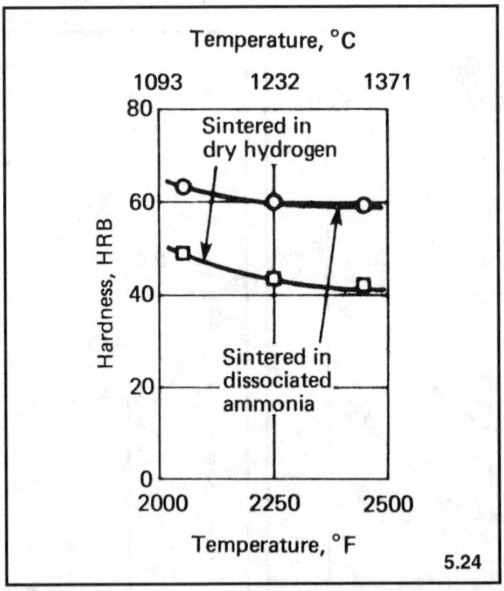

Figure 5.22, 5.23 and 5.24. Effects of sintering temperature on tensile strength, yield strength and apparent hardness of type 316L stainless steel powder pressed to 6.85 g/cm^3 and sintered for 30 min in various atmospheres. Courtesy of Powder Metallurgy, Applications, Advantages and Limitations, American Society for Metals.

show you the trends of increasing sintering temperature from 2050°F to 2450°F, and to compare atmospheres of hydrogen with dissociated ammonia, for 30 minutes at temperature. Figure 5.20 shows the shrinkage. Note that hydrogen at 2450°F cause shrinkage of about 1.5%, linearly. Armed with the handy approximation above, the parts would shrink about 3 × 1.5% in volume or 4.5%. The density goes up by a factor of 100%/(100% - 4.5%) or 4.7%. If you are puzzled why the 4.5% decrease in volume gave a 4.7% increase in density, think about the effect of a 50% decrease in volume, which will increase the density by 100%! (Note the analogy in this calculation to the one up above to get at the size of a die that compensates for dimensional change during sintering of the part.) The elongations of 10 to 25% in Figure 5.21 are some of the highest that we ever see for a sintered steel part. In Figures 5.22, 5.23, and 5.24, note how the D.A. atmosphere makes the stainless steel stronger and harder. This is nitrogen alloying: about .2 to .6% nitrogen from the D.A. dissolves in the stainless steel and strengthens it. You can have some fun by taking these data and comparing them with the N2 and L grades for 316 as published in Standard 35 and see what kind of job the standards committee did!

Sometimes sintered or presintered parts are repressed for higher density or closer dimensional control, as in Chapter 6. The sintered parts have to more or less fit in the tools and co-author WW lives and "dies" on calculations of die size!

CHAPTER 6

Repressing

Sintering is certainly a crucial step in the P/M process as you have just learned. Good sintering techniques produce many parts that are ready for use as sintered. The bad news is that there is some size change and distortion that takes place during sintering.

The size change is predictable but does depend on density. For materials that grow in sintering, lower density results in less growth. For materials that shrink in sintering, lower density results in greater shrinkage. If we have a green part that is very uniform in density, we can have a sintered part that is very uniform in size change. Unfortunately we don't live in that ideal world so many close tolerance parts need secondary operations after sintering. There are also many parts that require higher densities than can be obtained by conventional pressing and sintering.

One method for solving these situations is repressing. The ASTM B-9.01 Subcommittee defines re-press, "to apply pressure to a previously pressed and sintered compact. Includes coin, re-strike, size and double press double sinter." All cause plastic deformation and close pores. Restriking reduces volume, thereby increasing density and improving strength, hardness and dimensional accuracy. Another version of repressing can also be used to modify the surface shape such as embossing [coining]. Sizing primarily improves dimensional accuracy with little increase in density or strength.

Restriking is usually done with ferrous materials and probably the easiest one to work with is F-0000 pure iron. As sintered it is quite ductile and therefore can be repressed to a higher density without putting too much stress on the tooling. Densities of 7.3 gm/cc can be attained at pressures of 35/40 tons per square inch (tsi). There is a pressure versus density curve somewhat similar to that of compacting, so higher densities dramatically increase tooling loads. Figure 6.1 is a pressure vs. density curve for North American Höganäs Distaloy AB. Because this is a partially alloyed material, repressing pressure is higher than that required for pure iron.

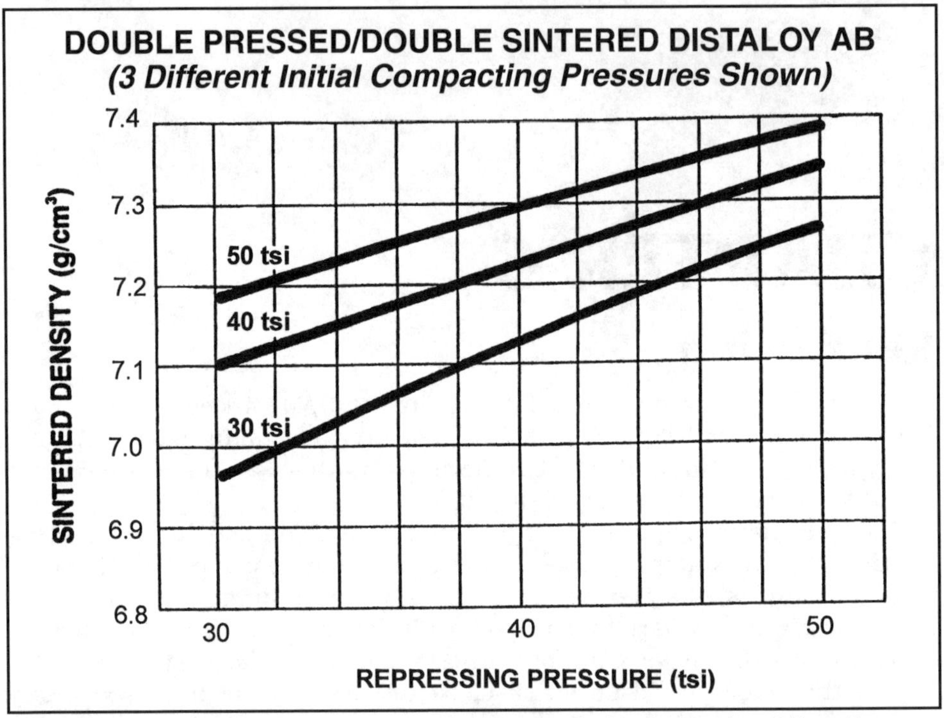

Figure 6.1. This figure is a pressure vs. density curve for North American Höganäs Distaloy AB.

Copper steels and nickel steels can also be restruck but the pressures are higher than those necessary for pure iron. This is because they are less ductile and have higher yield strengths. Ferrous materials with other alloying elements such as manganese, phosphorous, etc. usually require an additional process step when repressing is required. These materials are pre-sintered at temperatures of 1300 to 1550°F. This results in softening the iron and creates a low state of sintering of the iron in the part. It does not begin the alloying process where the carbon dissolves in the iron. The parts can then be restruck to increase density and then fully sintered to obtain the high strength and hardness of alloy material. The disadvantages are increased cost and some size variation during the final sinter. This process is called Double Press, Double Sinter for obvious reasons. As we mentioned in the chapter on Compacting, there are new techniques being developed which can eliminate the necessity for repressing many parts which require densities up to 7.4 gm/cc.

Sizing usually involves non-ferrous materials but we routinely size ferrous materials such as FC-0205, FC-0208 and FC-1000. Sizing was first used in the manufacture of bronze bushings and is still necessary today for those applications. Properly sintered bronze can distort considerably and its size change is greatly affected by density. Most bronze bushings need close tolerances because of the nature of their applications and therefore sizing is necessary. They are usually used to support steel shafts turning at medium to high speeds and loose fits would cause vibration and excessive wear of the bushing and shaft.

As we mentioned earlier, pressures are low and sizing has little effect on density. Tool wear and breakage are greatly reduced compared to restriking and the process is usually much faster.

Tooling for repressing is subjected to both shock and abrasion. Carbide dies and core rods are used wherever possible for abrasion resistance but dies which have thin sections may result in breakage and steel may have to be used for these applications. Small carbide core rods are subject to breakage and therefore coatings on steel such as titanium nitride are frequently used in these applications. High-speed steel cores are also used but can wear out quickly and must be carefully monitored for wear. Punches are usually made from medium carbon chrome steels such as S-7 or A-9 hardened and tempered to a hardness of Rc 56-58 to improve toughness.

The dies and core rods usually have a lead-in section which is .002/.004 larger than the finished size of the die cavity and .002/.004 smaller than the core rod finished size. This allows the part to fit into the tooling and be properly aligned before pressure is applied. Lead-in areas are usually ¼ to ½ inch in length and then blend smoothly to the finished size of the tool component.

Most parts are tumbled before restriking. This process eliminates sharp edges and enhances the loading process. Care must be taken to make sure no abrasive remains on the part from tumbling and any liquids must be removed from pores. Liquid removal is usually accomplished by cycling the parts in a heated spin dryer. Sizing usually does not require lubrication but repressing usually does. There are several lubricants marketed specifically for P/M repressing and some parts makers have their own “magic” formulas that may include metallic stearates in a liquid carrier or oil mists. Once again liquids cannot be allowed to fill pores because this would create a hydraulic system when pressed and result in tool breakage.

As in compacting, there are several different types of presses used for restriking. Most presses are mechanical but there are some hydraulics, particularly in the larger sizes. Figure 6.2 shows a small [12 ton] sizing press widely used for small bronze or iron bushings. Parts are fed automatically from a vibratory bowl feeder down a track. At the bottom of the track is a nest area on a pivoting arm that retains the part but is open on one side. The pivot arm is actuated by a cam on the press. Figure 6.3 shows the track and the part in the nest of the pivot arm. The cam moves the end of the arm to the center of the die and a core pin with a lead in moves up and through the hole in the part that is at the end of the pivot arm. The pivot arm then moves away from the die area to pick up the next part at the end or the fixed track. The die and upper punch are mounted in the top ram of the press and they move down to capture the part. Figure 6.4 shows a schematic diagram of the tooling movements that take place during this cycle. Friction between the part and the die forces the part down over the full size portion of the core rod and finally the part is compressed slightly between the top and bottom punches. The

Figure 6.2. 12 ton automatic sizing press.

part height is usually reduced about 5% and this plastic deformation confined by the die and core rod qualifies the size of the hole.

The press ram then starts upward and the part remains in the die as it pulls off the core rod. After the part is off the core rod the upper punch ejects the part out of the die. This cycle is then repeated and very rapid cycle times are possible. Speeds as high as 3000-4000 parts/hr. are common.

Figure 6.5 shows another type of press used for restriking. It uses a dial feed that indexes through 360 degrees and is driven by gearing from the press crankshaft. Parts are loaded in a nest that is shaped to control the orientation of the part. This dial has 12 stations and parts are loaded at the front or six o'clock position. Loading is either by hand or from an automatic feeding device depending on part quantity. As the dial indexes around counterclockwise,. sensors check to make sure there is a part in

Figure 6.3. Track and part in swing arm and on core.

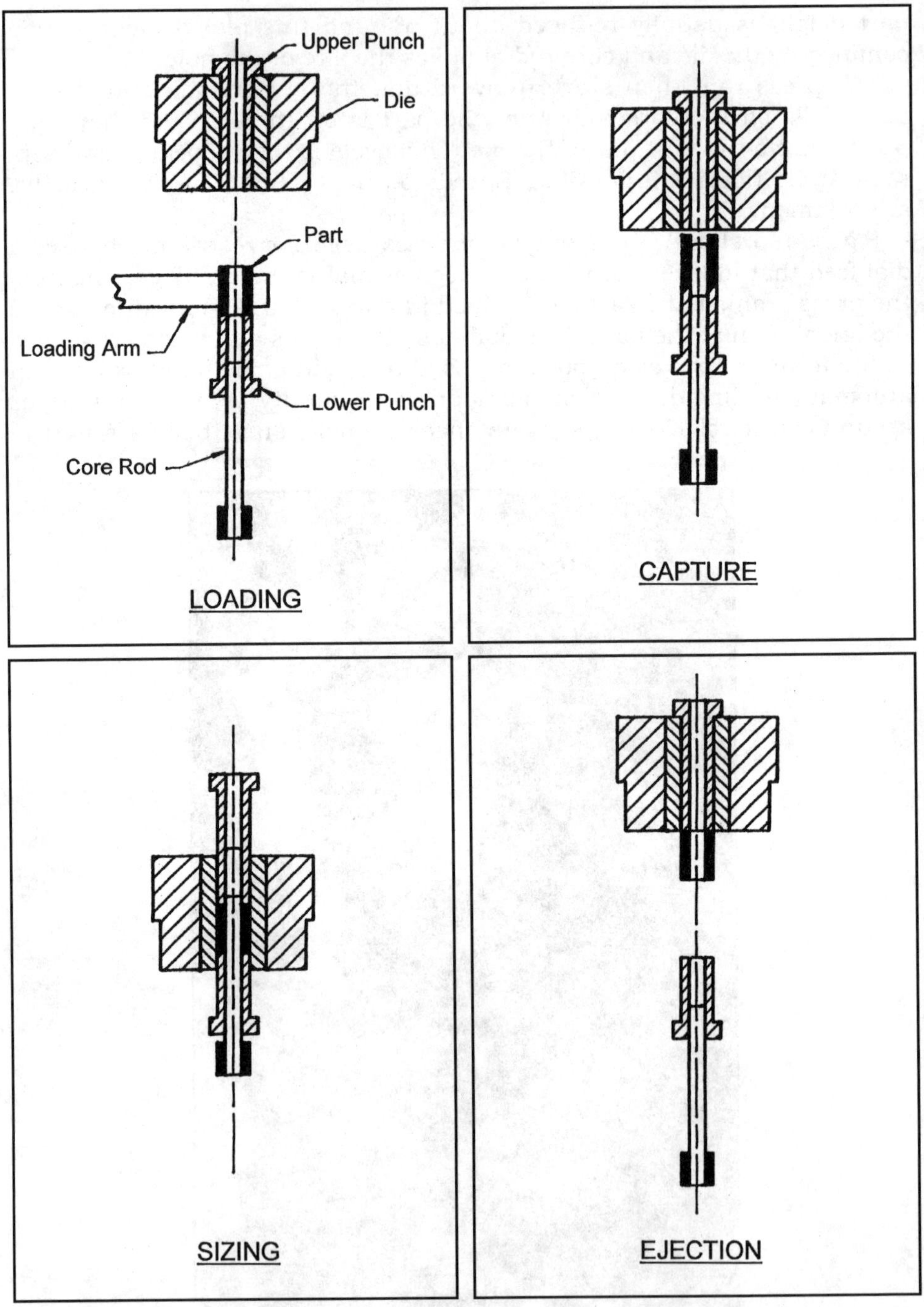

Figure 6.4. Automatic sizing cycle.

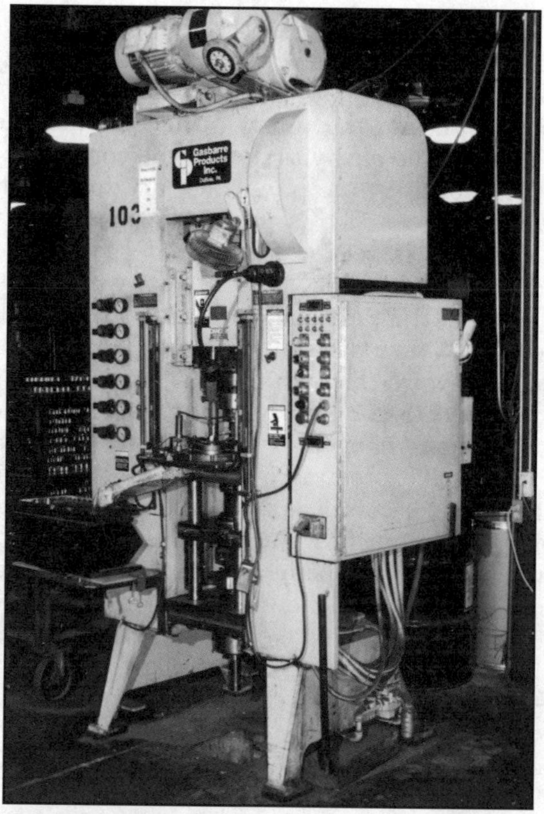

Figure 6.5.a. 30 ton dial feed press.

Figure 6.5.b. Close up of dial.

each station and that each part is properly seated so that it will feed smoothly into the die. At the twelve o'clock position, the station is over the die cavity and the part falls into the lead-in section of the die. The upper ram comes down and the upper punch pushes the part from the lead- in section into the die cavity and onto the core rod.

The upper punch continues to push the part down until the part contacts the lower punch face. At this time, deformation of the part begins and the part thickness is reduced to the desired dimension thereby increasing the density. The upper punch then withdraws and the lower punch pushes or ejects the part back up into the station in the dial. The dial then indexes and a new part falls into the lead-in area. The finished parts are ejected downward through the dial by an air cylinder located at about nine o'clock. Speeds for this type of press depend on the size and complexity of the part and can range from 300 to 1200 parts/hr.

A third type of restrike press is a single cycle hand feed press. This press is generally used for complex shapes that are difficult to feed automatically or lower part quantities that do not justify automation.

The die is usually mounted in the top ram of the press, similar to the swing arm press. The operator loads the part on the core rod lead-in and then pushes dual buttons to engage the clutch. It is necessary to hold down a switch button with each hand in order to engage the clutch. This prevents reach- in accidents and if either button is released early, the clutch will disengage.

The upper ram moves down and the part is forced into the die and over the core rod by friction and lower punch contact. As the ram continues downward, the part is squeezed between the punches and confined by the die and core rod. The upper ram moves away and usually the part stays in the die because there is more frictional force against the die wall than on the core rod The upper punch hits a stop and the die continues upward thereby ejecting the part.

Figure 6.6 shows some parts for the sporting firearms industry that have been repressed. Some are double pressed double sintered because they are made from alloy steels. These parts require repressing because they need high density for strength, ductility, hardness, size control and surface finish.

Figure 6.7 shows some parts for the business machine industry that have been sized. Some are bronze and others are iron or steel. The main reason they have been sized is for dimensional control. Tolerances after sizing for these parts are typically +/- .0005 in. on diameters up to 0.500 in. Length tolerances are typically +/- 0.003 in. on lengths up to 1.000 in.

Restriking costs vary considerably depending on the part size and complexity. Small simple parts such as bushings can cost as little as

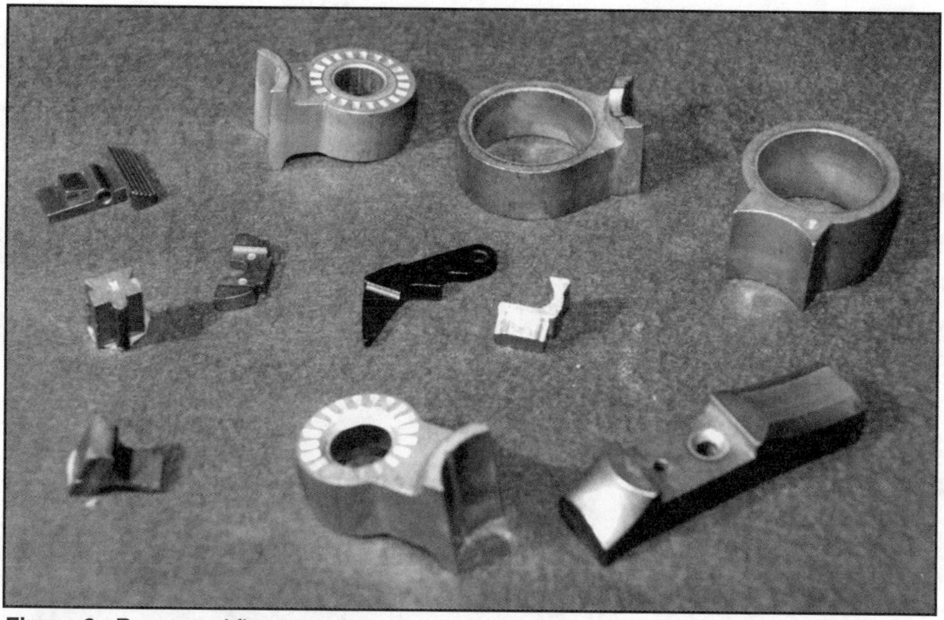

Figure 6. Repressed firearm parts.

Figure 7. Sized business machine parts.

$12/hr. if they are automated and the operator can run more than one press. Larger more complex parts which must be hand loaded can cost as much as $40 to $50/hr. Production speeds for complex parts can be in the order of 200/300 parts/hr. and small simple parts can be sized at rates of 2000/3000/hr. The end result of repressing which is a stronger, more precise part, usually justifies the additional cost. This, however, is an area where discussion with the end user sometimes can result in redesign, thereby eliminating the necessity for this operation.

CHAPTER 7

Secondary Machining

One of the obvious goals of making a part by the P/M process is to achieve net shape as pressed and sintered. There are, however, many parts produced which require varying degrees of secondary machining. The conventional molding process cannot produce parts with cross holes, undercuts or threads so these features must be added later by machining. Machining may also be required because of sections that are too thin to be molded or where there are multiple levels requiring more punches or molding functions than are available or practical. The value added by machining in many instances is greater than the value of the P/M blank and secondary machining can be a considerable profit generator.

There are some parts that are machined "green" but because of low strength in the green state, this occurs only on rare occasions. A paper presented at the 1997 International Conference on Powder Metallurgy and Particulate Materials by Gerald Gegel and others from Caterpillar Inc. investigated drilling of green parts. They used samples made from FC-0208 material that were warm compacted to a density of 7.22 gm/cc. The green strength of these samples was 3715 psi. That is very high and a major factor in the success of green machining. It appears from their results that speeds of 205/290 ft./min. and a feed rate of 7.5 to 13 ins./min. produced the best results. It is important to note that these samples were warm compacted to achieve high density and correspondingly high green strength.

Most parts are machined after sintering and before heat treating. The exception to machining after heat treating is for parts which are ground. Machining P/M parts presents some special problems. Ask anyone who has tried it and you will hear a myriad of complaints ranging from poor tool life to poor surface finish. Those who have had experience machining wrought parts know that interrupted cuts cause problems with tool wear and surface finish. The porosity of a P/M part causes constant interrupted cutting on a microscopic scale, unless the part density is extremely high. There are methods available to combat this problem. We have had excellent results with secondary machining when the following techniques are employed.

One method used to improve machinability is resin impregnation. A detailed description of this process can be found in Chapter 9. The pores of the P/M part are filled with liquid resin which is then cured. Not only does the resin eliminate porosity, but it also provides some lubrication which stops tools from overheating, a major cause of failure. Resin impregnation works extremely well when used in stainless parts. Wrought stainless is difficult to machine and P/M stainless can be a real "bear". We have seen improvement in tool life of 10 times after resin impregnation of stainless. Table 7.1 shows a table from a paper on The Machinability of P/M Materials as determined by Drilling Tests. This paper was presented by Howard Sanderow at the 1997 International Conference and is found on pgs 15-125 through 15-143 of the 1997 Conference Proceedings, "Advances in Powder Metallurgy & Particulate Materials".

This table ranks the machinability of P/M steels with and without additions of manganese sulfide and with and without resin impregnation. Look at the difference in performance of FC-0205 material in the "plain vanilla form" compared to resin impregnated. The results show the same number of holes drilled at 25% of the drill force with 50% less wear on the drill cutting edge! This improvement certainly justifies the cost of resin impregnation.

Another method of eliminating porosity is copper infiltration and that was discussed in Chapter 5. This is an expensive process. When we copper infiltrate, we mold a slug or disc from copper infiltrating powder which must then be placed on or under the iron or steel part to be infiltrated. The slug weighs from 15% to 30% of the iron or steel part and infiltrating powder costs 3 - 3.5 times the cost of iron powder. The molding and loading of the slug, and a slower sintering rate, added to the copper cost, make this an expensive process. The result is that copper infiltration raises the part cost between 35% and 50%. For this reason, it is not used solely for machinability improvement. Increased strength, sealing for pressure resistant applications and plating preparation are other common reasons for copper infiltration. Infiltrated parts machine extremely well yielding both good tool life and excellent surface finish on the part. Once again, the elimination of porosity and lubrication from the copper are the reasons for success.

Sometimes additives are used in the powder that aid machinability. Lead can be added to brass in amounts of 1% to 2% by weight. This addition makes the material much easier to machine, just as it does in wrought brass. Because of health concerns connected to lead, powder producers are working on other additives and we hope they will soon "get

TABLE 7.1. RANKING OF P/M STEEL MACHINABILITY (AVERAGE)

Material	# Holes Drilled	Ax Drill Force	Cutting Edge Wear
FD-0208	6.5	664	
FN-0208	22	648	
FC-0208	27	624	
FL-4605	28.5	667	
FL-4205	42.5	637	
FD-0405	53.5	535	
FD-0208+.5MnS	67.5	394	
FD-0405+.5MnS	81	364	
AISI 1045	122	479	.017
FL-4405	128	496	.017
FLN-4205	134.5	548	
FLN-4205+.5MnS	139	292	.016
FN-0205	178	494	.016
F-0008	192	488	.011
FC-0205	192	469	.014
FD-0205	192	416	.007
FC-0208+.5MnS	192	391	.012
FL-4205+.5MnS	192	387	.013
F-0005	192	377	.016
FN-0208+.5MnS	192	369	.013
FL-4605+.5MnS	192	365	.013
FC-0205+.5MnS	192	336	.008
FN-0205+.5MnS	192	305	.015
FD-0205+.5MnS	192	304	.015
F-0008+.5MnS	192	300	.013
F-0005+.5MnS	192	231	.011
FC-0208+IMP	192	193	.008
FC-0208+.5MnS+IMP	192	161	.007
FN-0208+IMP	192	153	.005
F-0008+IMP	192	140	.004
FC-0205+IMP	192	123	.008

the lead out". Small percentages of manganese sulfide [.5% to .7%] can be added to ferrous materials and considerable improvements in machinability have resulted from these additions. Some tests have shown that iron parts with certain types of MnS additions tend to rust more than those without MnS so there is a downside to this approach.

We referred to lubrication in our discussion of resin impregnation and copper infiltration. This is an important issue in machining P/M, just as it is in machining wrought materials. Keeping the cutting tool cool leads to longer tool life and better surface finish. Air is sometimes used when liquids can't be and there are some new methods used to deliver cold air to the cutting area. Ideally, one uses water soluble cutting fluids; these do not cause any special problems if the part porosity has been sealed. If sealing has not been done, fluids will be absorbed by the part and may have to be removed after machining.

These fluids can sometimes be removed by using a heated spin drier, sort of like your clothes drier at home. (We do not recommend trying this at home.) A more costly but successful method is to re-sinter the parts at lower than normal temperatures and shorter times. This will eliminate fluids but care must be taken not to change size or hardness. A third method is vapor degreasing that uses solvent vapors to dissolve trapped fluids. This method is not widely used because of ecological concerns.

A good summary of how machinability of P/M steels is affected by factors we have discussed is found in the conclusions of Sanderow's paper referred to earlier in this chapter. These conclusions are as follows.

For P/M steels the machinability is affected by:

A. Carbon Content-0.5% much better than 0.8%
B. Composition
   Plain carbon steels best
   Pre alloyed steels less machinable than copper or nickel steel premixes
   Diffusion alloyed steel machinability strongly dependent on carbon content
C. MnS-0.5% optimum for best machinability
D. Resin Impregnation-most effective for improving machinability
E. Density-not a significant factor [6.7-7.2 gm/cc range]
F. Special base irons
G. Reduced iron very poor machinability
H. QMP 29M superior to Domfer MP37

The machinability of P/M stainless steels is very poor, much lower than P/M steels. Resin impregnation dramatically improves the machinability of P/M stainless steels. The P/M copper base alloys [bronze, leaded brass] exhibit very good machinability, much better than the P/M steels.

We have already discussed improving machinability by additives such as MnS or resin impregnation. Another approach to improving machining performance is to make changes to the cutting tool. Walter Ball (now retired from Zinc Corporation of America) collaborated with W. Shelton of Hertel NA and Chuck Delong of Symmco, Inc. on one approach to machining problems. Walter presented a paper concerning the machining of nonleaded brass at the 1997 MPIF technical conference. We mentioned the concerns about leaded brass earlier in this chapter. This paper focused on changes in tool geometry and composition to improve the machinability of nonleaded brass.

They tested carbide sculptured-edge three fluted (SE)(TF) drills vs high speed steel conventional chisel point drills. They drilled both leaded and unleaded sintered brass plates. The conventional drills were totally unsatisfactory for drilling nonleaded brass. Holes were egg-shaped and chipping of the edges of the holes was common. Drills also exhibited very rapid wear. The carbide sculptured-edge drills produced over 3,000 holes, in tolerance and at a much faster production rate. Figure 7.1 shows the geometry of the drills which were used and Figure 7.2 and 7.3 show the feed rates and numbers of holes drilled. These results are certainly dramatic and our experience with production parts confirms this improved performance.

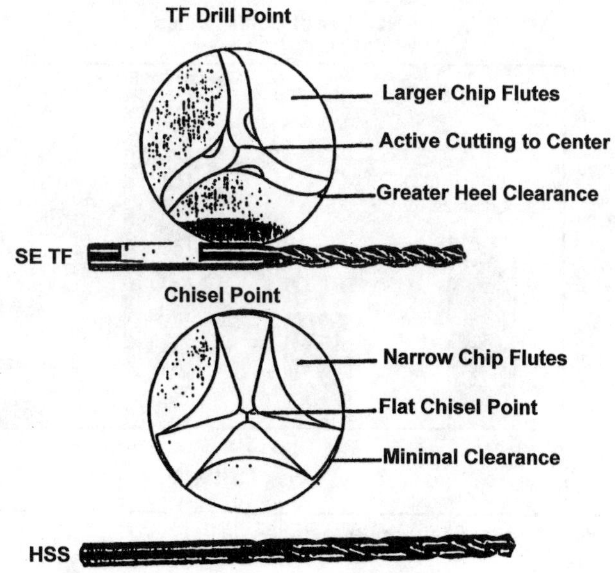

Figure 7.1. Drill geometry.

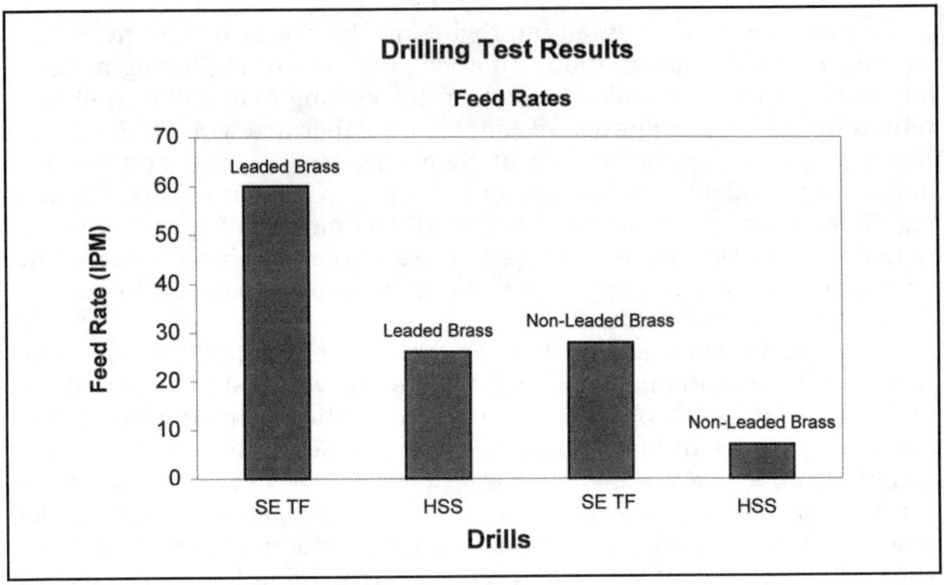

Figure 7.2. Drilling test results-feed rates.

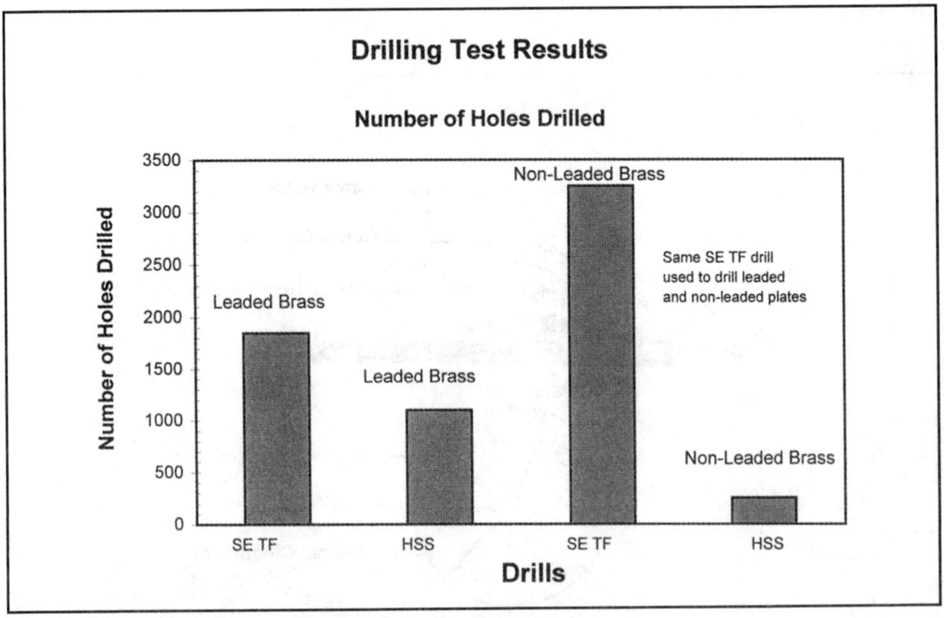

Figure 7.3. Drilling test results-number of holes drilled.

It is important to note that the cost of the SE TF drill is about 10 times that of the conventional drill. The SE TF drill can be re-sharpened from 10 to 20 times at a nominal cost while the conventional drill is usually discarded because of its low cost. Another critical factor is that the SE TF drills must be used in a very accurate machine like a CNC machining center not a drill press. This is because they must run at high speed and have excellent alignment. Carbide is quite brittle and poor alignment or vibration will cause breakage.

Robert Causton of Hoeganaes Corp. reported on Machinability of P/M Steels at the 1995 MPIF Conference on Powder Metallurgy & Particulate Materials. He investigated machining several P/M steels including Hoeganaes Ancorsteel 85HP, 2%NI, .5% graphite. This is a high strength material which has an apparent hardness of 80HRB as sintered. It is very difficult to machine for those reasons. The results of his drilling tests illustrate this. Table 7.2 shows data from the drill tests and we have included his comments on these results. "They (the results) indicated that individual improvements to high speed steel, flute form or coating had relatively little effect on drill life under the test conditions employed. However, when combined, drill life increased significantly."

At National Sintered Alloys, we perform some machining operations on high volume production parts made from the same material (Ancorsteel 85HP, 2%Ni, .5% Graphite). We have to remove significant amounts of material in a turning operation. Some surfaces have recesses which cause an interrupted cut and one finished diameter has a tolerance of plus or minus 0.0004 in. The machining is done in a robotics loading CNC lathe and a rough cut and a finish cut are necessary in order to attain surface finish and tolerance requirements.

A special grade of carbide is used for the rough cuts and cubic boron nitride (CBN) material is used for the finish cut. The tool inserts are triangular and can be rotated when worn so each insert has three cutting

TABLE 7.2. EFFECT OF DRILL TYPE UPON DRILL LIFE WHEN CUTTING ANCORSTEEL 85HP:2% NICKEL, 0.5% GRAPHITE

Drill Steel Type	Coating	Flute Form	Point	Drill Life
Base HSS	None	Standard	135°	2
M42	None	Standard	135°	2
Base HSS	TiN	Standard	135°	1
HSS	None	Parabolic	135°	1
Cobalt HSS	None	Parabolic	135°	3
M7HSS	TiN	Parabolic	135°	44
Cobalt HSS	TiN	—	Split	85

edges. We initially tried using carbide for both cuts and were only getting 400 parts per edge. By switching to CBN we are able to machine 3,000 parts per edge, a dramatic improvement! We cannot use CBN for the rough cut because the interrupted cut puts too much shock load on the tool insert and it chips.

P/M parts present special machining problems but we have shown some of the methods currently used as solutions. The following are general guidelines for performing various machining operations [from "The Powder Metallurgy Design Manual"]. The recommended speeds and feeds vary with the material being cut. Softer materials usually produce better results with faster feeds and higher speeds vs. harder materials.

TURNING-Sharp pointed tools are recommended. Carbide is used for softer materials and titanium nitride coated carbide or cubic boron nitride for harder materials. Cutting angles on single point tools should be 15 degrees, back rake about 10 degrees, and reliefs about 10 degrees. See Figure 7.4 Cutting speeds of 0.8 to 1.6 m/s [175 to 350 sfpm] and

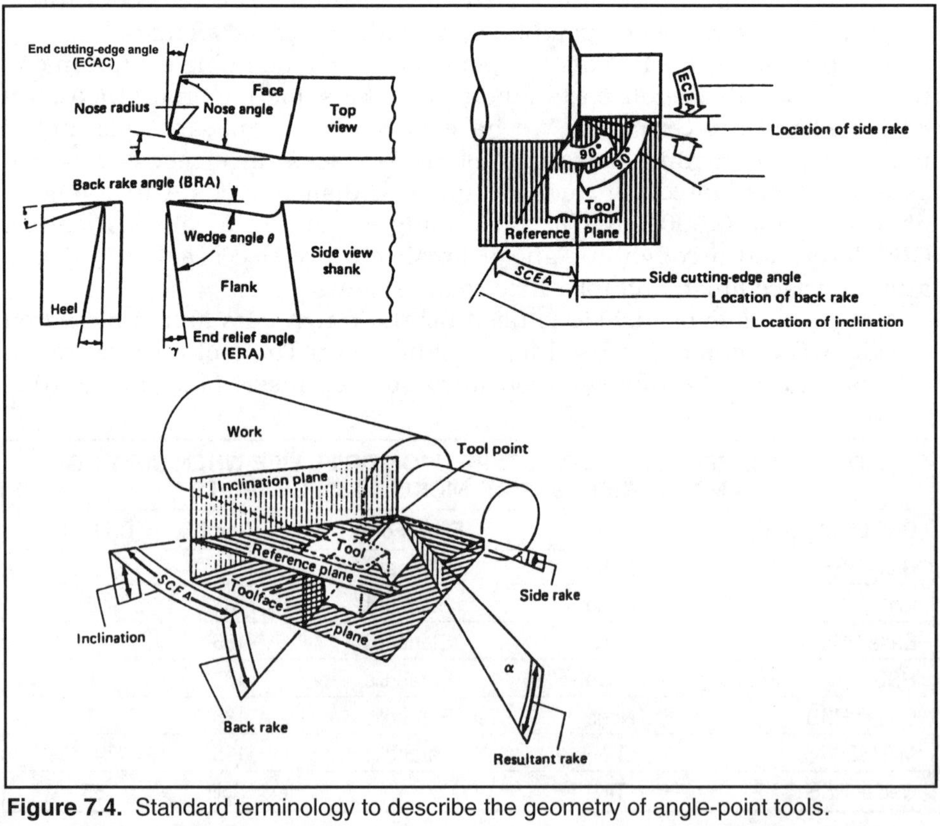

Figure 7.4. Standard terminology to describe the geometry of angle-point tools.

feeds of 0.05 to 0.1 mm [0.002 to 0.004 in.] are normally used. Depth of cut is preferably at least 0.1 mm [0.004 in.] to avoid surface rubbing, which work hardens the material and accelerates tool wear. When closing of surface microporosity must be avoided, slow speeds of 0.5 to 0.75 m/s [100 to 150 sfpm], lower depth of cut of 0.05 mm [0.002 in.] and very sharp tools are used. Figure 7.5 shows a typical turning operation.

MILLING-Cutters of high speed steel and carbide are recommended. Helical cutters with axial rake are preferred so that chips are sheared on an angle. Recommended speeds are 0.36 m/s [70 sfpm] for high speed steels and up to 1.52 m/s [300 sfpm] for carbide tools. Feeds should be 0.05 to 0.13 mm [0.002 to 0.005 in.] per tooth for roughing and 0.03 to 0.05 mm [0.001 to 0.002 in.] per tooth for finishing.

DRILLING and TAPPING-Speeds and feeds used for drilling P/M components are usually 80 to 85% of those for wrought metals of the same composition. Nitrided steel, high speed steels and carbide drills are recommended for long life. When holes in the direction of pressing must be drilled, they can be spotted or located for drilling by pressing in a small conical depression so that the drill has a starter hole. This is usually done when the hole is too small to be molded or located too close to an edge or another hole. If we tried to mold the hole, the tool would be fragile because of the thin wall section. Holes can also be partially formed or

Figure 7.5. Turning.

chamfered in pressing the component, thus simplifying the drilling operation. Internal threads are produced by conventional tapping methods. Figure 7.6 details a P/M part being tapped. Premium grade steel and nitrided taps produce excellent life. Titanium nitride coated taps may also be used on harder materials. Roll form tapping is also used, especially in blind holes. Roll form tapping works best with ductile materials such as F-0000 straight iron and some of the brasses. The material is not cut as in conventional tapping but displaced to form the threads. This means that material from the largest or major diameter of the threaded hole is pushed inward in the hole to form the root or minor diameter of the thread. It's sort of like pushing a fork down on a piece of cookie dough. Dough under the fork tines is depressed and some flows up through the spaces between the tines.

GRINDING-Grinding operations similar to those for wrought products apply to P/M components. However, where surface microporosity is required, grinding may be unacceptable because it tends to reduce or close microporosity. For hardened materials and high precision, cubic boron nitride grinding wheels are very effective.

Machining almost any material results in a burr or raised sharp area where the tool starts or finishes its cut. These burrs are caused by pressure from the cutting tool which moves material aside. At the start and

Figure 7.6. Tapping.

finish of the cut, the material is not well supported and therefore protrudes from the part surface. P/M materials are not an exception to this fact and the greater the ductility, the larger the burrs. Tapping usually causes larger burrs than drilling because of the tool shape and resultant forces on the part. Burrs usually present problems because they cause oversize conditions on the outside surface of parts and under size conditions on the inside surfaces. P/M does offer a design advantage; because, for a small tooling charge, a burr relief slot can be provided in a hole. Figure 7.7 illustrates this slot. This slot does not eliminate the burr but causes it to form in a space which does not interfere with the hole surface.

If a burr relief slot cannot be used because of adverse affects on part function, the burr must be removed. Countersinking or chamfering the hole is the most common method of burr removal on outside surfaces. Burrs which protrude into holes can usually be removed by reaming the larger hole. There are also chamfering tools which are spring loaded flexible blades. These rotating tools pass through the drilled hole and expand outward with spring force. This allows the tool to cut a chamfer on the inside of the hole when the tool is pulled back through the hole.

Ball sizing is commonly used in place of reaming for hole finishing. A steel or carbide ball, lubricated with oil is pushed through the hole. This process can be done manually using an arbor press or small air press or

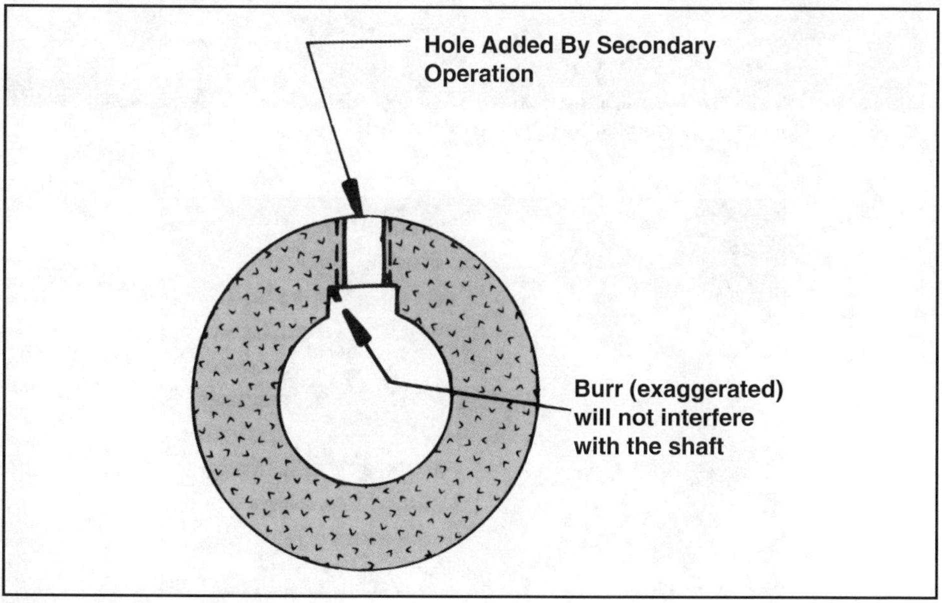

Figure 7.7. Burr relief slot.

automatically if the part quantity justifies. Material displaced depends on hole size and can range from .0005 in. to .0020 in. for small holes to as much as .0010 in. to .0030 in. for larger holes [½ to 1 in. dia.] Finished tolerances of .0005 in. to.001 in. are possible depending on hole size. Excellent surface finishes result because the ball burnishes the porous surface as it passes through the hole. Figure 7.8 shows a typical hole surface before and after ball sizing.

An example of a part which cannot be molded to net or final shape and therefore requires machining is shown in Fig. 7.9. It is a shifter cap for a ratchet wrench and the material is FX-2008-60, copper infiltrated steel.

The three ribs are on the top of the part as it is molded. The largest diameter section is too thin to be molded to finished size. That section is molded thicker and then turned in a lathe to finished size. The stem of

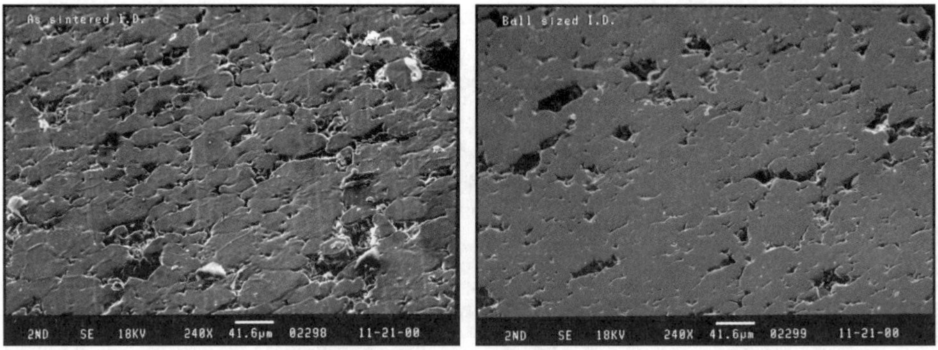

Figure 7.8. Ballsizing (surface shown before and after).

Figure 7.9. Shifter cap.

the part has a slot across it and a blind cross hole. These features are undercuts and as we mentioned in the chapter on Compaction, cannot be molded. The hole is drilled and the slot milled. After machining, the part is tumbled to remove burrs and then bright zinc plated. Remember what we said about copper infiltration. It is expensive, but here is an example where it strengthens the part in difficult to mold, low density areas, helps machinability and seals the part for plating.

Figure 7.10 shows another copper infiltrated part which has several machining operations performed on it. There is a cross hole which is drilled. The inside diameter is bored out larger for a length of about ¼ inch. A series of semi-holes are machined around the edge of the inside diameter using a small end mill cutter which is similar to a drill. A large thread is then machined on the inside of the part, beginning just past the semi-holes. You may wonder why some of these features cannot be molded in the part, which is a very logical concern. The tools would be too fragile and part density could not be attained in some areas, if these features were included. We hasten to point out that the P/M process does provide the square through hole and the series of spherical depressions at the bottom inside of the part. We are also able to provide the depressed lettering on the top of the part. Before this part was converted to P/M, it was machined from a solid chunk of steel and the cost savings offered by P/M were considerable. Once all of the machining is done, the part is heat treated, centerless ground to provide a very smooth finish on the outside diameter and finally, chrome plated.

Figure 7.11 is a part made from SS-316 stainless steel at a density of 6.4 gm/cc. The inside diameter is bored out and the top surface is milled to produce the step on top. These operations are necessary for the same reasons cited in the last example i.e.. thin sections which cause

Figure 7.10. Torque Limiting Tool Components

Figure 7.11. Stainless lock part.

tool breakage and uneven density. We mentioned the benefit of resin impregnation and H. Sanderow also mentioned this in his conclusions, especially concerning the machinability improvement with stainless steel. This part is resin impregnated before machining, resulting in greatly improved machinability, better finish and longer tool life.

CHAPTER 8

Heat Treating of Powder Metallurgy Parts

Hardening iron based materials

We imagine that everyone is familiar with the idea that steels can be heat treated or hardened, to increase their strength and wear resistance. In a general, way you probably realize that the blade in your Swiss Army Knife was somehow heated and quenched in a liquid to develop its high hardness. Most of the P/M parts exiting the sintering furnace have not been quenched fast enough to be thoroughly hardened, and secondary heat treatment is needed to develop maximum strength and hardness. Unfortunately, in order for you to properly understand the process of hardening the steels, we will first have to force a little more metallurgy on you!

When we sintered the iron and graphite mix in Chapter 5, we told you that the graphite dissolved in the iron to form steel. Well, just how does that happen? Do the atoms of carbon jump off the graphite flakes and attach themselves to the iron? Maybe some do, but it seems more likely that the furnace atmosphere helps. There is a little oxygen on the iron, and it can react with the carbon in the graphite to form CO. The carbon monoxide, CO floats around and can decompose on the iron, leaving carbon atoms on the iron surface and some residual CO_2 in the atmosphere. Or, some water, H_2O, or hydrogen could react with graphite and carry carbon to the iron. If the iron is hot enough when this happens, the carbon will diffuse right into the iron and we form steel. Hot enough means over 1333°F, and since we sinter iron at 2050°F, the carbon moves in very readily. Depending on the temperature, iron can exist in one of three different crystal structures. Above 1600°F, Iron has a high temperature crystal structure, called face centered cubic (FCC). This means the iron atoms sit at the eight corners of a cube and also in each of the 6 faces, as shown in Figure 8.1. Strangely, this arrangement of iron atoms, called austenite, is more open than the form that exists at room temperature. This means that the small carbon atoms can quite

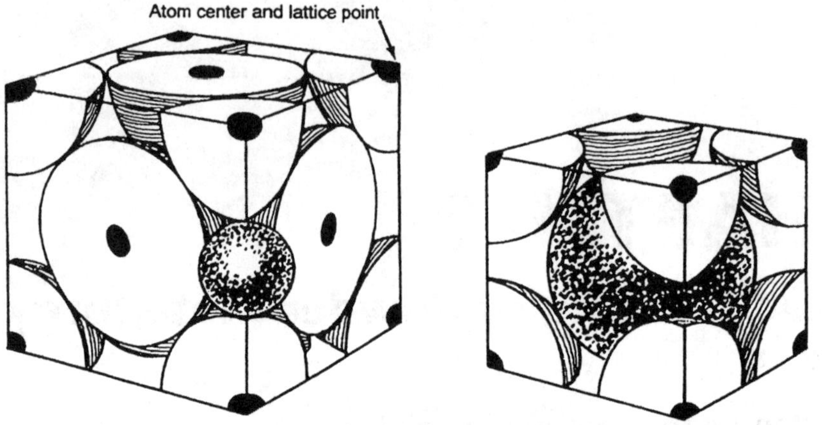

Figure 8.1. Face centered cubic structure of the high temperature form of iron, Austenite. The space between the iron atoms is larger than in the low temperature, ferrite, body entered cube shown on the right. In the FCC we see a small atom of carbon sitting neatly in among the larger atoms of iron. (Courtesy of Metallurgy for the Non-Metallurgist).

easily sit in between the iron ones, and so carbon atoms can diffuse in among the iron ones. The temperature that must be reached to have austenite depends on the carbon content of the steel. With no carbon, the austenite forms at 1670°F. With .8% carbon in the steel, the austenite forms at only 1333°F. This change in temperature with carbon content is neatly shown in the iron carbon phase diagram, Figure 8.2. Remember, a phase diagram shows you, for any given temperature and chemistry (%C in Fe), what phase or crystal structure exists. It is very handy to have when you want to see what happens when you make a mix of two or more elements and heat them up!

The low temperature form of iron is ferrite or alpha phase and also has a cubic arrangement of iron atoms. Again, there is one at each of the eight corners of a cube, and one atom right in the center of the cube, touching each of the eight corner ones. This body centered cubic (BCC) arrangement has the atoms a little closer together than in the high temperature form. So, when we cool a steel with .8%C in solution in austenite down through 1333°F, something has to give. There is not enough room for .8%C dissolved in the low temperature ferrite. Ferrite will hold only .025%C, as you can see by looking at the phase diagram. The carbon has to get out of the array of iron atoms. It does this by forming a completely different crystal, Fe_3C, called cementite or iron carbide. So, the carbon exits the austenite, and forms Fe_3C. This reduces the carbon in the austenite to only .025%, and it cheerfully reverts to ferrite below,

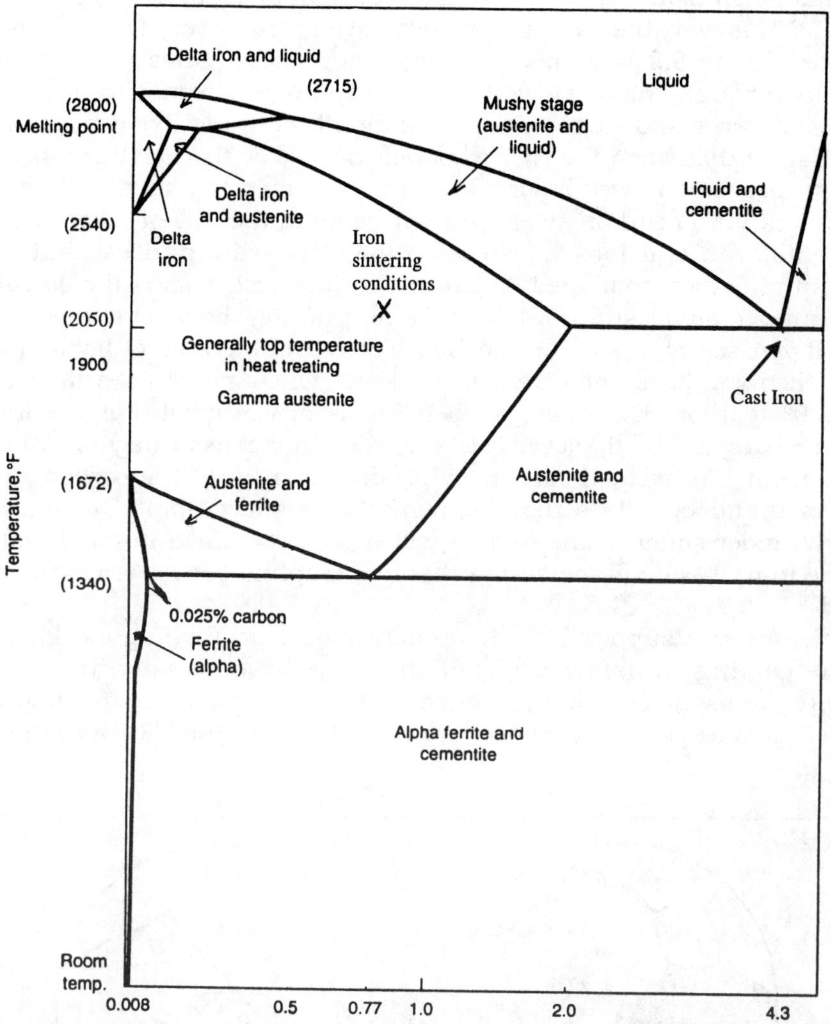

Figure 8.2. Simplified iron carbon phase diagram. Courtesy of "Metallurgy for the Non-metallurgist". A phase diagram shows what phases would be present for each mixture of iron and carbon, if you held it at any particular temperature. For instance when we sinter iron with .8% carbon at 2050°F, see the X on the diagram, we would see only one phase, high temperature iron, called austenite. On cooling slowly to room temperature we would see a mixture of two phases, low temperature alpha iron (ferrite) and iron carbide (cementite).

1333°F. If we look at the result of this transformation (austenite >> ferrite + carbide), what do we see? Is it a big lump of ferrite and another lump of carbide? No, not in our lifetimes. As the transformation takes place, we see that small plates of ferrite are formed, right along side small plates of carbide. This way, the carbon does not have to travel very far to form the carbide. Figure 8.3 is a sketch to show how these plates form. A cross section through them, viewed at 1000X shows alternating layers or stripes of ferrite and carbide, that look for all the world like a fingerprint! We imagine that when the earlier metallurgists saw this mixture, the carbide stripes weren't well resolved, and they named this mixture pearlite. For ordinary iron carbon steels without much in the way of other alloying elements, you can look at the amount of pearlite and estimate the amount of carbon combined as carbide in the steel. This is the so called lever rule. In simplest form, if you see all pearlite, the steel has .8% carbon. If you see half pearlite and half blocky white grains of ferrite (pure iron), then you have half of .8% C or .4% C. Of course 0% pearlite would mean 0% carbon. Everything in between is proportional. For the iron - 2%Cu + carbon mix, the lever rule still works quite accurately to estimate the percent of *combined* carbon. What does it mean if the carbon is *not combined* and how does that happen? if you add graphite to iron and grossly under-sinter, some carbon will not be combined in, or dissolved in, the iron, it will still be sitting there as graphite particles, i.e. uncombined.

But, this was supposed to be a chapter on heat treating steel. Right! In the cooling transformation of the .8% carbon austenite to form pearlite, we assumed there was time for the reaction to occur. That is, a few minutes elapsed on cooling from 1333°F to say 1000°F. But suppose

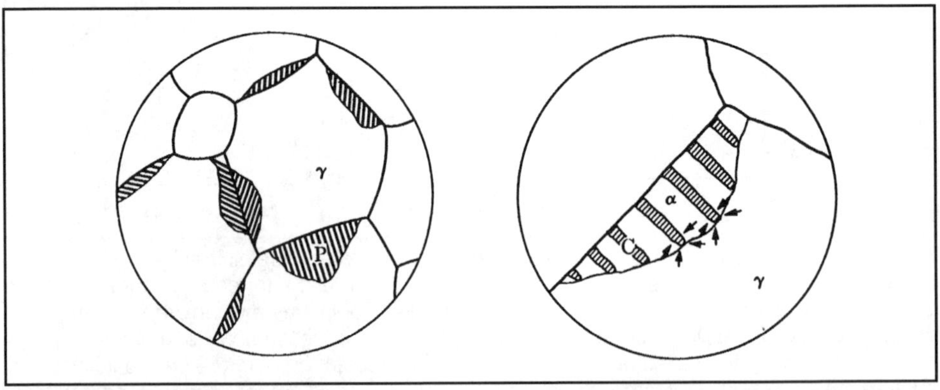

Figure 8.3. Transformation of austenite to pearlite, on cooling below 1340°F. The alternate platelets are nucleated at the austenite grain boundaries on the left. On the right, we see that the carbon must leave the .8% C austenite and move to form the 6.4% C cementite platelets, leaving behind ferrite iron platelets that have about .025% C. This fingerprint-like structure is called pearlite. The transformation from one solid phase to a mix of two solid phases is called a eutectoid decomposition.

we took the material at a temperature above 1333°F and quenched it in oil at 140°F. We are not going to give the carbon atoms time to leave the austenite and form iron carbide. The material becomes severely stressed as it is cooled, because there is no room for the carbon atoms. At about 500°F, something dramatic happens. Some of the highly stressed austenite instantly transforms into another crystal form, martensite. This is called the martensite start temperature. The martensite itself can hardly accommodate all the carbon in between the iron atoms, and it is too highly stressed. It is a very hard structure. It is hard, because with all those carbon atoms pushing on the iron, dislocations can not move through the crystals, and we can not easily deform the martensite. As the temperature gets lower and lower, more and more of the retained austenite transforms to martensite, until the austenite is all used up and we have all martensite. That is the martensite stop temperature often below 0°F.

The martensite is very hard, but it also is very brittle. We need to relieve some of the internal stresses by tempering the steel. If we heat the martensite up to 300 to 400°F, some of the trapped and overstressed carbon atoms can leave the martensite and form fine precipitates called epsilon carbides. These carbides are not large enough to be seen in the optical microscope. Reducing the internal stress in the martensite lowers its hardness a little but dramatically increases the toughness. It also increases the ductility of the steel and increases the tensile strength. Interestingly, if a steel is too brittle, and you pull on it in tension, it will fail prematurely. The first little crack will carry right across the part. So, lowering the hardness can actually result in a stronger steel. It is our practical experience that as quenched P/M parts and test bars are never as strong as tempered ones. Also most P/M steels with 10 or 20% of a softer phase than martensite will be stronger than one that has all martensite.

To see the effect of hardening, consider a 6.9 g/cm^3 sintered nickel steel FN-0205: as-sintered, it has a tensile strength of 50,000 psi and 2.5% elongation at fracture. After hardening and tempering 1 hr. at 500°F, the tensile strength jumps up to 120,000 psi and the elongation drops to <.5%.

Hardenability

Hardenability is the ease of hardening a steel. It is related to how quickly austenite transforms to pearlite on cooling below 1333°F. If the pearlite tends to form very quickly, then we have to quench very fast in order to avoid it, in order to form martensite. If we give the low hardenability steel half a chance on cooling, it will form some amount of pearlite in the range from 1333 to 1000°F, and not much of the austenite will be left to survive all the way down to the martensite start temperature. Which steels have this difficult characteristic of not hardening easily? The plain iron carbon steels are the most difficult, and among these, the lower the carbon content, the more likely that some amount of pearlite and ferrite will form at 1333 to 1000°F. Metallurgists have time temperature transformation diagrams to remind them of how quickly they have to quench a given steel to avoid forming the softer pearlite.

Figure 8.4 shows a T-T-T diagram for a 1080 steel. That is essentially plain iron with .8% carbon in it. Notice that the time scale is logarithmic. The different areas on the graph show what constituents would be present if you could cool very quickly down to a given temperature, like 1000°F and then hold the work for the times shown. At 1000°F you can see that in about 1 second, the part would begin to change from austenite to a mixture of austenite, ferrite and carbide. The ferrite + carbide would be in the form of pearlite. By the time 5 seconds had elapsed, the austenite would be all gone, and you would have all pearlite (ferrite + carbide). So, if you don't want any pearlite in your hardened steel, you need to cool so quickly that by the time 1 second has elapsed, you are down past 1000°F, or, as metallurgists like to say, down past the pearlite nose in the T-T-T curve! If you do cool quickly past the nose, and then slow the cooling rate so that you enter the δ ferrite + carbide zone, the δ phase does not transform to the fingerprint-like pearlite you are used to. Instead, it makes a feathery mixture of the two, called bainite. Once you reach the martensite start temperature of 500°F, ferrite plus carbide will never form; the only possible transformation of the austenite is to martensite. It starts to happen at 500°F and it finishes at about -60°F. If you never cool all the way to -60°F, then you will always have a little austenite left over in your martensite. You should remember that each alloy steel and carbon level has its own T-T-T curve, and some are more complicated than ordinary .8% carbon steel (1080 steel).

If ordinary carbon steels are not easily hardenable, what is? Alloying elements like molybdenum Mo and Ni are a great help, as is 2% Cu. Low alloy P/M steels often have up to 1% Mo and 2% Ni to increase the hardenability. Other useful elements are manganese, chromium, tungsten and vanadium. We like Ni and Mo because they do not oxidize too easily in our sintering furnaces. Greater care is needed in sintering alloys with Mn and Cr. We need steels that do not readily transform to pearlite on cooling so that when we quench a real part, even the inside, which does not cool as fast as the surface, will still cool fast enough to form substantial amounts of martensite.

There is a quantitative way to describe the hardenability, so that you can find the material with the best hardenability. The Jominy end quench test gives us the number by which we judge how hardenable the

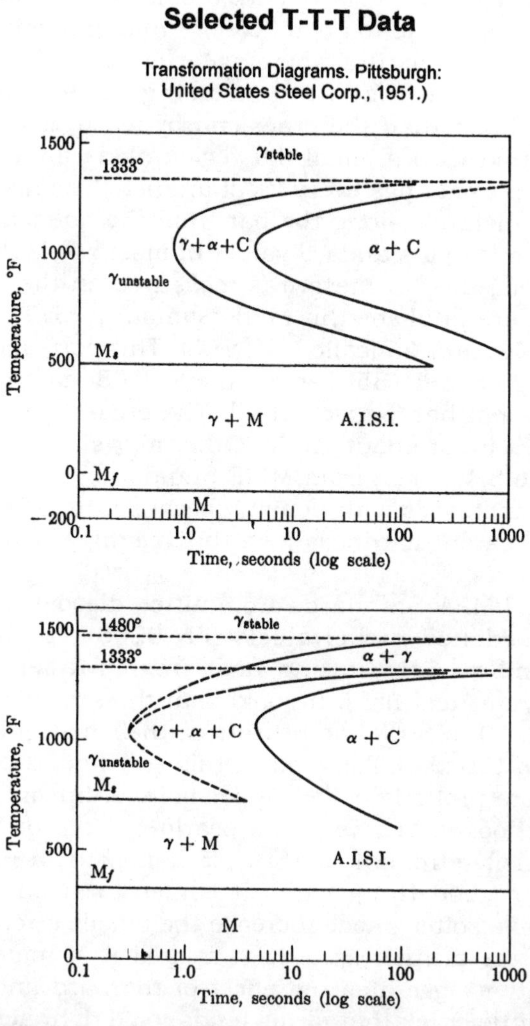

Figure 8.4. Isothermal transformation diagrams from "Atlas of Isothermal Transformation Diagrams", U.S. Steel, 1951. The data are for steel with .8%C (called AISI 1080) and .4% C (called AISI 1040). To develop this data, the steel was first heated above 1333˚F. It was then quenched to a lower temperature, 1000˚F and held for a defined time, 1 second. It was then quenched to room temperature, thus preserving whatever phases were present. The steel could then be examined to see what had been present at the elevated temperature. Four our choice of time and temperature, the investigator would have seen that all austenite, γ gamma, had been present. Repeat this process for other times and temperatures and you generate maps like these. For 1080 steel, if you want to have all austenite left, to eventually transform to martensite, you had better cool this steel down below 1000˚F in less than one second, or some of the other phases will start to form, i.e. α ferrite and carbide in the form of pearlite.

steel alloy will be. In that test, a 1 in. diameter by 4 in. long bar of the sintered steel is heated up to about 1550°F in a protective atmosphere. It is then dropped into a fixture where water is sprayed on the flat face at one end. You can see that the end being hit by the water will be quenched a lot faster than the other end by which it is suspended. The bar is then ground with a small flat region along its length, so that we can take hardness readings every 1/16 of an inch. The hardenability number is then the distance along the bar from the quenches end at which the hardness drops to a standard value. In most steels this is Rockwell C 50. In P/M steels, the hardness was measured on the Rockwell A scale and those data are published in MPIF Standard 35. The hardness level chosen was Rockwell A 65 (called HRA 65). This corresponds to HRC 52. As an example from MPIF 35, Table I. the F-0005 material has a depth of <1/16 in., i.e does not harden well at all. The prealloyed FL-4605 hardens to 2.5 units of 1/16 in. or about .16 in. Other alloys are even better, as you can see in Table 8.1, taken from MPIF Standard 35. Note that the hardenability depth increases with density, because the hardness measuring method is an apparent hardness test, that averages pores along with the solid steel.

In 1981, the PMPA Standards Committee discovered an interesting thing about hardened nickel steels (FN-0208). In those days, each committee member sintered and heat treated a series of identically pressed 6.85 g/cm^3 test bars. In looking at the strength, Figure 8.5, the companies with 10 to 30% fine pearlite in their martensite had the best strength (105 to 110 ksi UTS). (Today, this very fine, two phase, platelike structure is more properly called upper bainite, but in 1981 we saw, in the SEM, what looked like very fine pearlite). More or less fine pearlite reduced the tensile strength to 75 to 85 ksi. Thus, a little fine pearlite, 10 to 30%, is a good thing when it comes to strength. Why should adding a little of a softer phase increase the tensile strength? We think it has to do with the brittleness of heat treated P/M materials, where the residual porosity exaggerates the effect of the hard and brittle martensite. The least little crack that forms tends to go right across the test bar. But, if you add some HRC 40 fine pearlite (upper bainite) to some HRC 59 martensite, the cracks tend to be stopped or blunted or deflected by the softer material. This is also the logic behind the nickel steel and the diffusion alloys being so strong: the centers of the nickel-rich regions are soft and tough, and blunt the crack movements. This beneficial effect of the softer phase becomes less important when the density approaches the pore-free density, as in a powder forging or an MIM part. It also becomes less important when the pores are really round and not like a bunch of small notches. That happens in sintering at 2400°F for press sinter parts or in MIM parts.

TABLE 8.1

HARDENABILITY DATA
Depth to 65 HRA

Material System	Material Designation	Density g/cm³	Depth in 1/16 inch units J_{65}
Iron & Carbon Steel	F-0005	6.65	<1.0
		6.87	1.0
		7.03	1.0
	F-0008	6.78	1.5
		6.91	2.0
		7.06	2.0
Iron Copper & Copper Steel	FC-0205	6.50	<1.0
		6.82	1.5
		6.96	1.5
	FC-0208	6.40	1.5
		6.81	2.0
		7.15	2.5
Iron Nickel & Nickel Steel	FN-0205	6.90	1.5
		7.10	1.5
		7.38	2.5
	FN-0208	6.88	2.0
		6.97	2.0
		7.37	3.0
Low Alloy Steel	FL-4205	6.75	2.5
		7.00	3.5
		7.20	3.5
	FL-4405	6.64	2.0
		6.94	3.0
		7.20	4.5
	FL-4605	6.76	2.5
		6.99	5.0
		7.12	7.0
	FLN-4205	6.68	2.0
		7.00	5.0
		7.29	6.0
	FLN2-4405	6.71	7.5
		7.11	10.5
		7.22	10.5

HARDENABILITY DATA
Depth to 65 HRA

Material System	Material Designation	Density g/cm³	Depth in 1/16 inch units J_{65}
Low Alloy Steel (cont'd)	FLN4-4405	6.72	8.5
		7.10	14.5
		7.23	17.5
	FLN6-4405	6.79	13.0
		7.15	18.0
		7.30	26.0
Sinter Hardened Steel	FLC-4608	6.63	26.0
		7.06	36.0
		7.14	36.0
	FLC-4908	6.72	8.5
		7.08	9.5
		7.16	10.5
	FLN-4608	6.82	22.5
		7.08	36.0
		7.27	36.0
	FLNC-4408	6.65	9.0
		7.06	11.0
		7.22	15.5
Diffusion Alloyed Steel	FD-0205	6.98	2.5
		7.24	2.5
		7.32	4.5
	FD-0208	6.78	4.0
		6.97	9.5
		7.29	12.0
	FD-0405	6.70	2.0
		7.13	4.0
		7.26	10.0
	FD-0408	6.70	3.0
		7.08	8.0
		7.21	15.0
Copper Infiltrated Iron & Steel	FX-1005	7.40	2.0
	FX-1008	7.39	2.5
	FX-2005	7.38	<1.0

Courtesy of MPIF Standard 35, "Materials Standards for P/M Structural parts", 2000 edition.

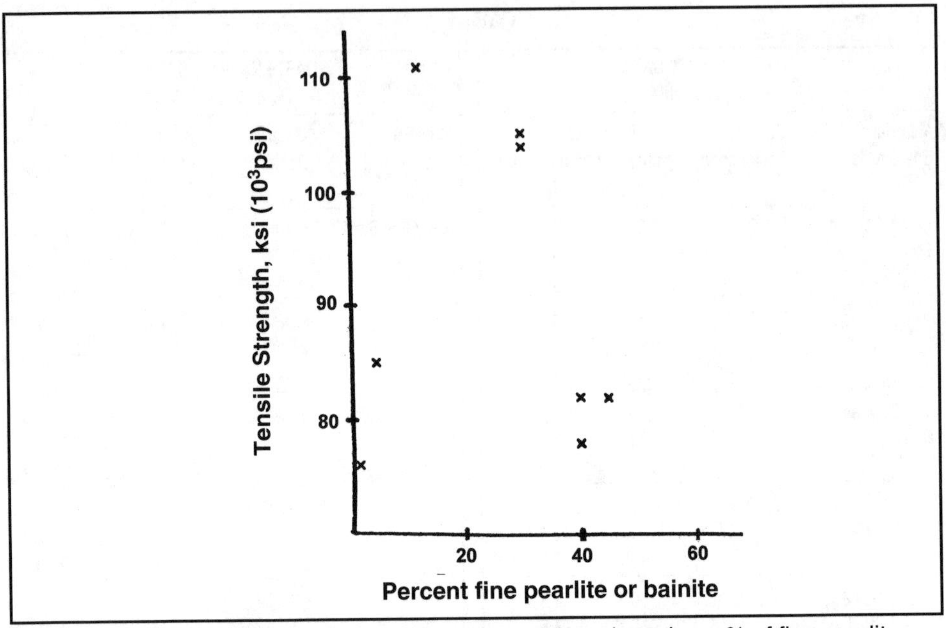

Figure 8.5. Tensile strength of FN-0208 heat treated, vs the volume % of fine pearlite (upper bainite) in the structure. There appears to be an optimum amount of fine pearlite for this sintered steel which was processed by eight different companies. Each point represents a different P/M parts producer, average of three tensile bars.

For years, companies who make the tool steel punches and dies used to compact our powders have known about and used "air-hardening" tool steels. The have so much in the way of alloys added to the iron, that you hardly have to quench them at all, to cool fast enough to form all martensite. The formation of pearlite is very delayed. Toolmakers would wrap tooling in stainless steel foil and heat the assembly in air in a tool hardening furnace, to 2000-2200°F. After an hour or two at temperature, take the bundle out of the furnace and sit it on the floor, to cool in air. Even at that slow cooling rate, the result was all martensite. There are only a few P/M houses that sinter parts made from tool steel powder, but they see the same effect: they come out of the furnace fully hardened. The parts sinter-hardened rather than air hardened. Up until about 1980, there were only a few P/M houses who quietly practiced the art of sintering the low alloy steels while having them harden at the same time. We now have a standard for that material, FLC-4608, and you can see its hardenability in Table 8.1. At 26 sixteenths of an inch, it is 10 times more hardenable than prealloyed steel, FL-4605. As a bar 2 x 8 x 1 in. thick, this material can pass through an ordinary sintering furnace with ordinary cooling and emerge with 80% martensite and 20%

fine pearlite, even at the center. That is a good imitation of the hardening process in Figure 8.5 that gave the highest tensile strength! Its tensile strength will be 90,000 psi, with <1% elongation; properties are in the middle between a fully hardened structure and an as-sintered structure. Further, it has nearly twice the impact strength of the FN-0205- HT, at the same density. The advantage of the process is less distortion than is found in oil quenched parts, and no entrapped heat treating oil to be removed. Remember, P/M houses like to limit tempering of oily parts to 350°F to prevent fires in the tempering furnaces. Without oil in the part, you can temper at whatever temperature gives you the desired balance of strength and impact resistance. Not content to leave well enough alone, those clever P/M parts producers are buying sintering furnaces with accelerated cooling zones. Then they can make a cheaper part with less expensive alloying, cool them quicker, and still get the same hardness! Further, the lower alloying in the powder comes with the benefit of easier compressibility, and so they can offer parts at even higher densities and better properties. At the same time, during 1998 - 2000, powder producers now offer other low alloy steels that harden more easily than FLC-4608, and are more compressible.

Equipment for Hardening and Tempering P/M Steels

The sintered parts are usually heat treated in controlled atmosphere furnaces that have an internal oil quenching system. Figure 8.6 shows a photo of a typical heat treating furnace and Figure 8.7 shows a furnace with a basket on a loading platform, ready to enter the front door. The parts are loaded in baskets outside the furnace, the furnace door is opened and the parts are pushed directly into a chamber at 1500 to 1600°F. The parts can be dumped in the basket if they are not too critical with regard to distortion. Better dimensional control is accomplished by careful arrangement and spacing. Sometimes the parts are hung on wires so that the quenching oil will strike the part in a preferred direction to minimize distortion. The furnace atmosphere can be endothermic gas (40% hydrogen, 40% nitrogen and 20% carbon monoxide with a small amount of natural gas (methane) or it can be nitrogen base with 10% hydrogen and some added methane. The gas composition is adjusted so as to be in equilibrium with the desired carbon level in the part. That level is often .5% to .8% C. Equilibrium means that carbon will not leave the part and enter the atmosphere, nor will carbon be picked up by the part from the atmosphere. Sometimes the P/M part has lost a little surface carbon during sintering. Instead of being .8% C at the surface, it is only .3%. The atmosphere in the heat treating furnace will supply carbon to the part (i.e. carburize it) and bring it back to the desired level.

Figure 8.6. Integral quench atmosphere furnace courtesy of Ipsen International. The parts inside the furnace are surrounded by an atmosphere neutral to the desired carbon content, e.g. .8% or .5% carbon. After 30 minutes at 1550˚F, the parts are lowered from the hot zone into rapidly circulating oil. This quickly reduces the part temperature below 400˚F and converts the high temperature form of steel into hard martensite.

Figure 8.7. Integral quench furnace courtesy of Lindberg Div. Notice the loading dolly in front of the furnace which is holding a typical metal tray. The parts can be carefully arranged in the tray for uniform cooling during quenching, or simpler shapes can be loosely packed atop each other. The tray slides in through the front door, into the heated interior chamber.

After about 30 to 60 minutes, the parts are uniformly heated and the carbon is at the right level. The tray of parts is pulled to a position above the quenching oil and dropped down into the oil. The oil is circulated at a high rate, rushing in between the parts and hopefully uniformly hardening all the surfaces. The oil temperature is controlled, often at 140°F. If the oil tends to get too hot, it is deliberately cooled back down, and vice versa. The oil quickly lowers the temperature of the austenite, whisking it down through that temperature range of 1333° to 1000°F where the softer pearlite might form. At the M_s, martensite begins to form, and by room temperature, the part is pretty well all converted over to martensite. The parts are then washed to remove excess oil and tempered as soon as practical after the hardening.

We have been talking about the possibility of distortion occurring on quenching. How does it come about? Picture dropping a nice flat steel ruler from a 1550°F furnace into the 140°F oil. The wide flat side that the oil hits most effectively will start to contract, and the ruler will bend in the direction of the contraction. When the other side gets cooled and contracts, the bend may not get completely reversed. We picked the example of the ruler, because the quenching stress could bend it a lot easier than a 1 in. thick steel bar. Smart people will either buy their rulers already hardened, or will quench them by having the long narrow edge hit the oil first, hopefully uniformly quenching the opposite flat faces. Trick question: if the ruler is dropped in so that the numbered side is down toward the oil, which way does it bend, toward the numbers or away from the numbers. Strangely, the top, un-numbered side is the direction of bending. The side facing down traps a few bubbles of gas and cools slower than the top side, so the top side cools faster and the curl is toward it! (Lessons we learned at our heat treater's knee.)

Well, what else can go wrong? If the carbon content of the furnace atmosphere is too high, e.g. >1%, the surface of the part becomes overloaded with carbon. Once there is more than .8% carbon in the part, at temperature, massive Fe_3 C carbide can form in the austenite grain boundaries. This intergranular carbide embrittles the part. Also, excess carbon can result in not all of the austenite transforming to martensite. Some of the austenite is retained as little triangular regions in among the martensite needles, leading to the term "retained austenite". People with very precise parts worry that at low temperature, say -20°F, that austenite will suddenly transform to martensite. Martensite is lower density than ferrite or austenite, and the part may then grow a little. Further, the martensite is untempered, so it is brittle. These people will refrigerate their parts to -80°F, transform all the retained austenite to martensite, and then temper them. Other people think a little retained austenite is pretty nice. It is a soft and ductile phase mixed in with those hard and brittle martensite platelets, and this increases strength and toughness.

Assuming the parts do not distort, do they shrink or grow on hardening? That depends on the hardenability of the alloy and whether you are looking at OD's or ID's. If a part forms all martensite, since martensite is lower in density than ferrite and pearlite, it will grow uniformly, pretty much like thermal expansion. That is, both the ID and the OD grows. That probably describes small parts with sections less than ¼ inch. Once you get to larger parts, the interior does not form all martensite, and so it does not grow as much. Picture a ring with a ¾ in. wall, of material of low hardenability, e.g F-0008 with only 1½ sixteenths in.=.094 in. depth of hardening. The OD transforms to martensite and grows outward. The interior does not harden much and stays the same size. The ID forms martensite and wants to expand outward in the direction of the OD. The interior says "I'm not moving, do what you need to do!" The ID does the only thing left and expands into the hole's ID, reducing the diameter of the hole, i.e. the ID shrinks.

What do real people do when faced with this dilemma? Folks who have been in the business a long time have made enough hardened parts with enough problems on dimensional control that they start by quoting wider tolerances. Then they reason by analogy and experience of a part that they once made that looks something like this and make a guess as to how much the size change will be. When they are wrong, they fix the problem by machining the part or building a new set of tools. "But", you ask," what about finite element analysis and expert systems?" Maybe some day.

Case Hardening

Case hardening is a very common heat treatment for wrought steels. A low alloy material of low carbon content, e.g. .2%, is selected. It is heat treated in an atmosphere neutral to .8 to 1.0% carbon. The surface picks up .8%C to a depth of .010 in. On quenching, the surface is easily hardenable and forms martensite on cooling. It cools a little faster than the interior. When the martensite forms, the surface layer expands outward a little. The interior is more slowly cooling, and contracting back toward its original size, unchanged. As the interior continues to cool, it pulls back on that martensite layer, trying to drag it back to a smaller size. That puts the martensite in compression. This is a good arrangement. First, the surface is now a nice hard wear resisting martensite. Second the interior is soft and tough. Third, the martensite surface being in compression is now much stronger. When someone wants to break that exterior martensite by pulling on it in tension, it must first overcome the compressive stresses on it. Only then, can the external force go on stressing it, to try to reach the ultimate strength of the martensite in tension.

The lessons of case hardening have not been lost on the P/M world. Their application, however, is something of a problem. If we take some nice FN-0200-20 nickel steel at 7.0 g/cm^3 with a tensile strength of only 35,000 psi, we are tempted to case harden the surface to a tensile strength of 145,000 psi, typical of FN-0208 heat treated. The problem is that during the 30 to 60 minutes of carburization in the heat treating furnace, the carbon not only diffuses in through the iron, its' carrier gas (CO or CH_4) also moves easily in through the pores. For a part with a 1/4 in. thick section, you probably now have enough carbon all the way to the center, to fully harden it. There will be no actual case. It is not practical to use very short carburizing cycles like 5 to 10 minutes, because the work is not uniformly heated. One way around the problem is to get the density up to 7.3 g/cm^3 so the pores will not be connected, and a definite case can be formed. It is also possible to close the pores by copper infiltration of low carbon iron. Since most low carbon irons grow during copper infiltration, it is necessary to use a sponge iron to hold down the growth. (Remember Harold Harrison!)

We have done some work with vacuum carburizing, using the C. I. Hayes vacuum carburizing furnace. Here, the work is heated in vacuum to the austenitizing temperature. Because the work is in vacuum, no carbon is picked up. Once the work is at thermal equilibrium in the furnace, small amounts of methane are introduced. The methane breaks down on the surface of the parts and supplies carbon for carburizing. The carburizing time can be as short as desired, e.g. 10 minutes. The work is then drawn back into a separate chamber and oil quenched. We have formed definite cases on pure iron at a density of 6.8 g/cm^3 by this method.

How do you measure case depth? In the powder metallurgy section of the ASTM Book of Standards, B02.05, there is a test method, B 721, "Microhardness and Case Depth of Powder Metallurgy (P/M) Parts" or in MPIF Standard 52. You need a microhardness tester and a properly polished section through the sample of interest. Figure 8.8 shows a microindentation hardness tester and Figure 8.9 shows 100g Knoop microindentation hardness marks in a polished as-sintered P/M sample. The sample must be properly prepared so that we can see all of the porosity in it. (The ASM Handbook "Powder Metal Technologies and Applications", explains how to polish a porous material.) Using the microscope on the microindentation hardness tester, you look at the polished section and select a region to test for hardness. At 200X magnification, it is relatively easy to find a solid spot in between the pores. That way, you get the hardness of the pore-free material and don't average in any pores. You need to pay attention to what phase or microconstituent you are checking. At 200X in a heat treated steel you could see martensite and fine pearlite that we have earlier described. You could also see

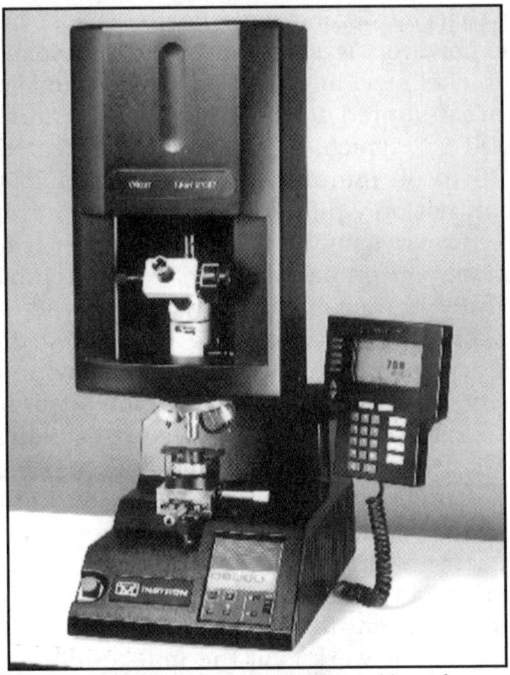

Figure 8.8. Microindentation hardness tester with closed loop force control. By revolving the turret, one can select magnification, load, or indenter. Interfaces with a computer for acquisition of images, stage movement, data collection, and analysis. Photo courtesy of Wilson®, a trademark of Instron®, Corporation.

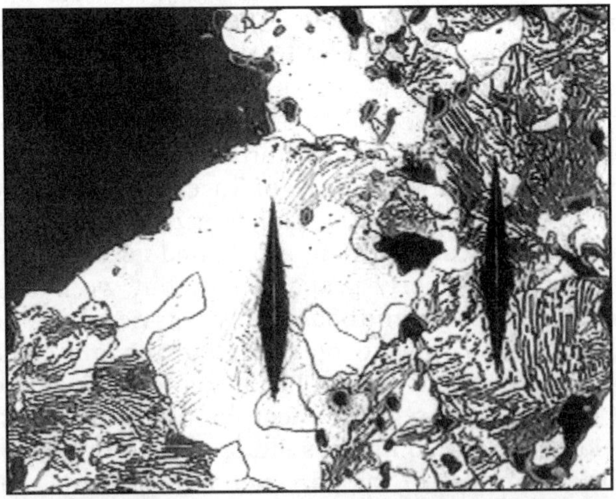

Figure 8.9. 100g (force), Knoop indentation in as sintered steel. The mark on the left landed on white ferrite and under-etched pearlite, the one on the right is on a pearlite colony, the finger-print-like material. The left hand mark is .08mm long implying a hardness of only Rockwell B85-90 or $Knoop_{100}$ of 221. Properly hardened steel would be $Knoop_{100}$ 650-750.

the light colored centers of nickel-rich regions, or copper in a mix with greater than 2% copper. It is customary not to land on the copper or the nickel-rich regions. It is OK to land on the martensite or the fine pearlite but you need to list which one you "Knooped".

With the target selected, you release the machine to cycle. It carefully lowers a loaded diamond point down onto the surface of the sample. The applied force can be varied, but 100g is typical, The diamond point can be elongated (Knoop, preferred) or a square base, making a small square hole, (Vickers, not preferred). The eyepiece is used to measure the length of the elongated diamond shaped indentation, often to an accuracy of .1μm. The Knoop hardness is calculated in terms of kg/mm^2. Hard steels are often 550 $KHN100_{100}$ (Knoop 100) or about 50 HRC. You can plot the Knoop 100 data vs distance from the surface, and when it drops below about 550, we say we have reached the depth of the case. Case depth hardnesses are often specified as HRC 50, i.e. when the hardness falls to HRC 50, you have found the edge of the case hardening. How do you convert your Knoop data to HRC? What you don't do is look at your Wilson wall chart for apparent hardness and Knoop hardness and read the HRC value that corresponds to the Knoop 500g value. The 500g load is too high, and even with pore-free materials, you don't get the same result when you use a 100g load vs a 500g load. The answer is to make your own conversion chart. You take some of your old Wilson HRC standard test blocks and cut out some small pieces. Then mount them in epoxy resin, oriented so that you can do microindentation hardness on the face that was standardized. Lightly polish them to remove scratches and then lightly etch them. For each block take 5 readings with the 100g load and note the average length of the indentations in micrometers or filar units. (Filar units are the arbitrary scale units that you see in the eyepiece of the microindentation hardness tester.) Finally, make your own conversion graph of HRC vs filar units or micrometers.

Carbonitriding

We have just seen that carburizing is carried out by admitting a little natural gas CH_4 or propane (C_3H_8) into the heat treating furnace atmosphere. The gas breaks down on the iron and the carbon is released to move into the iron and harden it. In carbonitriding, you first carburize the P/M material, and then, late in the heat treat cycle, the nitriding is performed. You admit about 2% of ammonia gas (NH_3) into the heat treating furnace for 30 minutes. It mixes with the endothermic gas in the furnace and then breaks down to form individual (nascent) atoms of nitrogen on the surface of the steel. The process takes place at 1550 to 1600°F. Nitrogen diffuses into the austenite, just like the carbon, and forms a shallow case. On quenching, the nitrogen strains the iron crystals, as did the carbon, and the effect is additive to the carbon. The martensite that forms is even harder than when carbon is used alone, about HRC 65, as quenched but untempered. The nitrogen also increases the hardenability of the steel, i.e. it hardens well even under less than effective quenching conditions. This means that the surface of the part will show fewer soft spots from slack quenching. The nitrogen embrittles the steel to some degree and so carbonitriding is best done on parts that will see abrasion or rubbing, but not high impact. To help reduce brittleness, parts should be tempered for one hour at 350 to 400°F.

You will remember that good hardenability really means that the austenite is quite stable and does not rush to transform to soft pearlite on cooling below 1333°F. It holds back and then transforms to martensite at the M_s temperature. Well, the nitrogen stabilizes the austenite. It does such a good job that some of it never does transform to martensite, i.e. it stays as retained austenite. This austenite is relatively soft, but tough, and helps overall toughness of the part. On the other hand, it lowers the overall hardness a little.

Ferritic Nitrocarburizing and the Lindure Process

Wrought steels can be heat treated in a nitrogen containing atmosphere to develop a hard white nitride layer (Fe_2N) at low temperatures, e.g. 950°F. The wrought steel should contain elements like aluminum or chromium that tend to react with nitrogen to form hard nitride phases. The steel is first heated to 1550°F, oil quenched, and then tempered at 900 to 1000°F. It next goes into a pit type furnace or retort furnace where it contacts a mixture of 70% ammonia plus nitrogen and hydrogen, at 950°F. The cycle can take 24 hours. At the end of the cycle, the furnace is cooled and purged with nitrogen. The parts are not quenched. The structure will be a hard nitrided surface on top of a very well tempered core. The surface can have a microhardness of HRC 68 to a depth of .004 inches. In the microscope we see that the surface has the white layer, and the underlying tempered martensite has the nitride precipitates in it. The advantage of the process is that not much distortion occurs during the nitriding cycle. (We suppose all the distortion occurs during the initial quenching, and must be corrected by machining prior to nitriding.) The Fe_2 N white surface layer, (.0002 in. to .002 in. thick) is relatively brittle and is generally removed by machining or by chemical means.

This is a lot to teach you about nitriding wrought steel, only to say that it is not commonly done to porous P/M parts! For one thing, we don't usually make low alloy steels with Al or Cr, to make them easily nitridable. That said, it is possible to form a wear resistant epsilon (ε) nitride layer $Fe_{2-3}N$ on P/M parts by ferritic nitrocarburizing. Ferritic is the key word. We don't heat the iron or the FC-0208 high enough to form austenite, so on cooling, there are no distortions of the type that happen when quenching austenite from 1550°F. The process takes place in one - two hours in a closed retort at 1060 to 1165°F while the iron is ferritic. The atmosphere gas is ammonia with some other hydrocarbons. To keep the gas from penetrating into the interior of the part, and causing both expansion and embrittlement, the parts need to be 92% of the pore-free density, i.e. 7.25 g/cm^3. In the patented Lindure version of the process, a lower density P/M part can be steam blackened to fill the pores with Fe_3O_4 (see steam blackening information later in this chapter). It is important that only a thin oxide layer be used to avoid a reaction with the nitrogen that hurts the nitrided layer. The oxide keeps the nitriding gases out of the pores and results in a well defined case with only .001 in/in expansion from sintered size. After the nitriding step, the parts are oil quenched to avoid precipitation of other nitrides that lower the fatigue resistance. Processing of up to 1000 lbs. of parts costs about $150. per furnace hr. and the floor to floor time is about 3 hrs., or $.45/lb. Some thought must be given to the arrangement of the parts so that one part does not touch or cover another, which would keep the nitriding gases away from the part's surface. This is also true of many other heat treating processes. In addition to the nitriding cost, one must add the cost of steam blackening for parts that are in the density range of 6.8 to 7.15 g/cm^3, to seal porosity, about $.06 to $.10/lb. Lower carbon parts are more suitable for this process than higher carbon, because the carbon interferes with the nitrogen diffusion.

The epsilon nitride white layer produces high hardness, (it resists a HRC 55 file hardness test) and enough lubricity to reduce coefficient of friction by 50%. The layer is porous on a fine scale and holds lubricating oil to increase the part's surface lubricity. This white layer does not have to be removed, because it is a different chemistry from the one discussed above for wrought steel. Because the nitrided layer is in compression, a 7.1 g/cm^3 FC-0205 material had its fatigue limit raised from 18,000 to 39,000 psi. If you look in Standard 35, you will see that FC-0208, when fully hardened and tempered, has essentially the same endurance limit at this density. The process works well on parts that are subjected to sliding wear or fretting wear. It does not provide high indentation hardness resistance. In one test on synchronizer hubs that had been ion nitrided in Europe, the ferritic nitro-carburizing process produced better life and hardness.

Ion Nitriding

No, we did not leave out the "r". People really do bombard "iron" with nitrogen "ions" and form hard nitride layers on the surface. This happens in a vacuum chamber partially back-filled with nitrogen gas and hydrogen gas. The pressure is between 1 and 10 Torr. (As a reminder, normal atmospheric pressure is 760 Torr or the 30 in. of mercury that you see on the weather channel. One Torr would be about .1% of atmospheric pressure. And if you have already forgotten what a Pascal is, there are 132 of them in a Torr). At 700 to 1200°F, a voltage of 200 to 1000 volts DC is applied between the vessel walls and the work pieces. The nitrogen molecules, N_2, begin to break apart into nitrogen positive ions, N^+. They migrate (stream) toward the P/M parts that are negatively charged. When they get near the parts, a negative electron from the steel part jumps over to the positive N^+ and neutralizes the + charge. When the electron lands on the N^+ ion, the ion emits a photon (a small amount of light). This light creates kind of an eerie glow inside the chamber. The "neutered" nitrogen atom continues on its track toward the P/M part and slams into it. It then diffuses or moves into the part. If we could slam the nitrogen molecule, N_2, into the hot iron, not much would happen. The molecules are too big to dissolve in the iron.

Now we have to describe the condition of the part which is under siege by this nascent N. Usually, the part has already been hardened and tempered, probably at 700°F or higher. If the part is going to form nitrides, as we noted above, Al and Cr are useful elements. Failing those elements, (and most prealloyed P/M alloys don't have much Al or Cr in them), some Mo will do, as in FL-4600 or FL-4900. The nitrogen forms a compound layer on the surface (a so-called white layer). The layer is .0003 in. thick and is called gamma prime (γ') Fe_4 N, iron nitride. The layer has a file hardness equivalent to an HRC 52 file. But, the N diffuses in well past this layer. First, it dissolves in the iron and hardens it. Then it forms very fine nitride precipitate particles with the iron and the other alloying elements. The precipitate particles strain the tempered martensite and make it more difficult for planes of atoms to slip by each other. The hardest region of this diffusion zone is within .004 in. of the surface, but some beneficial effects persists to .008 to .012 in. The hardness at the surface depends on the alloy being nitrided: a low alloy steel might be HV (Vickers) 800, a nitriding steel might be 1000, and a cold work die steel might be 1300. By comparison, our heat treated and tempered FC-0208 as shown in Standard 35 would be about HV 640, the equivalent of HRC 57.

In one example of the cost to have this process done on a P/M part by an outside heat treater we learned the following information. For a 2 in. diameter part .150 in. thick, weighing .12 lbs., the cost was $.30 each. The parts must be carefully arranged so that the nitrogen ions can easily make their way to all the important surfaces.

Steam Blackening

If you are not familiar with this term, you probably think of the steam coming from a tea kettle and its ability to remove unused postage stamps from junk mail return envelopes. How could it blacken anything? Well, steam or water vapor is a useful process gas at elevated temperatures, like 950 to 1100°F. The reaction is simple :

steam + iron >>>> iron oxide + hydrogen (at 1000°F)

$$H_2O + Fe >>>> Fe_3O_4 + H_2$$

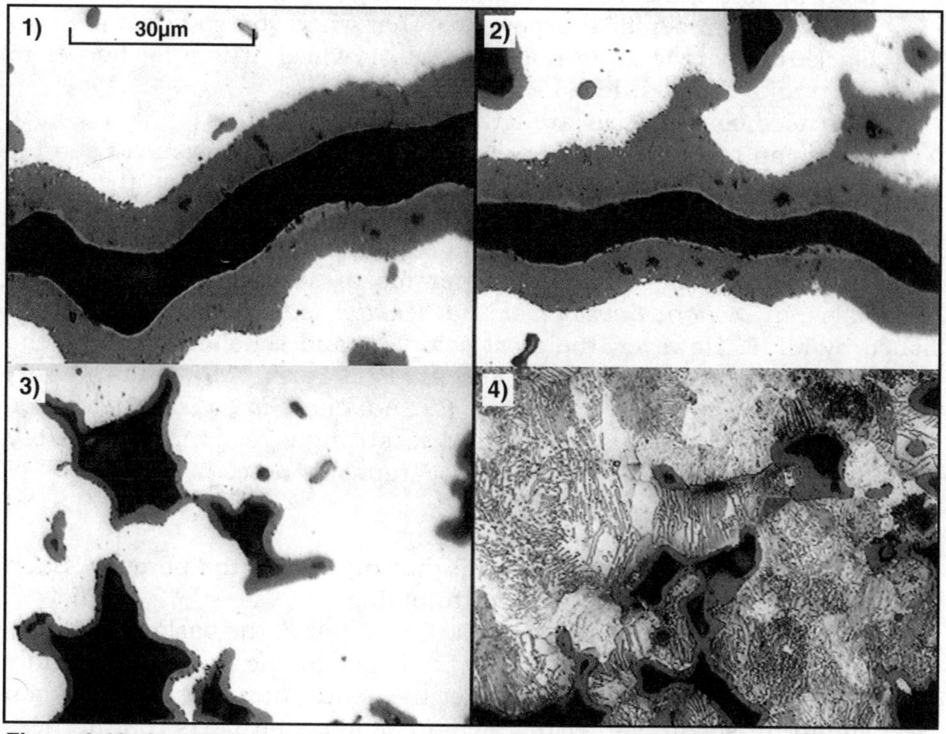

Figure 8.10. Images 1 and 2 show the $Fe_3\,0_4$ layer about 7 micrometers (.0003 in.) thick that has formed on the surface of a pre-existing crack in a P/M part. Image 3 is in about .2 in. from the part surface and the oxide layer is thinner. Image 4 shows the etched P/M steel structure with pearlite (fingerprint-like) and the gray oxide on the walls of the pores.

The Fe_3O_4 is black, hence the name, "steam blackening". In the microscope, the Fe_3O_4 looks gray, and at 200-1000 times magnification, we see a gray layer on the surface of the part and a gray layer on the inside of the pores, going down into the part for some distance. This is shown in Figure 8.10.

Why do people like to fill the pores of their P/M parts with Fe_3O_4? The survey says:

1. Visual appearance, a neat black color. And it doesn't reflect light like chrome.
2. To seal the porosity and make the parts pressure tight, so that they could be used, for instance, in a refrigerating compressor. It is the cheapest way to seal a P/M part. Steam blackening costs about \$.06 to \$.10/lb. You can also electroplate the sealed parts.
3. For increased hardness and wear resistance. For 6.2 g/cm^3 plain iron, you can drive the hardness from HRB 37 up to HRB 88. The Fe_3O_4 itself has a hardness of HRC 50. Everyone says it increases the wear resistance, but we have never seen any data. We can say, however, from personal experience, that when you go to cut a steam blackened part with an abrasive cut-off wheel, it is a lot harder to cut than a fully hardened P/M part!
4. Increased compressive strength
5. Corrosion resistance. We have no data on how corrosion resistant it is, but it is certainly better than plain iron. And dipping the blackened parts in a rust-inhibiting oil dip also helps.

The survey noted above was done in conjunction with a paper we published in 1988, "Mechanical Properties of steam Blackened P/M Materials", in *Modern Developments in Powder Metallurgy,* p. 275, published by MPIF. However, the most complete and scholarly work on the subject was "Steam Treatment of Ferrous P/M Parts, A State of the Art", by Dr. G. F. Bocchini in Volume 4, p. 77, *Advances in Powder Metallurgy and Particulate Materials,* 1992, and is strongly recommended. Very helpful, and practical as well, Howard Ferguson's notes in "Volume 7" of the ASM handbook, pp 650-651.

Here is one way to do steam blackening:

1. Start with clean parts, free of oil that may be in the porosity. Such oil can come from machining or tumbling.
2. Preheat the pit type furnace to 600°F and put in the baskets of parts.
3. Let the parts continue to heat in air until the load is uniformly heated throughout. (Failing to get the center nice and hot and then admitting steam will let the steam condense on parts that are less than 212°F and form red rust.)
4. Introduce superheated steam at line pressure of 5 to 15 psi and purge out any air for 15 minutes.

5. Increase the furnace temperature to 950 - 1060°F and hold for 2 to 4 hours, depending on whether you want a light blackening for appearance, or lots of Fe_3O_4 to seal the pores.
7. Reduce temperature to 600°F.
8. Purge furnace with nitrogen to remove any remaining hydrogen (safety requirements).
9. Unload the hot parts into air, to cool.

This will give a layer of Fe_3O_4 about 5-7 micrometers thick. You will need to devise a pressurizing test specific to each part, to be sure that the parts are leak tight under your required conditions. The steam blackening can lower certain mechanical properties, and you may want to use a destructive static or impact test to be sure that the parts are as strong as you hope they are. In our 1988 paper we showed that the tensile strength properties of F-0008 and FC-0208 in the sintered density range of 6.2 - 6.8 g/cm^3 were reduced by 12 to 58%. This means that the tensile strengths we measured were 13 to 25% below the mandatory .2% yield strength in MPIF Standard 35. For pure iron, the tensile properties were increased a little. For all three materials, the elongations were reduced from 30 to 80%. This suggests that steam blackened parts should be used in compression.

CHAPTER 9

Secondary Operations

Oil Impregnation

If you don't count ancient Egyptians and electrical brushes, sintered bronze bearings are the earliest P/M application, dating back to 1921. The bronze by itself has good compatibility running against steel, particularly hardened steel shafts. But, it is even better to have the pores of the bronze bearing filled with lubricating oil. When the bearing heats up a little, some oil runs out of the pores to lubricate the shaft. The question is, how do you get the oil into the bearing in the first place? If you don't have a vacuum system, you can run it in by capillary action. You put the bearings in a bucket of oil and heat the oil up to 180°F and hold it at temperature for 4 hours. The viscosity of the oil is lowered and the oil runs into the bearing, like water into a sponge. You can then turn off the heat, and let the bearings cool down in the oil, or you can quickly transfer them to cold oil and let them cool that way. You cool in oil because if you just took the bearings out of the hot oil and dried them while hot, as they cooled, some open and unfilled pores would occur at the surface. That is because on cooling, the hot oil shrinks more than the bronze. This method is described in ASTM B 328 and is mainly intended as a laboratory method to be used in the absence of a vacuum system.

In the real world, companies have specially constructed vacuum systems to add the oil to bearings. In the preferred embodiment (as they say at the U.S. Pat. Off.), the bearings go into a metal basket that goes into a vacuum chamber. The chamber is pumped down to -28 in. of mercury (7 kPa) to pull most of the air out of the bearings. This takes about 10 minutes. The bearings can then be lowered into a tank of oil, or oil can be admitted through a valve to enter the vacuum chamber and cover the bearings. Once the bearings are covered, the chamber is returned to atmospheric pressure. This applies a net pressure of about 14.5 psi or 93 kPa or about 0.1 mPa on the oil. It forces the oil into the pores in about 30 minutes. The bearings are often centrifuged to drain off the excess oil and make them easier to handle. Of course, thereafter, the bearings are always stored in non-porous containers, to keep the oil from wicking away.

In another technique, the bearings and basket are immersed in the oil in the vacuum chamber. The vacuum is then drawn above the oil and the air comes bubbling out. And does it ever come bubbling out! When you watch this in a transparent vacuum chamber in the lab, it seems like the bubbling will never stop, and the bubbles seem very large and numerous compared to the volume of the bearings. Then you realize that the room pressure air in the bearings is expanding about 15X as it is pumped out at -28 in of mercury. We prefer the technique of pumping the air out first, because there is no oil in the way to impede the exiting air. In any event, having filled the bearing with oil, you need to know whether you got the right amount into the pores. This is where you make use of ASTM Test Method B-328. A bronze bearing at 6.4/6.8 g/cm^3 should have 19% by volume of oil. To do this, you first need to know the volume of the bearing. This is numerically equal to the (weight in air minus the weight immersed in water) divided by the density of water (about .9975 g/cm^3). Then you need to know the weight of the oil. This is the oiled weight of the bearing minus the dried weight without oil. You remove the oil by repeated solvent extraction (Soxhlet extraction), or, if you have a handy protective atmosphere furnace running at 1400 to 1550°F, you pass it through there for a few minutes. Once you know the weight of oil, you need to get its volume by dividing by the density of oil, often .885 g/cm^3. Now you have the cm^3 volume of oil as well as that of the bearing. Divide the oil volume by the bearing volume and multiply by 100 to express the results as a percent. Hopefully you exceeded 19%! The test methods reduce this to a routine procedure, but we thought that it would be useful, at least once, to think about what the method is doing!

If we have just determined the oil content, what is the porosity of a P/M bearing? It is the volume % oil that you can add under specified conditions. The conditions are specified in ASTM B 328. Basically, it says get rid of all the oil that the bearing has in it at the beginning. Then impregnate it with a certain oil (100-330 SUS viscosity). That is done by immersion in oil, vacuum of -28 in mercury for 30 minutes, and return to room pressure for 10 minutes. Now weigh the bearing and see what it gained for weight, and calculate the volume of oil. Use the weight in air less the weight in water technique to get at the volume of the bearing. Calculate the volume % porosity as the oil volume divided by the bearing volume, x 100. Does that capture and measure each and every pore in the bearing? NO! But it gives you an agreed upon way to estimate the open porosity. Any pores that are isolated and not connected to the surface will be overlooked. Also, the vacuum never pumps out all the air, so the pores with air in them will not accept the oil.

Is there any way to measure all the porosity, connected or not? If you knew the pore-free density of the bearing, you could calculate it:

[(pore free density - dry density) / pore free density] X 100.

There is no standard method to measure the pore-free density of a material. If you are talking about bronze, you could accept a value from the ASM handbook for wrought bronze, 8.9g/cm^3, and make a pretty good estimate. By the way, we were curious to know why you wanted to know the total volume of porosity in the bearing.

Now you know a lot about how to oil and test sintered bearings. But what kind of oil do you use, and what about surface finish and the mating shaft? Norbert Arnold produced and tested bearings at Keystone Carbon Co. from 1939 to 1985 and offers you some guidance. You really must look at all the elements in the system: the bearing itself, the kind of oil used, and the shaft (the alloy, the hardness, and the surface finish). The harder the shaft, the better, and surface finishes of 8 RMS work well. If shafts are not hardened, there can be adhesion against the bearing. After grinding the shaft, treat it kindly; do not nick and scratch it. You may not find a nick with a surface finish tester, but it will pierce the oil film and create noise and earlier failures. The turbine quality oils like Terrestic 68 (300 SUS) and 77 (400 SUS) work well in most applications. For low temperature service, near or below 0°F, you need a synthetic oil that will not be too high in viscosity on cold mornings. Under ordinary conditions, you want an oil which is viscous to develop a boundary layer that will separate the running shaft and bearing. If it is too viscous, friction will overheat the oil and degrade it to carbonaceous sludge. The bearing should be carefully sized to get a very smooth surface finish, with nice flat lands between the pores. Sizing is needed because as sintered, the particle surfaces are relatively rough, and could poke through the oil film.

Resin Impregnation

As much as metallurgists like to admire the shapes of the pores in P/M parts, and bearing makers like to fill the pores with oil, there are times when they are a nuisance. The first instance that comes to mind is the case where you want the part to be pressure tight. That is, it is a component of a device in which there will be gas or fluid pressure against it, and you don't want the gas or fluid to leak through, e.g. in a power steering pump. Then there is the matter of food handling and sanitation. One can imagine what a fine home the pores in a stainless P/M food processor part would make for food bacteria. And finally, there is the problem of electroplating P/M parts. The cleaning agents, acids and alkalis, that are used to prepare the surface for plating tend to stay in the pores. When you plate over this entrapped material and wait two or three days, your nice shiny parts will have sprouted a funny white crust. The solution to these three dilemmas is to fill the pores with a plastic resin.

The methods of getting the resin into the pores are very similar to what we described for impregnating bearings with oil. In the wet vacuum method, the parts are immersed in resin and the vacuum pumps the air out through the resin, i.e. it bubbles. A vacuum of 1 to 5 millibar will extract the air in 5 to 10 minutes. (What's a millibar? A bar equals atmospheric pressure, something we all live with. A millibar =.001 bar, so at 1 millibar, we have removed 999 millibars of pressure or 99.9% of the air.) The vessel is then returned to atmospheric pressure, which forces the resin into the pores. The process of encouraging the resin to enter the pores can be improved by pressurizing the vessel to 85-100 psi. When the impregnation cycle is done, the basket of parts is spun to remove some of the excess resin. The parts are then washed to remove the rest of the surface resin. One can imagine that not cleaning off the resin would allow the parts to become glued together in a most unsatisfactory way. Also, lumps of resin on the surface hurt surface finish and dimensions. But rinsing is slightly more than just getting even. The rinsing can actually remove a small amount of resin from the pores. Normally this is not a big deal, but if you are trying to seal against or across this surface, the rinsed- out surface pores can create a leakage path.

Another method of resin impregnation is the dry vacuum method. It is like the "preferred embodiment" mentioned above for bearings: using a vacuum chamber, the air is sucked out of the parts. Then, the resin is flooded over the parts. On return to atmospheric pressure, the resin gets forced in. Sometimes elevated pressure is used to really force the resin in. This method is used with the more viscous resins and with high density parts with small pores. Finally, for automotive cylinder liners, a fixturing method is used. The cylinder liner is a hollow cylinder, They clamp covers on each end and feed resin in through a valve in the cover. The cylinder liner fills up and then the resin is forced radially outward until it exits on the outer surface. That leaves no open pore unfilled!

The most commonly used resins are anaerobic. They harden down inside the pores, in the absence of air. To help cure the surface pores, which have their resin exposed to air, the parts are dipped in a catalyst solution that hardens the resin in the surface pores. The cycle time for impregnation and rinsing is about 20 to 40 minutes. Job shops charge $.20 to .60/lb to resin impregnate, but this depends on the density of the parts and the labor costs at the site. As you can imagine, nice high density parts don't soak up much resin at all, and the converse is low density which uses up a lot of resin. The anaerobic resins are blends of organic molecules: methacrylate monomers. When they are stored prior to use, air must circulate through them to prevent their hardening. When hardened, they can resist temperatures up to 400°F and pressure up to 10,000 psi. This means that porous P/M parts can be resin impregnated and used as components of pumps.

Plating

In order to electroplate porous P/M materials, the surface pores must be closed off. This avoids the bleedout problem from cleaning agents noted above. The pores can be closed by resin impregnation, copper infiltration, or steam blackening. Double pressed and sintered steel parts at 7.3 g/cm^3 have the their pores well enough isolated that they require no added sealing. This also applies to powder forged (P/F) parts and MIM parts. The steam blackening is the least expensive at \$.06-.10/lb. but you really have to do it well. Also, it changes the mechanical properties, i.e. impact strength and elongation are reduced, while hardness is increased. For that reason, resin impregnation is usually the method of choice. The parts have to be rinsed well to get rid of surface resin, as the resin is an insulator that can interfere with plating.

The "Powder Metallurgy Design Manual", 3rd Edition, published by MPIF is an excellent source of information, and it offers us some practical suggestions on plating:

1. Copper plating - Acid copper plating produces the best results because of its excellent throwing power, (the copper atoms make their way into all the recesses on the part's surface). A clear lacquer coating will keep it from oxidizing.
2. Nickel plating - alone, or over a copper base, produces wear resistance and a bright corrosion resistant surface.
3. Chromium plating - decorative Cr is usually applied about .00001 to .00002 in. thick. For wear resistance, Cu and Ni plating are laid down first, followed by the Cr. This costs about \$.25/lb, once the parts have been resin impregnated.
4. Zinc and Cadmium plating- They provide corrosion resistance. Cadmium has greater throwing power, a lower coefficient of friction against steel, bonds well to solder and has a lower volume of corrosion products. It can not be used in food handling and cost 10X as much as zinc. The combination of resin impregnation followed by barrel zinc plating sells for about \$.32/lb., of which the zinc plating itself is \$.12 to \$.15/lb.
5. Phosphate coatings - can be applied to P/M for oil retention and initial wear. If the density of the part is greater than 6.8 g/cm^3, or steam blackened or copper infiltrated, the pores do not have to be sealed by resin impregnation. If resin sealing is required, the charge for sealing plus barrel phosphating is about \$.25/lb. Phosphate coatings are sometimes applied to P/M parts to help act as a lubricant during sizing or repressing operations.
6. Mechanical or peen plating refers to tumbling parts in a gas-driven stream of Ni, Cd, or Zn powder. The powders strike and adhere to the surface of the parts. They can be applied to ferrous P/M products that are greater than 6.4 g/cm^3 without having to fill the pores with resin. For some reason, parts made with reground iron powder seem to require a somewhat higher density to avoid entrapment and bleedout. Zinc plus a clear coat finish costs \$.30 to \$.45/lb. For salt spray resistance, zinc plus yellow coating costs \$.35 to \$.45/lb. The combination of tin and zinc is also good for salt spray resistance. Mechanical plating is advantageous because there is never a hydrogen embrittlement problem with high strength P/M parts. (With electroplating, some atomic hydrogen is set free at the metal surface. It diffuses into the hard steel and forms what amounts to very small pockets of hydrogen gas which can embrittle the steel. This is relieved by baking

the plated parts at 300 to 400°F) Prior to plating of heat treated parts that contain residual quenching oil, it is the usual practice to use a tempering operation, e.g. at 350 to 400°F to get rid of most of the oil.

7. Electroless nickel plating is used on P/M parts. (We have used it on steel parts without resin impregnation and found no bleedout problems but we are not sure if this is generally true). Job shops charge $10. per square foot of surface per .0002 in. of thickness to electrolessly nickel plated steel. The average thickness is .0005 in. and for good corrosion protection you need .0010 in.

Other Coatings

Allegeheny Coatings, Ridgway, PA supplies coatings like RTM 7020 for imparting 240 to 1000 hr. salt spray corrosion resistance per ASTM B-117. It does not require sealing of the pores in the P/M part. The coating is a dispersion of zinc dust and waterborne ceramic vehicle applied by spraying on clean parts and cured at 680°F. It resists service temperatures to 900°F. The coating is typically about .0002 - .0015 in. thick and the cost is $.25 to $.75/lb. of parts. It has the advantage of being free of chromium. This is important in Europe, where chromium interferes with recycling.

Joining and Welding of P/M Parts

Other than as decorative items, P/M parts are invariably used in assemblies of some sort. Some applications don't even seem like joining, for instance an idler pulley riding on a shaft, or a sprocket with a D-hole or key way to drive it. You just slide them into place, and secure them with a "C" clip or something that acts like a thrust washer. P/M parts are often joined to other things by bolts or screws, and this works very well. Forcing a P/M bearing into its housing is getting closer to a joining operation. An interference fit is a form of joining where you press one part into a hole in another part, and the elastic stress from the interference holds the two pieces together. But we like to think of "real joining" as bonding two different objects together. This would include methods like

Brazing
Sinterbonding
Welding: projection welding
gas metal arc, GMAW
high speed pulse welding
friction welding

If the P/M industry kept a list of FAQ's (frequently asked questions) welding would be high on the list. We hope the folks at the ladder company find our short review to be useful! The best recent summary of joining methods was done by Jack Hamill, Hoeganaes Corp., in "Volume 7" of the ASM Handbook.

Parmigiani and Kosco of Keystone Powdered Metal Co. conducted many joining experiments which they described in their paper, "Joining of P/M Components", SAE, 1999-01-0889, 1999. In the SAE paper, the authors used a simple but clever specimen for testing the weld or braze strength, see Fig. 9.1. After the parts are joined, to test them, you set the part on the outside step and press down through the center. This breaks the part in tension, at or near the weld area, or in shear, for the brazed specimen. The idea is that the tensile strength so measured ought to be as high as the that of the P/M part. If it is not, something went wrong. They tested the following P/M materials:

TABLE 9.1. MATERIALS SUBJECTED TO WELD TESTING

Code	Material	Density g/cm^3	Hardenability
A	SS-410-90HT	6.6	medium
B	F-0000	6.9	low
C	FN-0208	7.0	medium
D	FN-0208	7.2	medium
E	FLC-4608 mod	7.0	high
F	FN-0200	7.1	low
G	FL-4405	7.6	medium

The materials were welded to wrought 1045 bar stock (.45%C) except for the 410SS. That was welded to wrought 410 bar stock. The FN-0200 was welded to itself. There were three weld methods and one brazing method. The first method was friction welding, wherein one piece is rotated at high speed and then jammed into the other piece. The friction heating welds the parts together without melting. The second method was high speed pulse welding. Here, the P/M part would have one or two small raised circular rings on the face to be welded. It is clamped against the mating piece and a high electric current is discharged through the part. The current heats the raised rings and the pressure causes them to collapse, densify, and weld to the mating part. This all happens in .003 to .015 seconds. No melting occurs and there is only a very small zone near the weld that shows the effect of the heating. A second small pulse can be used to temper the weld.

The third method was gas tungsten arc welding, GTAW. It is also called TIG or tungsten inert gas welding. The welding current passes from a non-consumable tungsten welding tip to the work piece, through a protective gas layer, argon of helium. The electric current locally melts the P/M part and the mating part, joining them together. Since the P/M part densifies when it melts, a separate filler wire is fed into the melt pool to maintain the original volume of the two parts. The parts are not tempered after welding.

In brazing P/M parts, the two parts are placed close together with a gap of up to .010 in. between them. The gap can be as small as a light press fit, but in the present work, it was .002 in. The thinner the joint of brazing material, the better.

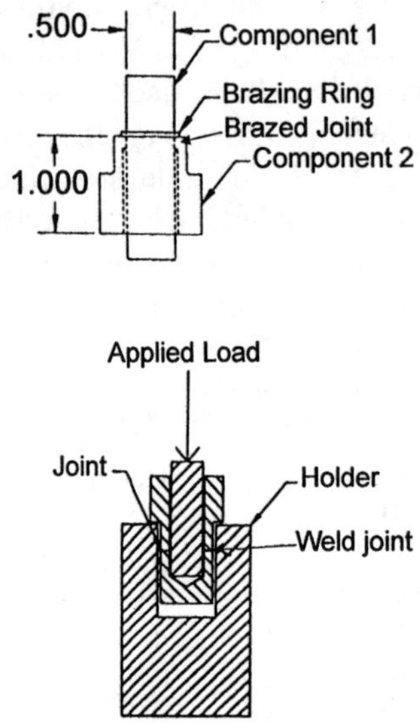

Figure 9.1. Arrangement of brazed and welded specimens for strength testing. The brazed joint is tested in shear and the welded joint is tested in tension. Devised by Kosco and Parmigiana of Keystone Powdered Metal Company.

The brazing material can alloy with the iron and be strengthened. During testing, the thin joint helps the steel on either side to support and strengthen the braze material. When the work is placed in a furnace with protective atmosphere, at a temperature above the melting point of the braze material, it melts and, under capillary action, it runs into the joint. The brazing material must be carefully chosen because most brazing alloys tend to run into the pores (capillaries) in the P/M part, and not stay at the interface where they are needed to bond the two pieces together. The usual choice for P/M is Hoeganaes Ancorbraze 72, a Ni-Cu-Mn alloy invented by Walter Knopp, an independent P/M consultant. (During 1999 about 80,000 lbs. of this material were used for brazing.) As this material starts to run into a pore, it dissolves a little iron, has its melting point raised, freezes, and goes no further into the pore. This keeps the brazing material at the joint, where you need it. In the present case, the brazing material in powder form was pressed into a little ring that sat near the joint. As it melted, it was sucked into the joint by the capillary forces. This brazing operation is done at 2050°F. Brazed parts can be hardened and tempered after the brazing operation.

Before looking at the results from the above test, consider what problems often occur in gas tungsten arc welding P/M materials that contain carbon and alloy elements. On cooling, the cooling rate can be fast enough to form some hard untempered brittle martensite in or adjacent to the weld. This can result in low fatigue strength unless the weld area is tempered. Sometimes the stresses from cooling and the formation of martensite can locally crack a weld. With that background in mind, here is what the above authors learned about the conditions of the welds:

TABLE 9.2. WELDABILITY OF P/M MATERIALS					
P/M Material And Code	**Mating Wrought Material**	**Friction Weld**	**High Speed Pulse Weld**	**Gas Tungsten Arc Weld**	**Braze**
SS-410-90HT 6.6	SS	Good	Good	Poor[2]	Poor[4]
F-0000 6.9	1045	Good	Good	Fair[3]	Good[5]
FN-0208 7.0	1045	Good	Good	Fair[3]	Good[5]
FN-0208 7.0	Self	Good	Good	Fair[3]	N/A
FN-0208 7.2	1045	Good	Good	Fair[3]	Good[5]
FLC-4608 mod 7.0	1045	Poor[1]	Poor[1]	Poor[2]	Good[5]
FN-0200 7.1	Self	Good	Good	Fair[3]	Good[5]
FL-4405 7.6	1045	N/A	Poor[1]	Fair[1]	Good[5]

1 High weld strength, with hard brittle (martensite) weld area, and no significant porosity in weld one.
2 Weld area was brittle (martensite) with porosity and cracking
3 High strength, no significant brittle martensite, but porosity and cracking at weld interface
4 Base metals, 410, became hard and brittle near joint
5 Joint strength likely to increase with smaller clearance. Most joints showed some porosity in the filler metal

The measured tensile strengths (shear strengths in the case of the braze) of the joint are as follows:

TABLE 9.3. TENSILE STRENGTH OF WELD AND SHEAR STRENGTH OF BRAZED JOINTS					
P/M Material And Code	**P/M Material UTS, ksi**	**Friction Weld UTS, ksi**	**High Speed Pulse Weld UTS, ksi**	**Gas Tungsten Arc Weld UTS, ksi**	**Braze USS, ksi***
SS-410-90HT 6.6	105.	100.	93.	75.	33.
F-0000 6.9	29.	31.	30.	29.	22.
FN-0208 7.0	62.	60.	57.	61.	31.
FN-0208 7.0	62.	60.	60.	62.	
FN-0208 7.2	75.	75.	74.	71.	32
FLC-4608 mod 7.0	100.	100.	76.	50.	44.
FN-0200 7.1	36.	36.	28.	32.	19.
FL-4405 7.6	120.	N/A	119.	100.	36.

The second column, P/M Material, shows the strength of the P/M material prior to welding. The other columns for the various welding methods are compared to it, to see how close the weld strength came to the parent P/M material. In the braze case, for comparison with the base P/M material, the Ultimate Shear Strength of the base P/M material should be approximated by .5 x the Ultimate Tensile Strength shown in Column 2.

We can see that the friction welding, high speed pulse welding and the brazing generally work well. A lot of questions surround the gas tungsten arc welding in this example. All of the samples with ordinary density show gas porosity at the weld. All but the 7.6 g/cm^3 sample show cracking near the weld. These defects don't hurt the tensile strength much, but they are reckoned to reduce the fatigue strength, in service. The cracking can be caused by the expansion when martensite forms in one of the hardenable alloys. That does not, however, explain cracking that occurred in the F-0000 and FN-0200, which should have shown little or no martensite formation. Jack Hamill in "Volume 7," shows an instance where low density, 6.2 g/cm^3 plain iron cracked at the weld to wrought 1010. There was a fairly large weld bead present. During cooling, the weld bead contracted quite a bit compared to the surrounding P/M. The low density P/M material was put in tension and cracked. A stronger P/M material might not have cracked. The lesson is that small weld beads are better, because, when they freeze and cool, they cause less stress on the abutting material.

In Gas Metal Arc Welding, GMAW, a filler wire carries the electric current to heat the weld region. The filler wire then melts along with the two base materials and forms the weld bead. This method, and also the GTAW described above, have the ability to control the heat input. Use as little heat as possible, and do this with the use of a small filler wire. This means that the region that will be contracting on cooling is smaller, and will pull less strongly on the surrounding P/M material. The use of an E70S filler wire to successfully weld 6.9 g/cm^3 FN-0205 to a 1035 steel is shown in Figure 9.2, from Hamill's work. He also shows a porous stainless steel weld, where he attributes the porosity to interstitial gases in the stainless P/M. He is referring to oxygen, nitrogen, or hydrogen as well as CO that can form from the oxygen and residual carbon in the stainless steel. Other ways to avoid cracking on cooling include:

1. Preheating the part to drive off moisture that can liberate hydrogen during welding. It will also lower the temperature difference and hence the stress between the base metal and the cooling weld pool.

Figure 9.2. Photomicrograph of a welded joint between porous 6.9g/cm^3, FN-0205 nickel steel with .5% carbon, lower left, and a solid rod of AISI 1035 (.35%C). The dark weld material was provided by a filer wire of type E70s, which filled in the space where the P/M materials densified on melting, and also added a weld bridge of material between the two parts. This was a successful weld done by Gas Metal Arc, GMA. Micrograph Courtesy of ASM Handbook, Vol. 7.

2. Post-heating immediately after welding reduces stresses in high hardenability materials that form martensite near the weld. It is sort of like tempering the martensite, a little.
3. Using austenitic filler metals which do not cool off to form hard martensite in the welds. They are also tougher materials.
4. Using a low heat input to the weld.

P/M parts that are to be welded need to be free of elements that are going to form gas porosity in the molten weld pool. That means the pores in the parts should not contain any lubricants, water or moisture, quenching oil, machining coolants, plating solutions, impregnating materials, cleaning and tumbling agents, free graphite or ash, or be steam blackened. Each form of contamination requires a different method of removal. Even if the parts are very clean, you need to be sure that the protective gas is working well during the welding. If you let air into the liquid weld, it reacts with carbon to form CO or CO_2. We once received a porous looking weldment and were asked to figure out why it welded so poorly. Our David Pease took out his trusty little Hobart MIG Welder, Handler 120 (DC Amps) and welded it perfectly on the first try. His best guess was that the other welder didn't have the gas turned on! If you are welding a hardenable P/M alloy and get a weld free of pores and cracks, but containing brittle martensite, all may not be lost. If the assembly is reasonable in size, you could consider tempering it for 1 hour at 350°F in air, just as you do other hardened P/M steel.

Not all P/M alloys are weldable by GTAW or GMAW. That includes the Fe Cu alloys with more than 4% Cu, copper infiltrated parts and the iron alloys that have .45% phosphorous or higher. However the high speed pulse welding will work on the phosphorous-containing materials. Figure 9.3a shows an example of high speed pulse welding an assembly of FN-0208 and an AISI 1010 low carbon stamping. The assembly is 8 in length and 4 in wide. Figure 9.3b shows a close up view of tooling in the welding press as well as some parts that were built up into an assembly, by welding. The equipment for high speed pulse welding is relatively expensive $200,0000 to $1,000,000, and even though you can weld 200 - 2000 parts/hr., in order to pay for a machine you would have to weld a lot of P/M parts. There is an older method, resistance projection welding, RPW, which is similar but less costly. Again, one of the two mating parts needs some small projections on it. The parts to be joined are clamped between two electrodes, with the projections from one part pressing on the other part. An electric current passes through the parts and the projections, heating and softening them. The pressure on the assembly collapses the projections and welds them to the other part. This results in a good weld, but the weld is limited to the areas where the projections were. The method works well on high or low carbon P/M parts at densities from 5.8 to 7.4g/cm^3. Normalizing or tempering the high carbon parts after welding will further increase their strength. Figure 9.4 shows an example of a projection welded part.

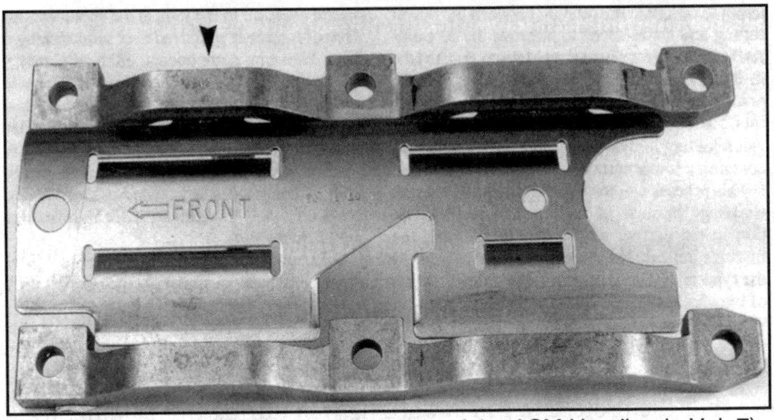

Figure 9.3a. High speed pulse welding (Courtesy of the ASM Handbook, Vol. 7).

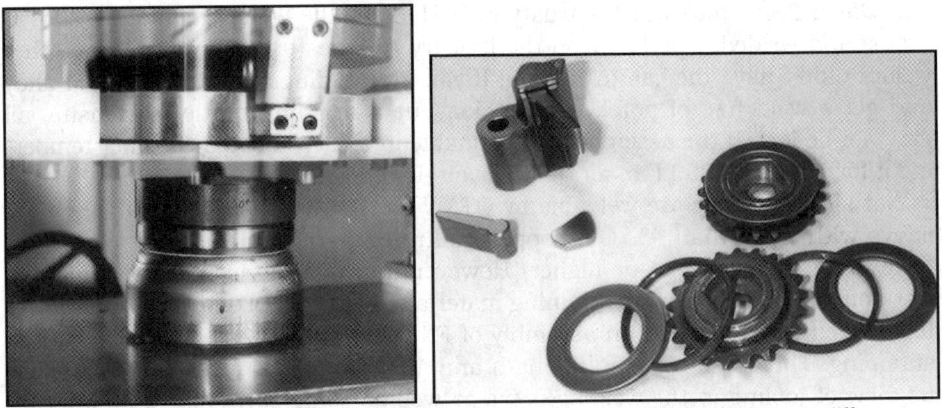

Figure 9.3b. Press for performing high speed pulse welding. On the right, two different assemblies built up by this welding technique. Courtesy of Process Equipment Company.

Figure 9.4. Resistance projection welding application joining an intermediate density P/M spur gear to a low carbon wrought steel stamping (Courtesy of the ASM Handbook, Vol.7).

Sinter-bonding

You can use sinter-bonding to join two P/M parts if they have the right shape and composition. In the green state, the outer member must slip over and fit closely around a projection on the mating part. If you look back at the brazing design in Figure 9.1, you will see an arrangement that could also be used for sinter-bonding. The arrangement for the welding trials, Figure 9.1 with the butting surfaces will not work, for reasons to be given shortly. The outer part is made from an alloy mix that shrinks during sintering, like FN-0205. The inner part is made from a mix that grows during sintering, like FC-0205 made with atomized iron. During sintering, one shrinks and one grows, bringing the mating surfaces into close enough contact for diffusion to occur and create a decent bond between the members. Figure 9.5 shows a sinter-bonded assembly courtesy of Norwalk Powder Metals. They achieved the differential growth between the mating members by making the inner one with iron plus 3% Cu and the outer one with only 2% Cu in the iron. The bond strength is usually not as high as might be achieved by brazing or welding. Sometimes folks apply the belt and suspenders technique and copper infiltrate the assembly of the two parts. They are trying to make sure that either the mating surfaces diffuse and bond together, or the two parts get brazed together.

Press Fitting P/M Parts

Press fitting occurs when you press a projecting region of one P/M part into a cavity in a mating part, and that cavity is a little smaller than the projecting member. The cavity is forced outward elastically and then perhaps plastically. (In the worst case it is forced out catastrophically and it fractures!) All of the gripping and holding occurs from the compressive elastic stresses. Expanding the

Figure 9.5. Sinter-bonded part courtesy of Norwalk Powder Metals, Norwalk, CT. Since there is no way to form the reduced diameter central section, two separate pieces are compacted. The male member is the reduced section with the gear on the right and a round extension. The extension fits into the I.D. of the larger gear, on the left. During sintering, the larger gear I.D. shrinks against the round reduced section's extension and bonds to it.

outer member beyond the elastic yield point gains nothing and courts disaster. For a 1 in. diameter round part being pressed into a mating cavity, an interference of .003 in. is reasonable. It is good practice to repress or size the inner and outer members to develop an O.D. taper on one, and a lead-in taper on the cavity, to prevent scoring and shearing at assembly. If the outer member is hardened and has thin walls, the .003 in. interference might be too much, and the outer member could crack.

Tumbling/Deburring

An attribute of P/M parts that is usually detrimental is that they have sharp edges. This condition results from powder extruding or flowing into the small clearances between punches and dies and punches and core rods. This causes burrs or "flash" as it is sometimes called. Obviously, tool wear makes the burrs larger and more objectionable.

Flash and burns are certainly not unique to P/M parts because most formed or machined metal parts come with burrs. The good news is that using the proper techniques, burrs can be removed successfully at minimal cost. Timing of burr removal is critical and is usually done before secondary machining because tumbling can damage threads or other machined features. Parts that require secondary heat treating are usually tumbled after heat treat. The tumbling operation cleans the parts after heat treat. If finishing were done first, there could be problems in heat treat caused by entrapped fluids from tumbling.

Mechanical finishing of parts in bulk is among the oldest known processes, dating back to the Roman empire. Mass finishing involves loading parts into a container usually with some medium and a chemical compound. The container moves and the parts and medium rub against each other. The result is that burrs are worn off and sharp edges are rounded. If you have ever found a piece of glass with smooth edges on a beach, you have witnessed the result of a medium (sand in this case) rubbing on sharp-edges and making them smooth. It takes a long time to polish sea glass and we need to process P/M parts more quickly and in a more metal friendly environment than the ocean. One common method (not as popular today as it was 25 years ago) is rotary barrel tumbling.

This process has its origins in the Middle Ages when casks filled with abrasive stones and pieces of armor were rolled around the courtyard for days until the desired finishing was obtained. (We don't know if knights did this at night also!). Today parts are placed in a special barrel along with the medium. A typical medium/part ratio is 3/1 by volume. Water, a cleaning agent or type of soap, and a rust inhibitor are usually included when tumbling P/M parts. These ingredients prevent abrasive from getting imbedded in the surface porosity of the P/M part. The rust inhibitor prevents corrosion that would take place as a reaction between the P/M parts and water. The barrel is rotated which causes the medium and parts to slide past each other thereby removing burrs and rounding sharp edges.

Only about the top fifth of the load in the barrel is sliding at a given time. The rotation of the barrel produces a "landslide"-action when the parts and medium rise and then fall due to gravity -see Figure 9.6. This means that the cycle time is fairly long. Increasing the rotational speed improves process time but can result in damaged parts because the "landslide" action is more aggressive.

Vibratory finishing is widely used today as a replacement process for tumbling. It is done in open containers that are either round or rectangular. These containers are loaded with parts, medium and water etc. just like the tumbling barrel. The container is vibrated at 900 to 3000 cycles/min. The entire load is under constant agitation and therefore cycle times are shorter than with barrels. Vibratory finishing also does a better job of finishing because of the improved movement of parts and medium. Another advantage is the ability to reach into the container at any time to inspect parts and check on progress.

We mentioned medium several times and rightfully so, because medium plays a critical role in finishing. The primary function of the medium is to cushion the parts so they don't damage each other while the medium is also cleaning and finishing. There are many compositions and shapes of medium and selection of the proper type is extremely important. Compositions can range from manufactured materials such as aluminum oxide or silicon carbide imbedded in resin or ceramic to natural materials like sand, granite chips or even hardwood pegs or

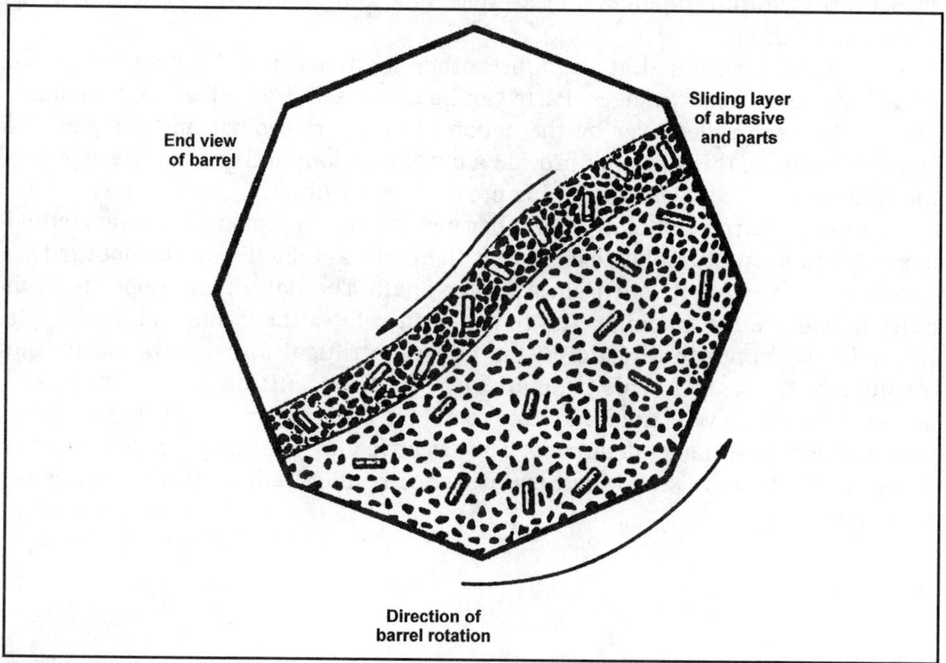

Figure 9.6. "Landslide" action in tumbling barrel.

nails. Shape and size of the medium must be carefully evaluated so that the medium will fit in holes without lodging. The medium must also allow for easy separation from the parts when the process is done. One cannot afford to hand separate parts from the medium or pry medium out of holes and slots where it has lodged during finishing.

Available shapes include triangles, stars, cylinders, rectangles and diamond shapes. Of course these shapes are only available in the manufactured materials and must be very uniform in size to prevent lodging and separation problems.

When the finishing cycle is done the parts and medium, are thoroughly rinsed: to remove residue. The parts and medium are then usually separated by using a screen. The screen allows the medium to fall through and parts stay on the screen. Most ferrous parts are then dipped in a chemical water displacer which forces water out of the pores of the parts. Drying in a heated spin dryer is sometimes the next step and finally most parts are dipped in a rust preventative. This coating dries in air, leaving a protective coating on the parts.

Vibratory systems are readily adaptable to automation. Many vibratory tubs allow the insertion of a ramp or dam when the deburring is done. The parts and medium climb up the ramp and the medium falls back into the container through a screen located just beyond the dam. The parts stay on top of the screen and move off to the side of the drum and into a basket or water displacing dip tank. The dam is removed and the next load of parts is added to the tub. This process improvement reduces cycle time and labor involved in handling parts and medium.

We should mention that there are other methods of deburring that have advantages and disadvantages. Parts can be processed dry and without medium. Sharp edges are rolled over by the action of one part against another part. As you can imagine, this does not provide a cutting action; and, parts are subject to nicks because there is no cushioning provided by medium.

Centrifugal barrel tumbling is another method used occasionally. This method is similar to barrel tumbling except that there are 2-4 small barrels mounted on a circular frame 2-3 ft. out from a center shaft. The barrels are opposite each other to balance the machine. The center shaft rotates the frame and the barrels rotate in the opposite direction. This adds centrifugal force to the parts and medium in the barrel. This dramatically increases cutting action. Increased action decreases cycle time and can result in process speeds 25 times faster than conventional barrel deburring. The negatives are that only small loads can be put in these barrels and high maintenance results from the high stresses on the equipment.

CHAPTER 10

Standards for Powder Metallurgy

There are two kinds of standards that are important to know about: on the one hand we have material specifications, and on the other, test methods. Material specifications tell you the properties of the material, like strength or chemical composition. Test methods tell you how to test the material or part, like how to measure its density. Another way to say all this is that material specifications tell you how strong it will be and test methods teach you how to break it to see if it is as strong as it is supposed to be.

Why bother to create a material specification at all? Why not let the company that wants to buy some P/M steel gears tell its suppliers what strength it wants? Prior to 1945, this was the situation, and P/M houses offered their customers their own proprietary mixes. Since we still have a P/M industry today, the system must have worked, but it did not work easily. If the company using the gears or bearings wanted an additional P/M supplier, it had to find someone who could match supplier A's properties. For sintered bronze bearings, this dilemma was first resolved by ASTM specification B 202 in about 1945. It specified the chemistry, radial crushing strength (K factor), density, and oil content. The companies that built little electric motors could list this material and the P/M houses could all quote the same thing. It leveled the playing field. The motor builder got his shipment of bearings. He decided he wanted to see for himself if he got the density, strength, and oil content, etc. that he expected. He turned to the ASTM test methods: e.g. B 328 to measure density and oil content, and B 202 (later B 438), to measure crushing strength .

ASTM (American Society for Testing Materials) consists of a group of professionals who develop standards. Their earliest non-P/M standards have been in use since 1898. The standards are referred to as "voluntary" and "consensus". "Voluntary" means that the ASTM committee develops them, not a government agency. "Consensus" means that the committee is not just made up of P/M parts producers. The members are classified according to their interest as producers, users and general interest. In a specific example, the producer is the P/M parts maker, the user works for the company that puts the bearing into the motors, and general interest includes all others, like one of the authors, L P. The committee must maintain balance among these interests, that is, the producers can not dominate the committee. The ASTM B-9 committee, Metal Powders and Metal Powder Products, has some consumers, but it is always in need of more end users of parts. (The committee maintains overall balance by having powder producers register as general interest on specifications that govern materials pressed and sintered by the P/M parts producers.)

So, if you show up at a B-9 committee meeting whom will you meet? What you won't find is a group of professional standards writers employed by ASTM! You will find a group of volunteer professionals, all interested in maintaining some semblance of order in the P/M industry. At a recent meeting, there were 35 members present, divided up according to 26 interested in ordinary pressed and sintered iron, steel and bronze, and 9 interested in refractory metals (for example tungsten and tungsten carbide). There are companies represented that make powder, that convert powder into parts, and those who wish to use P/M parts. A significant number of members are retired from the P/M industry, but choose to keep very active by helping with the standards work. ASTM headquarters is represented by the staff manager for the committee. This person keeps us on a straight and narrow - being sure that negative votes are fairly handled and that standards are periodically reviewed. New standards have to be balloted at the subcommittee level, the main B-9 committee level, and at the general ASTM society level: the staff manager helps to make sure this happens smoothly. What he does not do is offer technical advice, for example, on how to make P/M test bars. All technical matters are handled by the volunteer committee members.

How do the test methods come into being? Somebody comes to the committee with a problem like, "How should we measure microindentation hardness and case depth on heat treated P/M parts?" If it is generally recognized as a real need, some folks will step forward and volunteer to write the test method. Or maybe none of the committee members volunteer. That is when the DIY (do it yourself) system operates. If the proposer seriously needs a test method, he can have a copy of the ASTM Blue Book, (Form and Style Manual), take pen in hand, and write the test method himself. (He can also take computer keyboard in hand, and make use of a new electronic format that shows you what information needs to be filled in, and where.) The proposed method (whoever the author may be) gets reviewed (picked apart) at the next subcommittee meeting. Suggestions and revisions are offered. The proposer does draft #2 and it usually is sent out by the subcommittee chairman for voting at the subcommittee level. Members who disagree vote negatively and give their reasons. At the next subcommittee meeting the chair reviews all negatives and comments. If the subcommittee agrees with the negative vote, the test method must be changed by the proposer and resubmitted for ballot as draft #3. Eventually, subcommittee members, representing about 30% of all of the B09 committee members, are satisfied. The draft then goes for concurrent main B-9 committee vote, and overall society ballot. Sometimes it passes. Sometimes a B-9 member not on the subcommittee sees a problem. And, distressingly, sometimes a subcommittee member sees a problem he missed the first time! Will there ever be closure? Eventually yes, although once it required 14 drafts! The subcommittee has an informal award named to honor the persistence of that proposer, PT.

Well, now we have a method to measure case depth and microhardness. Actually, we *almost* have a test method. The last item in the test method is supposed to be the precision statement. The precision statement tells the test method user how closely the microindentation hardness value will be repeated if he repeats the tests himself on the same part. (That is called repeatability.) It also tells how closely two different laboratories will agree, if they test the same material. (That is called reproducibility.) As you can imagine, people agree more closely with themselves than with other labs, or repeatability is a smaller number than reproducibility. From whence cometh the repeatability and reproducibility values? They come from interlaboratory (round-robin) trials. For microhardness, a task group could send the same polished specimen to at least 6 different labs. They measure microindentation hardness following the procedure in the new test method and send the results back to the task force chairman. The task force uses software in ASTM E 691 to operate on all the microindentation hardness values. If any of the values are way out of line, they are flagged. The task force then deletes the peculiar (outlying) data and the software re-calculates I_R (reproducibility) and I_r (repeatability). These results are presented to the subcommittee for acceptance as a precision statement. In MPIF Standard 51, the statement reads as follows: "The mean hardness value was HK_{100} 701.1 with a repeatability of 22.4 and a reproducibility of 76.0. Duplicate Knoop hardness results from the same laboratory should not be considered suspect at the 95% confidence level unless they differ by more than 22.4. For the same test specimen, Knoop hardness results from two different laboratories should not be considered suspect at the 95% confidence level unless they differ by more than 76.0." This precision statement really only applies to heat treated materials where the Knoop hardness is in the range of 600 to 700 kg/mm^2. The 95% confidence statement means that if two labs differ by more than 76.0 units, that we are

95% sure that one or both are making an error in their measurement technique. Looked at the other way, if the two labs agree within 76 units, for instance if one laboratory found the hardness to be 663 and the other found 739, we would have to assume that they were both in agreement. This becomes interesting when there is a minimum hardness for example 701, and the buyer finds 663, and the seller finds 739. The buyer might want to say that the parts had failed the hardness test of 701 minimum. However, given the precision of the test method, the parts may very well have passed the 701 minimum. If you are offended by the high value of the reproducibility, and the apparent lack of precision in the test method, then you might want to join the ASTM B09.05 subcommittee, and show them how to do it right!

Having read how a test method is developed, we hope you will appreciate the next one you have to read! You may look at the precision statement and say "I can't believe that it is that bad, we always do better than that!" And, you probably do. And the committee felt the same way you do. It seems that no matter how carefully the test methods are written, when you let 6 or more reasonably competent labs run the test, the reproducibility number accurately represents the real differences that are seen. If you ever carry out one of these test methods and you notice a problem in the way it is written, let the ASTM B 09 or MPIF committee know. This way, the test methods get better and better. ASTM in West Conshohocken, PA can be reached at (610) 832-9500 and Metal Powder Industries Federation in Princeton, NJ at (609) 452-7700. Both organizations will cheerfully sell you their standards. And while you are on the phone, ask ASTM for a copy of the membership application form. Our editor would not let us put it in the appendix to this book! At MPIF, you can join the professional group interested in powder metallurgy, APMI International. Visit them on the web at "www.mpif.org" and sign up on line!

You notice that the test method speaks of both the precision and the bias. You now know a lot about precision but have seen precious little about bias. Bias is the error in the result given by a test method that is noticed when a test method is used on a standard material. The problem is that to determine bias, you need a standard material. Where do you get standard materials? The most common source is the National Institute of Standards and Technology, (NIST). The problem is that NIST does not have standard reference materials for all of our test methods. In fact, there are hardly any of our test methods for which there are reference materials. So, you won't be surprised to see that most of the test methods are unbiased! How could this remedied? Someone needs to convince NIST that it would be useful to create the standard reference materials. (It is a student exercise to count the number of soap boxes embedded in this section!)

Specifications

What properties could be found in material specifications? Tensile properties like ultimate tensile strength, .2% yield strength and percent elongation in 1 in. are common. You could also find chemistry, density, hardness, elastic constants, impact energy, and fatigue endurance limit, to name a few more. The list could be very long but it isn't. Specifications list two kinds of properties: the ones you must meet (required or mandatory) (in the U.K. man dat' ory) or informative (typical properties). How does the Standards Committee decide which properties are required and which are typical? Its really a matter of negotiation between buyer and seller. In the real world, the buyers wants as much guarantee as possible; however they really don't want to pay for all the work to guarantee the properties. (We suppose that in the aircraft industry they ask for and are guaranteed far more properties than in the industries served by press and sinter P/M). The press and sinter P/M industry has figured out that it can sell P/M parts to most end users by guaranteeing chemistry and minimum yield strength. Everything else is typical. What are the typical properties in MPIF Standard 35? Tensile strength, elongation, transverse rupture strength, impact energy, elastic constants E and ν, (Young's modulus and Poisson's ratio), apparent hardness, fatigue endurance limit, microindentation hardness, hardenability, coefficient of thermal expansion, corrosion resistance, magnetic properties, K_{Ic}, machinability and a general description of the microstructure.

What if an end user wants guarantees beyond chemistry and .2% yield strength? No problem. Real people just mark up their drawings with other things that they require, and, they are prepared to pay the extra cost (we hope). If you are an end user, you may be wondering what number to put on your drawing for a minimum density and apparent hardness. One answer is simply to use the typical values. But, now you are taking "typical" values and making them "required". Required means that parts producers will have to far exceed the typical density and hardness shown in Standard 35 so that *on a statistical basis,* the values never go below the minimum on your drawing. Unless you have some rational basis for turning typical into minimum, why not ask the parts producer what he would suggest for minimum density and hardness, consistent with the mandatory minimum yield strength? This idea is consistent with the handling of .2% yield strength in MPIF Standards. It has both a required minimum value, and it also is shown as a typical value. The typical value is always a little higher.

How does the parts buyer know that the gear or lever has a .2% yield strength greater than the minimum shown in the specification? He could machine a tensile test bar from the part and measure .2% yield strength. But that is almost never done, and when it is done, the bar must be the same size as in MPIF Standard 10. It is not done because parts usually don't have the right size and because it costs \$50 to \$100 to machine and pull a tensile bar. Companies selling parts made from wrought steel can supply data based on doing tests on the original steel bars that were used to machine the parts. The nearest P/M comes to that, is making some molded tensile bars (MPIF 10) and processing them with the regular parts: same chemistry and same average density. It works, and indeed, a very limited number of end users ask for this to be done. What do real people do? They follow the advice in MPIF Standard 35 on demonstrating yield strength. This means that when the first lot of parts is being processed, the end user tests them for durability or service life. If the parts are okay, the user or the producer works out a static or dynamic test that breaks the part through some critical section. The values from these tests are used to establish the minimum breaking values for future tests. The assumption is that the P/M parts have passed the minimum yield strength because the parts survived life tests. The mechanical break tests then keep things under control on future lots.

Of what use are the typical properties? They guide the designer in choosing a durable material. Having chosen the material and made the first run of parts, the state of the art is that these parts now require careful durability or life testing.

Where did the minimum and typical property values in MPIF Standard 35 come from? These data were generated under the guidance of the Powder Metallurgy Parts Manufacturers Association (PMPA) Standards Committee using funds from MPIF and the Navy MANTECH Program, through Concurrent Technologies Corp. (then Metalworking Technologies, Inc.) of Johnstown, Pennsylvania. The work began in 1992 and was pretty well finished by 1998. We would guess \$5 million went into the project. The PMPA Standards Committee chose the materials, densities, heat treatments and test methods. CTC bought the test specimens from three companies in the P/M industry, with all bars being sintered and heat treated under industrial conditions. In general, there were three test bars per condition. This means that for any one material like FC-0208-50, at 6.7 g/cm^3, the three bars that determined its values were sintered by one P/M house. A different chemistry like FN-0208, had bars that were sintered at another P/M house. The sintering companies were all PMPA members. After sintering and heat treating, bars of each alloy were checked metallographically for degree of sinter or heat treatment, and any other problem. The microstructures were recorded on film and saved. CTC and its subcontractors did all the testing. The data were given to the PMPA Standards Committee and also posted on a password protected CTC web site. The PMPA Committee Chairman, Howard Sanderow, evaluated all the data and presented the committee with the data plotted in graphical form. This was done by plotting a property like .2% yield strength, versus density. The committee studied the data and reached a consensus on the minimum yield strength values. For heat treated materials where yield strength is about equal to tensile strength, the tensile strength was set as the minimum property. The minimum values were not set by a statistical procedure such as average minus three standard deviations. As noted, they were consensus values, where the committee members believed that, on a statistical basis, their companies could produce material to meet the minimum .2% yield strength. It is expected that if any minima were set too high, that P/M houses will not hesitate to tell the committee. (And on the down side, if they were set too low, end users will not buy P/M parts!)

We talked about ASTM B09 as a balanced committee, representing end users, and not subject to undue influence of the P/M parts producers. By definition, the PMPA Standards Committee is composed only of parts producers. How are end users to view its findings? Our personal opinion is that they should have no concern. The data were based on bars tested by a third party laboratory (CTC) and the data were published on the CTC web site. Further, Howard Sanderow and several CTC professionals have published most of the findings at the MPIF/APMI annual technical conference. If you have any doubts, look at the data yourself. The typical values shown in Standard 35 are generally the average of three data points. They were not reduced to allow for them to be met on a statistical basis.

Many people wondered why, in 1984, MPIF went to a nomenclature system that did not include minimum density. Prior to that time, the nomenclature included a suffix letter which indicated a required density range. We already noted that density does not guarantee mechanical properties because of variations in sintering, or lack of sintering. But there is another reason. Not all iron powders have the same strength at a given density. Atomized irons press easily to 6.8 g/cm^3, but the same pressure on a sponge iron only gives about 6.3 g/cm^3. The significance is that the sponge iron is just as strong at the lower density as the atomized iron is at the higher density. The logic was: “Why pay for the extra weight?” For this reason, the PMPA Standards Committee decided that it was more important to deliver minimum yield strength than it was to deliver minimum density. The attitude was: do what you need to in terms of powder type, density and sintering as long as you meet the minimum yield strength.

Having told you all of this about Standard 35 with its sole requirements being chemistry and .2% yield strength, we have to confess that the red and blue colored 2000 edition of MPIF Standard 35 has more than one part to it. The folks who make bearings have their bronze colored version, and MIM folks have the red one, and the forgers have the teal blue one. These four different parts are all available separately. Because bearings, P/F parts, and MIM parts serve different markets, the end users have varying ideas about which properties should be required vs typical. Everybody agrees chemistry is important, but that is as far as agreement goes. The bearing folks have requirements for radial crushing strength (minimum and maximum), minimum oil content, oiled density range, and microstructure. (This author's opinion is that oiled density does not add much to the specification. What the consumer really wants is oil content and strength.) The MIM folks' standards are based on minimum ultimate tensile strength, minimum .2% yield strength, and minimum percent elongation in 1". Their typical values include all of those tensile properties, as well as density, apparent hardness, and microindentation hardness. They like to show their elongation, because their MIM parts are a lot more ductile than the press and sinter materials. That's because they use fine particles and sinter to a very dense condition with only a few isolated small round pores. The forgers are the latest to issue a version of MPIF Standard 35. They guarantee closer carbon control than the .3% offered in MPIF Standard 35 for press and sinter parts. They have a minimum density, set typically at .5 or 1.5% porosity. For P/F parts, density counts: the denser the better. Sponge iron is never used, as its useful limit by single pressing is 6.6 g/cm^3. Sponge iron also has oxides that detract from the forged properties.

Did you ever wonder how you name all of these mixtures? That is the duty of the standards folks who devised the nomenclature system. You are probably wondering why, when we number MPIF standards from 1 to 56, we don't just number the alloys starting with number 1. Well, the PMPA Standards Committee decided in 1965, that the designation for the mixes should help your memory: as they say, it should convey mnemonic value. Well the system does that: of course, you have to memorize the system! But it is not hard to remember. Take our perennial favorite iron plus 2% copper plus .8% carbon. Lets let the F stand for iron as in "ferrous". Let the C stand for copper. So lets begin with FC. Well how much is there of the major alloying element, copper? 2%! Okay make the nomenclature FC-02. Steels have carbon. How much carbon? Say .8%. Lets have the last digit be the tenth of a percent of carbon: FC-0208. Simplicity itself. What about nickel, lets use N. Okay - Fe + 2% Ni + .5% carbon will be FN-0205. You can see this is easy but don't get carried away and try naming some for yourself! Nomenclature only flows one way: from the standards committee down. Its not just jealousy that causes them to guard this perogative, its because they don't want you naming things for which there are no standard property data. (We suppose that some folks probably evade detection by using made-up names around their companies, but you can always tell the legitimate names by looking in MPIF Standard 35.)

This has been kind of a light hearted view of Standard 35 and its nomenclature system. We have copied the rest of the rules in Appendix 1, if you want to read them. The other thing to remember is the meaning of the digits that follow the FC-0208. It is a two or three digit number that stands for the minimum .2% yield strength in the as-sintered condition, or the minimum tensile strength in the heat treated condition. The digits stand for the thousands of psi's in the strength value. If nomenclature could have been frozen with the 1970 era materials, life would be simpler. But then we added the prealloyed steels with 1.8% nickel and .5% molybdenum like the AISI (American Iron and Steel Institute), 4600. Since end user engineers knew what a great hardenable steel this was, we grabbed their terminology and created FL-4605. "L" equals prealloyed steel, "05" equals .5% carbon. The mnemonic value came from the engineer already knowing what 4600 stood for! Ditto for an even lower alloy, FL-4205. Then the folks at Black and Decker, at their captive P/M operation in Hampstead, MD, found that 1-2% elemental nickel added to the .4% prealloyed Ni in FL-4205 gave better strength and impact resistance. Lets call it FLN-4205. The nickel is present in two forms: the 1-2% added elemental nickel, and the .4% dissolved in the prealloyed 4200 steel. Well, it's still mnemonic, (sort of).

Are you wondering why the author is perseverating on mnemonicism? As the pen is mightier than the sword, the MPIF nomenclature system is in the process of vanquishing all others. What all others? Well, the ASTM B-9 committee did not stop its standardizing work when it assigned K values, densities and oil contents to bearings in 1946. They went on to standardize the iron based materials such as plain iron, Fe-C, Fe-Cu-C, and Fe-Ni-C. They followed a format of types, grades and classes for example in ASTM B426-65. At that time, our FC-0208-40, typically 6.3 g/cm^3 was ASTM B426-65, grade 2, type II . There were other specifications for Fe-C, Fe-Ni-C, stainless steel, brass, bronze, and nickel silver. Each one had its own specification. We have to share a little secret: many of the same people serve on both the ASTM B-9 committee and the PMPA Standards Committee of the Metal Powder Producers Association Standards Committee. We noticed that we could simply hear the MPIF name for material and instantly tell you how to make it. With the ASTM multiple specifications plus types, grades and classes, we always had to look up the recipe. Mnemonic value was winning the war. In 1988, the ASTM B09.05 subcommittee took the MPIF Standard 35 nomenclature and mechanical values and created a unified ferrous specification, B 783. With permission of MPIF, the ASTM subcommittee took the entirety of MPIF Standard 35 (as they say on PEI, holus-bolus).

Where is this nomenclature system going? As this is being written, the PMPA Standards Committee continues on adding to its system, trying to maintain its easy to remember characteristics. It gets harder and harder. What did they do with the magnetic properties of the soft magnetic materials? (See the MPIF Standard 35, year 2000 edition for structural parts, but look at the special section on magnetic properties.) What will they do about the soft magnetic material made by pressing insulated powders and then using them as pressed? Maybe the fine young man from Connecticut who roots for the serially numbered chemistries is right! But his day has not yet come.

The North American participants in ISO TC-119, SC 5 have transmitted the MPIF Standard 35 "virus" to the material specification ISO 5755. Some important P/M folks in Europe have come to realize that the original ISO 5755 nomenclature system can't be remembered very well. When they think about a P/M mix, they think in terms of MPIF Standard 35 and good old FC-0208. Add to that, the huge data bank that MPIF 35 accumulated during the 90's, and the result is that ISO 5755 will probably look a lot like MPIF 35. The nomenclature will be close to, but not exactly the same (NIH), and the mechanical properties will be very familiar. To give credit where due, the North American automobile industry wants to use ISO 5755 on a worldwide basis and we guess they want it to look like their easily remembered friend, MPIF 35.

Material Selection

Standard 35, "Materials Standards for P/M Structural Parts" 2000 edition, lists about 150 different irons and steels. Figure 10.1 shows the page for copper steels. How does a designer choose which material to use? He has to sift through these and all the other pages to find the least expensive one that meets all of the required mechanical properties and the other service conditions. What are "other service conditions"? A non-exhaustive list includes:

Machinability
Pressure tightness = hermeticity
Controlled porosity
Corrosion resistance
Platability
Weldability
Need to be deformable enough to swage or stake to another part
Need for heat treatment or defined hard case
Need to be self lubricating
Magnetic properties, e.g soft magnet
Thermal expansion coefficient
Heat resistance and creep resistance

Iron-Copper and Copper Steel

P/M Material Properties

Material Designation Code	MINIMUM VALUES (A): Minimum Strength (A) (E) Yield, 10³ psi	MINIMUM VALUES (A): Minimum Strength (A) (E) Ultimate, 10³ psi	TYPICAL VALUES (B): TENSILE PROPERTIES Ultimate Strength, 10³ psi	Yield Strength (0.2%), 10³ psi	Elongation (in 1 in.), %	ELASTIC CONSTANTS Young's Modulus, 10⁶ psi	Poisson's Ratio	Unnotched Charpy Impact Energy, ft·lbf	Transverse Rupture Strength, 10³ psi	Compressive Yield Strength (0.1%), 10³ psi	HARDNESS Macro (apparent), Rockwell	HARDNESS Micro (converted), Rockwell	Fatigue Limit 90% Survival, 10³ psi	Density, g/cm³
FC-0200-15	15		25	20	1.0	14.0	0.25	4.5	45	18	11HRB		10	6.0
-18	18		28	23	1.5	16.5	0.25	5.0	51	21	18	N/A	11	6.3
-21	21		31	26	1.5	16.5	0.25	5.5	56	23	26		12	6.6
-24	24		34	29	2.0	19.5	0.27	6.0	63	26	36		13	6.9
FC-0205-30	30		35	35	<1.0	14.0	0.25	<2.0	60	35	37HRB		13	6.0
-35	35		40	40	<1.0	16.5	0.25	3.0	75	40	48	N/A	15	6.3
-40	40		50	45	<1.0	17.5	0.25	5.0	95	45	60		21	6.7
-45	45		60	50	<1.0	21.5	0.27	8.0	115	50	72		31	7.1
FC-0205-60HT		60	70		<0.5	16.0	0.25	2.5	95	57	19HRC	58HRC	27	6.2
-70HT		70	80	(D)	<0.5	15.5	0.25	3.5	110	71	25	58	30	6.5
-80HT		80	90		<0.5	18.5	0.27	4.5	120	86	31	58	34	6.8
-90HT		90	100		<0.5	20.5	0.27	5.5	135	95	36	58	38	7.0
FC-0208-30	30		35	35	<1.0	12.5	0.25	<2.0	60	40	50HRB		13	5.8
-40	40		50	45	<1.0	16.5	0.25	2.0	90	45	61	N/A	17	6.3
-50	50		60	55	<1.0	17.5	0.25	5.0	125	50	73		23	6.7
-60	60		75	65	<1.0	22.5	0.28	7.0	155	55	84		33	7.2
FC-0208-50HT		50	65		<0.5	15.0	0.25	2.5	95	58	20HRC	60HRC	25	6.1
-65HT		65	75	(D)	<0.5	17.5	0.27	3.5	110	72	27	60	30	6.4
-80HT		80	90		<0.5	18.5	0.27	4.5	130	91	35	60	35	6.8
-95HT		95	105		<0.5	21.5	0.27	5.5	150	105	43	60	40	7.1
FC-0505-30	30		44	36	<0.5	12.5	0.25	3.0	77	50	51HRB		17	5.8
-40	40		58	47	<0.5	16.5	0.25	4.5	102	54	62	N/A	22	6.3
-50	50		71	56	<1.0	17.5	0.25	5.0	124	58	72		27	6.7
FC-0508-40	40		58	50	<0.5	13.0	0.25	3.0	100	58	60HRB		22	5.9
-50	50		68	60	<0.5	16.5	0.25	3.5	120	63	68	N/A	26	6.3
-60	60		82	70	<1.0	18.5	0.27	4.5	145	68	80		31	6.8
FC-0808-45	45		55	50	<0.5	14.0	0.27	3.0	85	62	65HRB	N/A	21	6.0
FC-1000-20	20		30	26	<1.0	14.0	0.27	3.5	53	33	15HRB	N/A	11	6.0

Figure 10.1. 2000 Edition, Materials Standards for P/M Structural Parts. Approved: 1984 Revised 1987, 1990, 1994, 1997, 2000.

As an example of sorting for a minimum yield strength, take 50,000 psi. The closest ones, that is the ones with the lowest densities that are still strong enough, are:

AS SINTERED	AS HEAT TREATED
FC-0208-50 at 6.7 g/cm^3	F-0005-50HT at 6.6 g/cm^3
FC-0505-50 at 6.7 g/cm^3	F-0008-55HT at 6.3 g/cm^3
FC-0508-50 at 6.3 g/cm^3	FC-0205-60HT at 6.2 g/cm^3
FN-0208-50 at 7.4 g/cm^3	FC-0208-50HT at 6.1 g/cm^3
FN-0408-55 at 7.2 g/cm^3	FN-0205-80HT at 6.6 g/cm^3
FL-4205-50 at 7.3 g/cm^3	FN-0208-80HT at 6.7 g/cm^3
FL-4405-50 at 7.3 g/cm^3	FN-0405-80HT at 6.5 g/cm^3
FL-4605-50 at 7.35 g/cm^3	FL-4205-80HT at 6.6 g/cm^3
FLN-4205-50 at 7.05 g/cm^3	FL-4405-100HT at 6.7 g/cm^3
FD-0205-50 at 6.95 g/cm^3	FL-4605-80HT at 6.55 g/cm^3
FD-0208-50 at 6.75 g/cm^3	FLN-4205-80HT at 6.6 g/cm^3
FD-0405-55 at 6.75 g/cm^3	FLC-4608-50HT at 6.6 g/cm^3
FD-0408-50 at 6.70 g/cm^3	FD-0205-95HT at 6.75 g/cm^3
FX-1008-50 at 7.3 g/cm^3	FD-0405-100HT at 6.75 g/cm^3
FX-2008-60 at 7.3 g/cm^3	FX-1005-100HT at 7.3 g/cm^3
	FX-1008-110HT at 7.3 g/cm^3
	FX-2005-90HT at 7.3 g/cm^3
	FX-2008-90HT at 7.3 g/cm^3

Lets suppose that at a yield strength of 50,000 psi, you really did not need something heat treated for wear resistance. Lets eliminate the whole right column, and avoid paying about $.50/lb extra, and avoid the related distortion and oil pick up. The exception in the right column would be FLC-4608 which sinter-hardens and needs only to be tempered for 1 hour 350 to 400˚F after sintering. But why pay for tempering, if you don't need it.

Suppose that none of the "other service conditions" affect your part. Which mix do you choose? The least expensive is good old FC-0208-50 at 6.7 g/cm^3. The FC-0508 -50 at 6.3 g/cm^3 weighs less, why not choose it? The cost for the extra 5% Cu adds $.07/lb. and the weight saving of 6% only saves you $.025/lb. toward paying for the sponge iron. Also, sponge iron costs $.38/lb. and atomized iron for the FC-0208-50 only costs $.36/lb. And FC-0508-50 is less dimensionally controllable at sintering. You might wonder why FC-0508 is in the specification. The best answer is historical precedent, but you always could ask the committee to delete it and save a tree! The same logic applies to FC-0505-50 at 6.7 g/cm^3, only more so.

What about all the diffusion alloys, the FD-'s? The FD-0205 at 6.95 g/cm^3 would cost $.63/lb vs the FC-0208 at $.41/lb. But, that diffusion alloy has 2.5 times the impact strength of the FC-0208, 12 ft-lb vs 5 ft-lb, and might be chosen if the part sees more than average impact forces. Any drawbacks to FD-0205 -50 at 6.95 g/cm^3? It has partially alloyed regions around the 2% nickel that tend to have some martensite needles in them. Good for wear resistance, but good for wearing out the cutting tools that are used to machine it! The MPIF Standard 35, 2000 Edition and after has data ranking machinability. Similarly, FLN-4205-50 at 7.05 g/cm^3 has even better impact energy at 13 ft-lb. and it cost $.63/lb., but you need 1.5% more of it, because of the higher density. At that same 7.05 g/cm^3 higher density, the FD-0205 would have an impact energy of 15 ft-lb.

What about the nickel steels and low alloy steels that are in the 7.2 to 7.4 g/cm^3 range? They have terrific impact properties, (22 ft-lb.) but the cost penalty to double press and sinter would add 45 to 50% to the cost of the parts. (These high density materials are usually hardened to give very high tensile strengths.) What about the infiltrated steels at 7.3 to 7.5 g/cm^3? The cost premium vs FC-0208 would be 40%. The advantage would be pressure tightness or close to it, so that they could seal against air or gas pressure. The parts could be plated without need for resin impregnation.

What about machinability? The best choice is the FC-0208-50, with .5% MnS added. The extra .5% MnS at $5.00/lb. adds about $ .025/lb. to the parts. Controlled porosity is a characteristic of filters or bearings and they are usually lower in density than these examples with their 50,000 psi yield strength. The only way to get corrosion resistance is to steam blacken the parts (helps somewhat), or to resin seal and plate. Welding is aided by lower carbon and lower hardenability, and the FLN-4205-50 would be the best choice. See our chapter 9, for hints on welding and selection of materials. None of these materials are really ductile for swaging or staking, but the FLN-4205-50 is the best choice. It is hard to get something with a high yield strength and good ductility in a press and sinter part. The .45% Phosphorous materials have both a good yield strength and good ductility. They finally were standardized in MPIF Standard 35, for P/M Structural Parts, 2000 edition, but only after they had been in wide use for at least 15 years.

None of these materials can be heat treated to a defined case. You will need a 7.3 g/cm^3 density with .2%C or less, in the core. That will not give you the 50,000 psi min yield strength in the core. The FC-0208 at 6.7 can hold 12-13% oil by volume and is approximately covered by the MPIF bearing specification 35 as FC-0208-K40 at 6.4 g/cm^3. None of these materials make good soft magnets, because of their high carbon contents. Magnets don't usually need high yield strength, and so they are made from pure iron or Fe + .45%P mix. Interestingly, as noted above, it does have good yield strength and toughness.

As you can see, this was, as we used to say in school, "a non-trivial exercise"! What do real design engineers do, if they don't have all this detailed P/M background? We suppose this little primer helps, as does the excellent "Powder Metallurgy Design Manual" available from MPIF, in Princeton, NJ. But, once you have some basic familiarity with the P/M process, the most efficient thing to do is to work with the sales and engineering folks at a P/M parts producing house. This can be a very quick process, especially if the end-user engineer is careful to lay out all the "other service conditions", at the outset. The P/M folks have a kind of expert system in their heads that allows them to go though all the options in less than the time it took you to read the single example above.

Something that does not work well is having an end user design engineer say, dogmatically, "I want something like my cold rolled, low carbon steel that I can case harden." Did he design the part that way, or did he copy the specification that had been on similar parts for the last 50 years?" Cold rolled steel has very good elongation (20%) compared to most P/M materials of the same yield strength (33 ksi). Well, if he really has to have the identical elongation and impact strength, there is always powder forging and HIP'ing. But they have their penalties in terms of cost and nearness to net shape. Does P/M fail to make a sale at this point? Maybe, but it depends on the end user's willingness to try alternative materials. This suggestion comes from the fact that many parts are over-designed by virtue of being made from some readily available wrought material. The cold rolled steel bars were cheap, but the required machining was expensive. If shape permits, why not compact the least expensive, densest single pressed material that could be more or less case hardened. That could be FN-0205 or FN-0200 or a "made up" version with 2% Ni and .2% combined carbon, but with no actual MPIF Standard 35 equivalent. If the price looks OK, the parts are prototyped and tested in service. This could mean that hard tools are built and used to make prototypes, or, more often, prototypes are machined from blanks of the correct chemistry, density and mechanical properties. (See the "Powder Metallurgy Design Manual" for an unbiased view of the methods for prototyping) . It often turns out that the P/M material is better than would have been expected and works well in the application. The drawing could then reference the material, a mechanical test to be done on the part, and perhaps a case depth and hardness, for the particular example considered here. Or, maybe the single pressed part was not strong enough. What then? Can the application stand the cost of double pressing and double sintering to 7.3 to 7.4 g/cm^3, or single pressing and high temperature sintering on a ceramic belt? If "yes", the testing goes on. If "no", then "Thanks for calling, and keep us in mind in the future." P/M is not all things to all people. If we were, the industry would be a lot bigger than its $2-3 billion.

Failure Analysis

If an engineer specifies a material from Standard 35 or develops a proprietary material with a specific P/M house, and proves it out with field tests, does everyone "live happily ever after"? It depends a lot on the quality control system at the P/M house and how well the P/M house understands its process. You can have really nice ISO 9000 paperwork and not completely understand the P/M process! When a field failure occurs, the end user assumes that its own laboratory will be able to diagnose the problem and then approach the supplier with details about the cause of the problem to discuss ways to make sure it never happens again. End users can quite easily check apparent hardness or density or chemistry. When it comes to metallographically examining cross sections of the parts, the process is far more difficult than looking at wrought steel. One of us (LP) has published information on how to do this in "Volume 7" of the 10th edition of the "ASM Metals Handbook". The simple truth is that if you don't practice a skill regularly, it is hard to just pick up and do it, and this applies to metallography on P/M materials. This means that when there is a failure, the end user will be more than usually dependent on his supplier or a third party lab familiar with P/M.

If the P/M manufacturing process is under ISO 9000 control, how could it deviate in a way that would not be readily detected? Take compaction as an example. Mold a part to 7.05 g/cm^3 and everything is fine. Increase the green density a little to 7.15 g/cm^3 and some modest overpressing occurs, see Figure 10.1. You can't see it on the surface, you can't see it in X-rays, and if you are not a skilled P/M metallographer you may not see it in the microstructure. After all, conventional wisdom is that denser is better! Maybe the destructive break test did not stress the part perpendicular to the particle boundaries left locally by the overpressing and the test found nothing wrong. But service conditions occasionally do stress the part across these particle boundaries, and the part breaks. How do you cure this? Its not much fun but you do sectional densities to be sure that the density is not too high in one local area, or maybe you measure the local density by image analysis and check structure at the same time, and maybe you add a new destructive test to look for this particle boundary condition.

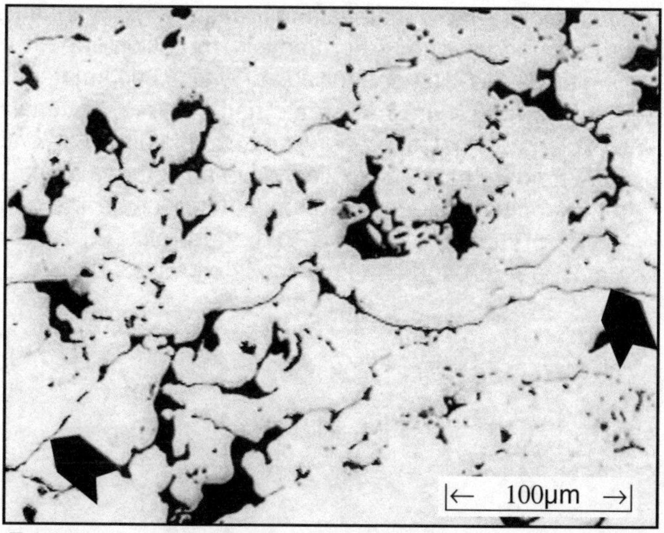

Figure 10.1. Example of overpressing in a P/M steel part. The pressure was high enough that on ejection from the tooling, some of the particle sprang apart vertically. Sintering did not re-bond the particles. The separations are the thin black lines shown by the arrows. There is no bonding across these lines, and very low strength. (250x)

Appendix 1

P/M Material Code Designation

The P/M material code designation or identifying code in the case of structural P/M parts defines a specific material as to chemistry and minimum strength expressed in 10^3 psi. For example, FC-0208-60 is a P/M copper steel material containing nominal 2% copper and 0.8% combined carbon possessing a minimum yield strength of 60 X 10^3 psi (60,000 psi) (410 MPa) in the as-sintered condition.

A coding system offers a convenient means for designating both the chemistry and minimum strength value of any standard P/M material. It is based on the system established by the industry with the addition of a two-or three-digit suffix representing minimum strength in place of a suffix letter indicating density range. The density is given for each standard material as one of the typical values.

Code designations in this Standard and revisions thereof apply only to P/M materials for which MPIF Standards have been adopted. In order to avoid confusion, the MPIF coding system is intended for use only with such materials and should not be used to create non-standard materials. The explanatory notes, property values, and other contents of this Standard have no application to any other materials.

In the coding system, the prefix letters denote the general type of material. For example, the prefix "CT" represents copper (C) and tin (T) which is known as bronze.

Prefix Letter Code

A	Aluminum
C	Copper
CT	Bronze
CNZ	Nickel Silver
CZ	Brass
F	Iron
FC	Iron-Copper or Copper Steel
FD	Diffusion Alloyed Steel
FL	Prealloyed Ferrous material except Stainless Steel
FN	Iron-Nickel or Nickel Steel
FX	Copper Infiltrated Iron or Steel
G	Free graphite
M	Manganese
N	Nickel
P	Lead
S	Silicon
SS	Stainless Steel (prealloyed)
T	Tin
U	Sulfur
Y	Phosphorus
Z	Zinc

Prefix and Four Digit Code

The four digits following the prefix letter code refer to the composition of the material.

In nonferrous materials, the first two numbers in the four digit series designate the percentage of the major alloying constituent. The last two numbers of the four digit series designate the percentage of the minor alloying constitutent.

For improved machinability lead is sometimes the third alloying element in a nonferrous alloy system. Lead will only be indicated by the letter "P" in the prefix. The percentage of lead or any other minor alloying element which happens to be excluded from the four digit nomenclature, is represented in the "Chemical Composition" which appears with each standard material. Illustration of P/M nonferrous material designation coding:

P/M Nickel Silver:

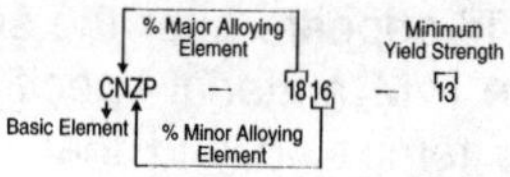

In ferrous materials, the major alloying elements (except combined carbon) are included in the prefix letter code. Other elements are excluded from the code but are represented in the "Chemical Composition" which appears with each Standard material. The first two digits of the four-digit code indicate the percentage of the major alloying constituent present.

Combined carbon content in ferrous materials is designated by the last two numbers in the four-digit series. The carbon content up to and including 0.3% is considered as zero; higher contents are included in ranges and coded as follows:

Carbon Ranges	Code Designation
0.0% - 0.3%	00
0.3% - 0.6%*	05
0.6% - 0.9%	08

*Exception: Carbon range for the FL-series is 0.4%-0.7%.

The range of carbon that is metallurgically combined is indicated in the coding system. Illustration of P/M ferrous material designation coding:

P/M Nickel Steel:

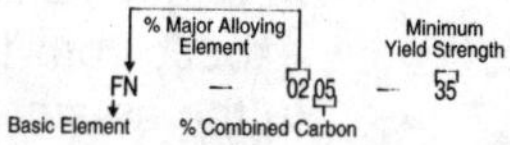

In the case of P/M stainless steels and P/M prealloyed low alloy steels, the four-digit series is replaced with a designation derived from modifications of the American Iron and Steel Institute alloy coding system, e.g., SS-316L-15, FL-4605-100HT.

When a P/M prealloyed low alloy steel powder is modified with elemental additions, a three or four alpha-numeric designator will be used, e.g., FLN-4205-40, FLN-4205-105HT.

Examples of P/M Material Designation Coding

Material	Compositions by Percent	Complete Code for Material, Composition & Minimum Strength (10^3 psi)	
P/M Bronze	Cu-90, Sn-10	CT-1000-13	
P/M Nickel Silver	Cu-64, Ni-18, Zn-18	CNZ-1818-17	
P/M Nickel Silver	Cu-64, Ni-18, Zn-16, Pb-2	CNZP-1816-13	
P/M Brass	Cu-90, Zn-10	CZ-1000-11	
P/M Brass	Cu-78, Zn-20, Pb-2	CZP-2002-12	
		As-sintered	**Heat Treated**
P/M Iron	Fe-99, C-0.2	F-0000-20	—
P/M Steel	Fe-98, C-0.8	F-0008-35	F-0008-85HT
P/M Copper Steel	Fe-96, Cu-2, C-0.8	FC-0208-60	FC-0208-95HT
P/M Nickel Steel	Fe-96, Ni-2, C-0.5	FN-0205-35	FN-0205-180HT
P/M Infiltrated Iron	Fe-78, Cu-20	FX-2000-25	—
P/M Infiltrated Steel	Fe-77, Cu-20, C-0.8	FX-2008-60	FX-2008-90HT
P/M Stainless Steel (austenitic)	AISI 316 (modified)	SS-316N1-25	—
P/M Stainless Steel (martensitic)	AISI 410 (modified)		SS-410-90HT
P/M 4600 Steel (prealloyed)	AISI 4600 (modified), C-0.5	FL-4605-45	FL-4605-140HT
P/M 4200 Steel (prealloyed+addition)	AISI 4200 (modified), Ni-1.5, C-0.5	FLN-4205-40	FLN4205-105HT

As with other P/M materials, the suffix number denotes the specified minimum strength value expressed in 10^3 psi.

In the case of soft magnetic alloys, the phosphorus containing irons are treated differently, since the amount of phosphorus is usually less than 1%. To indicate more accurately the nominal amount of phosphorus the code takes the nominal percent phosphorus, multiplies by 100 and uses this number for the first two digits in the code. The last two digits remain "00" since no carbon is required. For example, the iron-0.45% phosphorus alloy would be designated as: FY-4500.

Suffix Digit Code

The two- or three-digit suffix represents the minimum strength value, expressed in 10^3 psi, that the user can expect from the P/M material possessing that chemistry. In the as-sintered condition the strength is tensile yield; in the heat treated condition, it is ultimate tensile.

(See **Minimum Value** page 2 of MPIF Standard 35.)

Suffix Letter Code

When the code designation "HT" appears after the suffix digits it is understood that the P/M material specified has been quench hardened and tempered and that the strength represented is ultimate tensile in 10^3 psi.

In the case of the soft magnetic alloys the suffix does NOT designate yield or tensile strength, but rather the maximum coercive field (10 times the value in oersteds) and an alphabetic designator for the minimum density as follows:

Designator	Minimum Density (g/cm^3)
U	6.5
V	6.7
W	6.9
X	7.1
Y	7.3
Z	7.4

For example, a pure iron material at a minimum density of 6.9 g/cm^3 and coercive field of 2.3 Oe would be designated as F-0000-23W. The iron-0.45% phosphorus alloy at a 7.1 g/cm^3 minimum density and coercive field of 2.0 Oe would be designated as FY-4500-20X.

CHAPTER 11

Quality Assurance

We have covered many aspects of the P/M process in previous chapters and we hope you understand most of that material. You can now go forth and create P/M parts. If however, one of the process steps is out of control, the parts are going to be defective and no one will want them. How do we make certain that every part we produce is a good one? We put systems in place that monitor process variations. We can then control and reduce variation; that will assure quality.

In the "good old days", quality control usually consisted of a roving inspector checking certain dimensions and weight at the compacting press and at the end of the sintering furnace. These checks may not have been done on a regular basis. The inspector may have occasionally detected a bad part that was hopefully discarded. Nothing said, "Watch out, your process is out of control!" Statistical process control (SPC) is a tool which, when properly implemented, will achieve our desired results, i.e. all parts produced are acceptable and are of the highest possible quality.

Statistics is a method for communicating your knowledge of the product or process clearly to others. It is also a way that a process can communicate to you about what it is doing. Statistics can get very complicated and we are not going to get involved with a lot of complicated formulas. Let's just see how statistics can help us with P/M parts production.

The first thing we need to understand is that there is variation in every process and therefore no two parts are identical. Think about the height of people in your department. Their height varies and there is a range or spread in their height. If we measure and record everyone's height on a chart, we create what's called a histogram. It is a picture of the distribution of height (Figure 11.1). Distributions are very important because they show the pattern of variation. All patterns have three characteristics. The first is spread. That's just the width of the distribution. Spread is often measured by taking the difference between the highest and lowest measurement.

The second characteristic is the location of the distribution. This is measured by either the average or median. Adding the heights of all the people and dividing by the number of people determines the average or mean. In chart 1 the median is 5 ft. 10 in. because just as many people are taller than 5 ft. 10 in. as are shorter than that height. Therefore 5 ft. 10 in. is the median of the distribution. The true definition of median is the middle most value, or that value that has just as many values above as below. See figure 11.2. In general, the median is not exactly the same as the mean or the average.

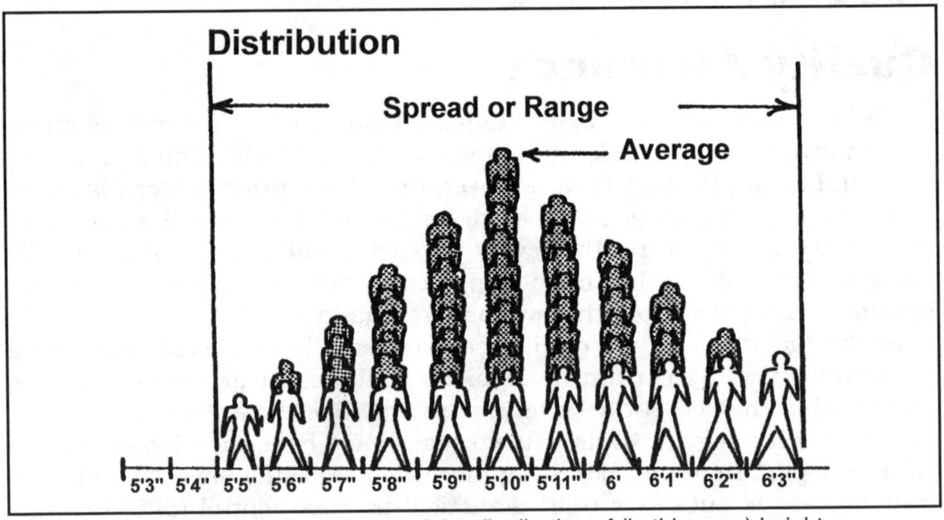

Figure 11.1. A Histogram or a picture of the distribution of (in this case) height.

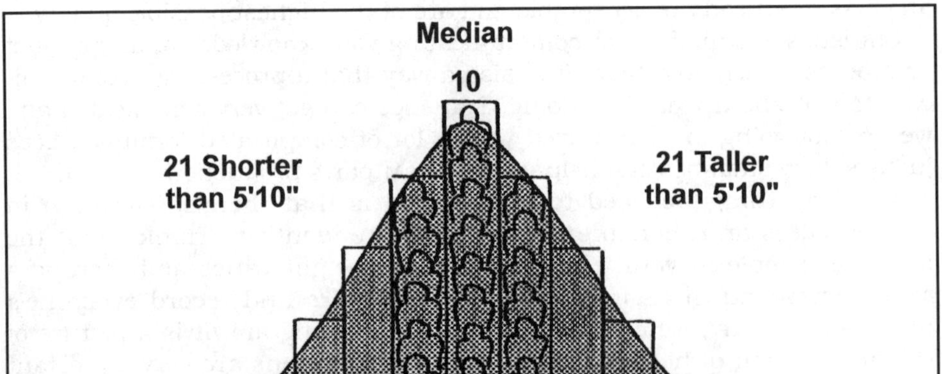

Figure 11.2. A bell shaped distribution showing the median or middle most value.

The third characteristic is the shape of the distribution. The distribution in Figure 11.2 is "bell shaped". This is called normal because the distribution of many measured variables exhibit this pattern. If for example, you are a member of the Presidential Honor Guard, your height chart distribution shape will not be normal. You are part of a select group that has strict minimum and maximum height requirements. This sets your group apart from the normal population and your distribution shape will reflect this special circumstance or cause. You will note that for a perfectly normal distribution, the average and the median are the same. You are now probably wondering what the height of your coworkers has to do with the Quality Assurance of P/M parts.

Many years ago, Dr. Walter Shewhart showed how data from a manufacturing process could be developed and interpreted using simple statistical methods. Given a sufficient number of samples, the output of any process operating solely under a set of common causes can be shown to exhibit a normal distribution. Take a look at Figure 11.3. Across the bottom of the charts you will see several notations of the Greek letter Sigma (σ). Statisticians use this letter to represent a measure of variation and call it "Standard Deviation". When we have a normal distribution, we can expect the values to fall within ± 3 standard

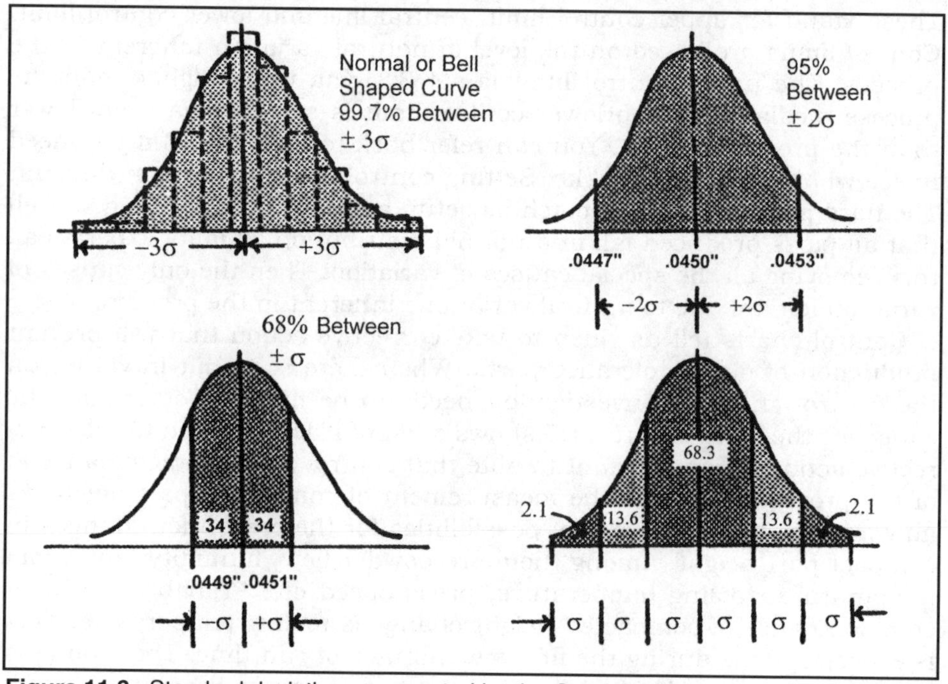

Figure 11.3. Standard deviation represented by the Greek letter Sigma σ.

deviations (3 σ) of the process mean in 997 out of 1000 cases. If a point appears outside the 3 standard deviation limit it is likely that a special cause has affected the process.

This system of sampling a process and tracking results generates charts that are known as Showhart control charts. These charts provide a tool to prevent defects by telling us when the process has shifted. We can then adjust the process before defective parts are produced. There are actually two control charts for each characteristic being tracked. One is the X-bar chart and the other is the R Chart. Figure 11.4 shows these charts. Each point on the charts represents a group of parts called a sample subgroup or "sample" for short. Samples typically consist of three to five parts and all samples for a given chart are the same quantity. The upper chart, called the X (X-bar) chart shows the history of the process average. Each point is the average of readings for the sample (3-5 parts).

The lower chart, called the R Chart (range chart) shows history of variation within a sample. If for example, the sample is 3 parts and the readings are 45, 45.5 and 46, then the average of that sample is 45.5 and the range is 1. By using the X and R charts, significant shifts in both the process mean and spread can be detected as an indication of a process out of control.

You will see the letters UCL, CL and LCL on the charts in Figure 11.4. These stand for upper control limit, central line and lower control limit. Control limits are based on the level of natural variation inherent in the process. The upper control limit is set 3 sigma units higher than the process median, and the lower control limit is set 3 sigma units lower than the process median. You can refer back to Figure 11.3 if you need to see what 3 sigma looks like. Setting control limits is not that difficult. The hard part is to get the machine setup and materials running so well that all parts produced fall in a normal distribution. Quality experts call this removing all the special causes of variation. Then the only causes of variation left are due to natural variations inherent in the process.

Control charts tell us when to take corrective action that will prevent production of out of tolerance parts. When there is a shift in values on the control chart, an investigation needs to be made to determine the cause for the shift. Figure 11.5 shows a list of rules on when to take corrective action. It is important to note that control charts indicate a trend of the process. Let's use the measurement of compacted part height as an example. There are several possibilities for the reason for changes in compact part height. Among them are powder flow, humidity, and room temperature, tooling temperature, press speed etc. Probably the most common cause of compacted height change is tooling temperature. This is especially true during the first few minutes of run time. The tooling is heating up due to frictional forces during compacting. When the tooling

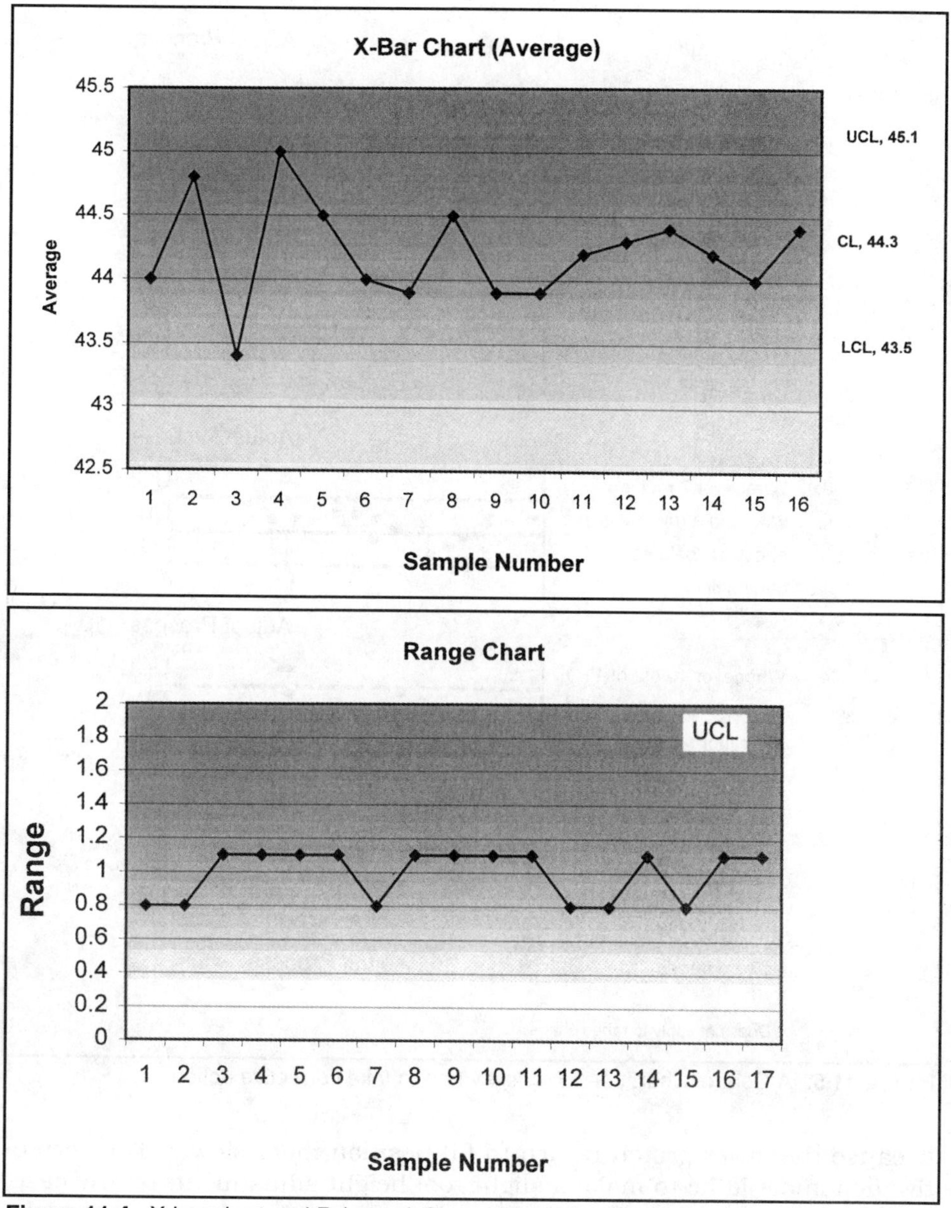

Figure 11.4. X-bar chart and R (range) Chart. The X-bar Chart shows the history of the process average and the R Chart shows history of variation within a sample.

heats up, the punches expand. The punch lengths expand, which causes the molded part length to shorten. Outward expansion of the punches causes the punch to fit tighter in the die. This changes the weight

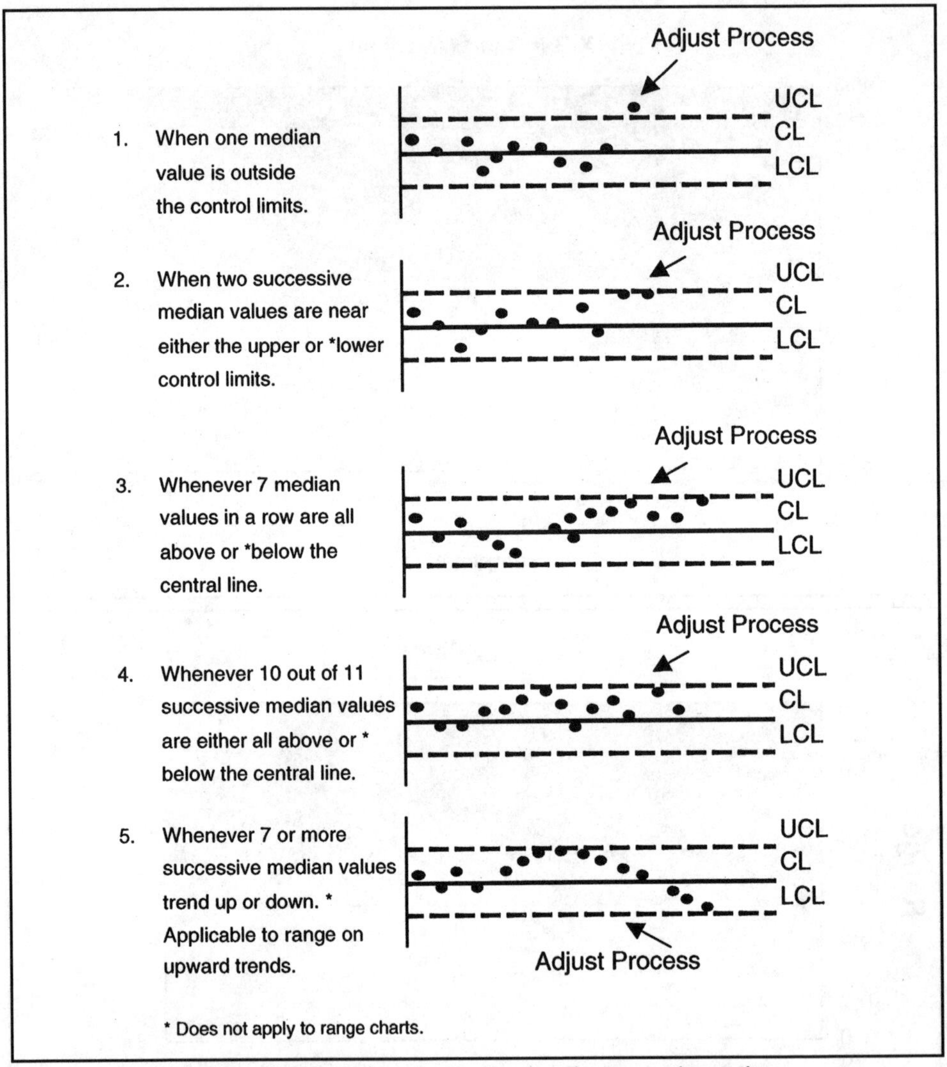

Figure 11.5. A Control Chart used to show when to take corrective action.

because the lower punch return to fill position more slowly. The corrective action could be to make a slight tool height adjustment. In any case, whenever the process is changed in any way, the change should be noted on the chart. This includes replacement of punches or other tooling and the change from one raw material lot to another. If the sample average is slightly above the UCL or slightly below the LCL the parts being produced usually will still be in print tolerance. This is because the UCL and LCL are not the tolerance limits of the part.

Control limits and tolerances must never be compared numerically and should not appear together on the same graph. Tolerances are limits on individual measurements. Control limits define the amount of variation to be expected in a process.

Once the process is in statistical control, it is common to measure the capability of the process. There are two measures, Cp and Cpk. Cp is the process capability index and it has a minimum value of 1.37 for 3 sigma deviation. Cpk is the process location index and has a minimum target value of 1.33. The process capability index, Cp, is the relationship between the engineering blue print tolerance and the process spread. It is calculated by the formula Cp=tolerance/process spread=tolerance/6 sigma. Sigma is the standard deviation of the process. If you are confused, take heart, you're not alone. You might want to look back at Figure 11.3 to redefine sigma. The second capability index Cpk, or process location index, is a measure of how well the process distribution is centered between the specification limits. It is calculated by Cpk=(S1-X)/3 sigma. S1 is the closer specification to X , which is the process average. Figure 11.6 shows a chart and capability calculations for a process that is centered and capable. The process location index Cpk is 1.33 and this is the minimum target for all processes. Higher Cp and Cpk numbers indicate a better level of statistical control and capability. Figure 11.7 shows examples of other process diagrams - some capable, some not. The last example is one that is capable, in control but producing output out of specification because the process is not centered. If the capability study shows that the process is not capable, then the following alternatives exist:

Change the process. Take engineering action to reduce variability of the process. This would involve changes in one or more of the following; materials, tools, methods or machines.

Change the specification. A relaxation in specifications could change a process from not capable to capable. This would probably involve changes requested from the customer.

Live with the situation. 100% inspection and sorting will be necessary to ensure meeting the specification limits. This is only acceptable as a short term solution. 100% inspection has proven to be ineffective.

Stop making the product. If economics do not justify the above actions, management should decide to stop production.

We hope you are feeling quite statistically capable now and ready to apply SPC to P/M parts production. How do we set up an effective quality system? Like any other successful methodology, the key is good planning.

Capability
(Quality Indices)

Case 1 - Centered Distribution

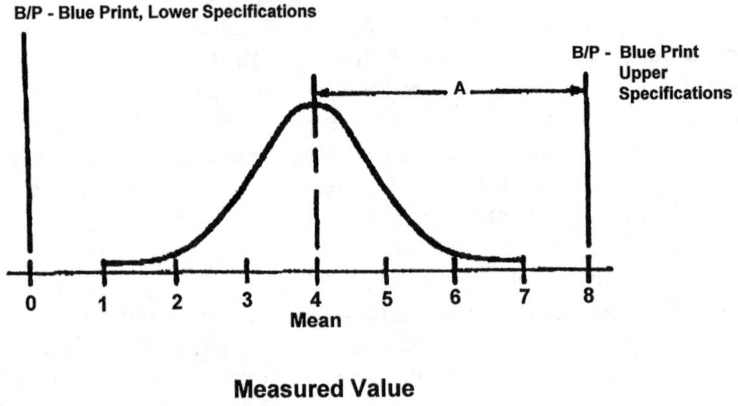

$$C_p = \frac{\text{Spec. Width}}{\text{Dist. Width}} = \frac{8-0}{7-1} = \frac{8}{6} = 1.33$$

$$C_{pk} = \frac{A}{3\text{ sigma}} = \frac{4}{3} = 1.33$$

Note: for figuring "A" when distribution is centered, either limit gives the same result.

A = Center - Low = 4 - 0 = 4
A = High - Center = 8 - 4 = 4

If not centered, the distance to closest specification limit must be used.

Figure 11.6. A Chart and capability calculations for a process that is centered and capable.

We first must have an overall quality plan; and, because each different part we manufacture is for a specific application we must develop a specific plan and methodology for that part. The plan development process includes discussion with the customer, ideally held before the design is finalized. This step gives the part manufacturer an understanding of the function and fit of the product. The discussion must also communicate a thorough understanding of the quality requirements including dimensional tolerances and other features. It is especially important that those discussions be held while there is still some flexibility in the design. Many times changes can be made at this point which accommodate the P/M process and can lower overall cost. These changes could include minor tolerance changes which would eliminate secondary machining.

Process Capability

A capability index (C_p) of 1.33 or more is considered good. The diagram below shows examples of processes that are in a state of statistical control and are capable (all output within specifications).

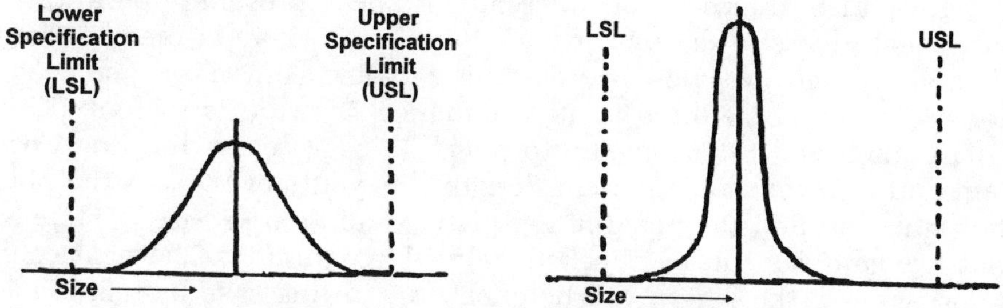

Processes in Statistical Control--And Capable*

The diagram below shows examples of processes that are in a state of statistical control and are <u>not capable</u> (output produced is beyond one or more specifications). (C_p is not good)

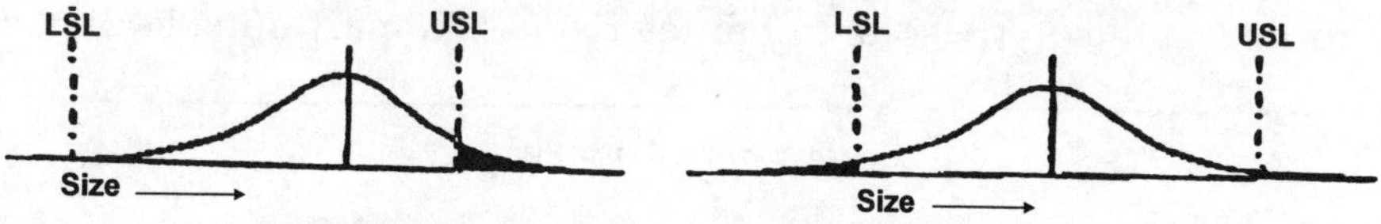

Processes in Statistical Control--And Not Capable*

Finally, the diagram below shows an example of a process that is in a state of statistical control, <u>is capable</u>, and is producing output beyond one specification because the process is not centered. (C_p is ok; C_{pk} is not good)

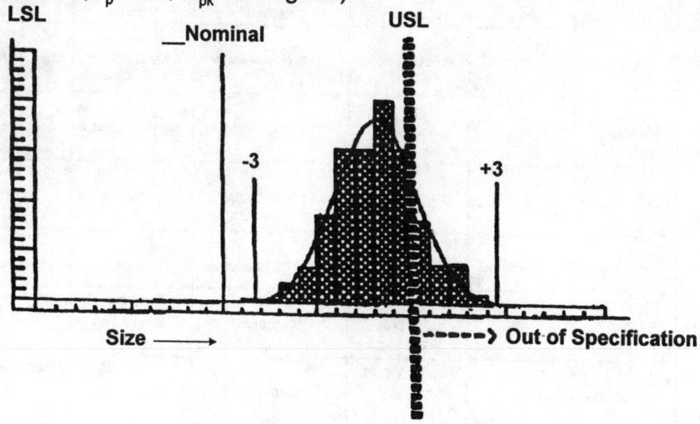

Process Not Centered

Figure 11.7. Process Capability.

Slight changes in shape can also be made which result in faster molding speed, lower tool cost or produce a stronger part.

After these discussions are concluded and the design is finalized, a control plan which includes gauging methods, control methods, inspection frequency and responsibility can be developed.

Figure 11.8 shows a control plan for the repressing operation of a part. Look across the columns of the plan until you come to Control Method. As you look down under this heading you will see some notations - X-bar and R chart and Frequency Distribution Chart. This is where our friend SPC comes into play. Go back to the heading Control Item and let's consider the item "length". As you move across the plan to the right, you find the heading Frequency and Sample Size. This tells us that five samples will be checked every thirty minutes The next column describes how the length will be checked - in this case by using a 1 in. micrometer. The last column describes what action will be taken if there is a problem with the length and who is responsible for the necessary corrective action. Now we can see how SPC can be used to monitor part length or other process variables.

There are some control items on the chart which are to be checked at setup and audited. These are characteristics like part outside diameter

Manufacturing Control Plan

Customer Part # | Part Name | Customer:
Supplier Part # | Date (Original) : | Date (Revised) :

Oper. Seq.	Process Description Process Purpose	Control Item & Specification	Dept. Resp.	Freq. & Sample Size	Measurement Method	Method of Recording Control Method	Problem Reaction Plan & Resp.
3	Coin	Length	220	5 pcs/30 mins	0-1" Micrometer	X-bar & R Chart	Press Adj. Sort, Rework, Scrap Operator
		Parallelism	220	2 pcs/ 1 hour	0-1" Micrometer	X-bar & R Chart	Sort, Rework, Scrap Operator
		Die-Controlled Features	220	At Set-up	Attribute Gages	Audit	Sort, Rework, Scrap Operator
		Density	220	At Set-up	Water Displacement	Audit	Press Adj. Operator
		No Cracks	225	5 pcs/lot	Magnaflux	Frequency Distribution Chart	Sort, Rework, Scrap Operator
		Appearance	225	5 pcs/lot	Visual Feel	Frequency Distribution Sheet	Sort, Rework, Scrap Operator

Figure 11.8. A brief example of a control plan for repress operation (Courtesy of Quality Control and Inspection of Secondary Operations-Pat Kenkel, John Enquist, Mike Blanton, Burgess Norton Mfg. Co.).

in the molding operation. This diameter is formed in a fixed tool member and will not vary in this operation. The feature should be checked at setup to make certain the die isn't damaged or worn and then only audited on occasion.

Let's pretend you are the press operator running this job. The Control Plan tells you that every 30 minutes you need to take five samples and measure them using a 1 in. micrometer. You will then determine the average of the five readings and record it on an X-bar chart (look back at Figure 11.4). If the X-bar chart tells you we need to take corrective action (review Figure 11.5) you will take the action described in the plan. You are now the master of your fate because the responsibility for the quality of the parts you are running is clearly fixed. No more waiting for an inspector to come around and tell you you're making bad parts. This may sound a little scary at first but if the process is in control and you monitor it according to the plan there is nothing to worry about.

Figure 11.9 is the print of a production part called a tube. Some information has been removed for confidentiality reasons. Figure 11.10 is the first section of the Control Plan for the subject part. There are four pages in the Control Plan but this one provides good examples of the specifics of a plan and you can relate the plan to the part print. The plan is slightly different than the one in Figure 11.8 and you will note that it covers several processes not just repressing or sizing.

One interesting characteristic which must be controlled is density. The density of the part is carefully checked when the initial samples are made using the method described in MPIF Standard 42. This method uses water displacement and is rather time consuming so we don't want to use it every 20/30 minutes. After the density has been checked initially, and found to be in specifications, limits can be calculated for weight and length which will maintain the density range. It is then possible to control the density by controlling weight and length during compacting and if necessary repressing.

One must also be aware that multilevel parts may require more stringent controls including a density check in each level. For example, the part shown in Figure 11.9 could have low density in the flange and still pass the specifications for overall density. This would be caused by the body density being too high which would compensate for the low flange density. This situation could cause cracks and low strength in the part, leading to field failure. Tool breakage can also result due to overloading if the density in one section is too high.

When SPC methods were first being used, the information was collected and recorded manually. As you can imagine, this was a very time consuming task. This may still be the case in some companies but there have been many technological advances in measurement, data collection and

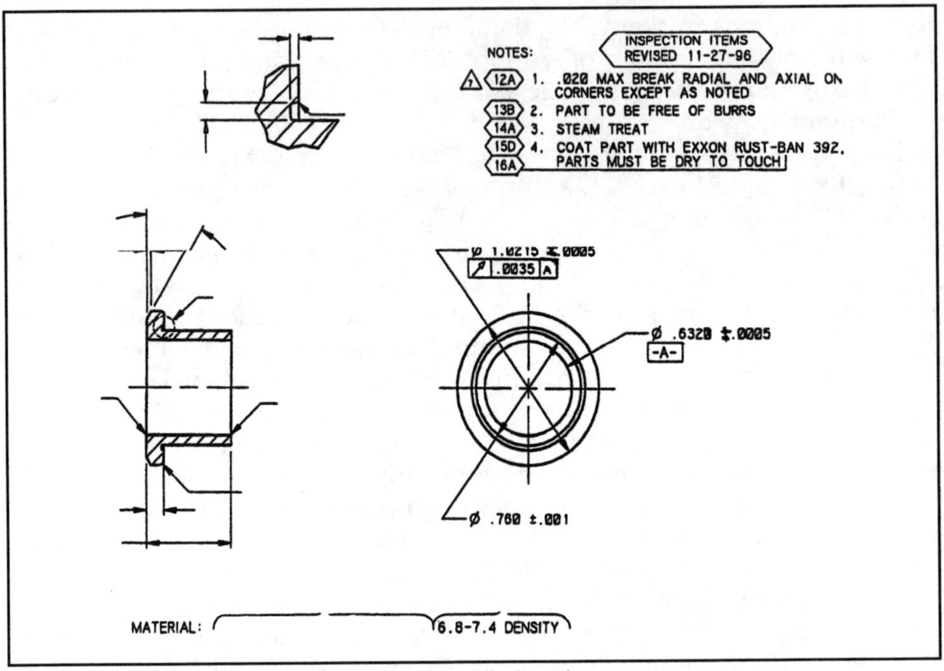

Figure 11.9. A print of production of a part called a tube.

chart creation. At NSA we have been using a PC based real time system for ten years. Work stations are located in different process areas within the factory. These work stations have a software package which includes SPC and other manufacturing control information. Measurement devices such as micrometers, electronic height gages and scales are hard wired to the PC's through interfaces. Figure 11.11 shows one of these work stations. All machine operators are trained to use this system and it is therefore much simpler to enter data in the system. As soon as the readings are taken, the X and R charts are updated and appear on the monitor. The corrective action situations (shown in Figure 11.5) are part of the SPC program and if action is necessary, the chart appears highlighted to attract the operators attention. The operator then takes appropriate action as noted in the control plan. The action taken is entered in the computer and initialed by the person who made the correction.

The workstations are networked with other PC's in strategic locations in the factory so it is quite simple to monitor the processing of any part or check the SPC history of a part. Some customers require X-bar and R charts from different processes with each shipment and that information is readily available with this system.

There are other more recent advances in P/M process control which

National Sintered Alloys Control Plan

Customer:
N.S.A. Job No.: 1970

Customer Part No.:
Title: Tube

Characteristic	Item	Specification	Total Tolerance	Inspection Tools	Frequency/ Sample Size	Inspection Methods	Reaction Program to out of Control conditions
Material	1D		N/A	Vendor Cert.	Cert. With Each Shipment	Receiving Inspection	Impound Lot/Contact Vendor for Resolution
Density	2D	6.8 / 7.4	0.6	Density Evaluation Part Weight Correlation	1 Pc at New Tooling 2 Pcs Every Twenty Minutes	Tool Sample Inspection In-Process Briquetting	Change Tooling Impound lot /sort / rework / remake/contact customer to request deviation
Inside Diameter	5C	.632+-.0005	0.001	Go No-Go Plug Gage	2 Pcs Every Twenty Minutes Full Sample	In Process Sizing Final Inspection	Trace to origin of discrepancy / isolate out of spec. parts / await disposition/correct process Impound lot /sort / rework / remake/contact customer to request deviation
Flange	6B	.120+-.003	0.006	Electronic Indicator	2 Pcs Every Twenty Minutes Full Sample	In Process Sizing Final Inspection	Trace to origin of discrepancy / isolate out of spec. parts / await disposition/correct process Impound lot /sort / rework / remake/contact customer to request deviation
Length	7C	.591+.000-.005	0.005	Electronic Indicator	2 Pcs Every Twenty Minutes Full Sample	In Process Sizing Final Inspection	Trace to origin of discrepancy / isolate out of spec. parts / await disposition/correct process Impound lot /sort / rework / remake/contact customer to request deviation
Outside Diameter	8B	.760+-.001	0.002	0 - 1" Micrometer	2 Pcs Every Twenty Minutes Full Sample	In Process Sizing Final Inspection	Trace to origin of discrepancy / isolate out of spec. parts / await disposition/correct process Impound lot /sort / rework / remake/contact customer to request deviation

Figure 11.10. The first section of the Control Plan for the subject part.

are noteworthy. Most press manufacturers can now provide presses with closed loop systems. These systems can measure compacted height and weight automatically. This information is used to adjust powder fill depth and part height dimensions without human intervention. As you can imagine, this equipment is very sophisticated and expensive. Continued refinements in these systems will probably result in lower cost in the future.

Most parts are subject to final inspection before shipping. This does not mean that every dimension and feature of the part must be checked. Once again, the control plan comes into play. If you go back to Figure 11.10 and look at the column headed "Inspection Methods", you will see that some characteristics require two locations for parts to be inspected. One is an in-process check and the other is at final inspection. The quantity of samples checked at final inspection is usually determined by Mil. Std. 105. When all the required characteristics have been checked at final inspection, a report is issued displaying the findings.

Figure 11.11. SPC workstation.

If a characteristic is found to be out of specification, the control plan describes the proper action. Sometimes the customer is able to approve a deviation for a dimension which is out of tolerance. Deviation requests are usually made in writing to the customer and require a written approval and a deviation number. The deviation number is shown on shipping paperwork, inspection reports and other related documents and protects both the supplier and the customer if a problem arises.

Many P/M parts users have eliminated incoming inspection of parts and rely on their supplier's quality systems instead. The part supplier must certify in writing that all parts in the shipment meet print specifications. As you can well imagine, the confidence level necessary for this type of relationship takes time and experience to develop. The supplier must pass a rigorous review of their entire operation. Once this has been accomplished, a program of reduced incoming inspections by the user is established. When a sufficient number of shipments has been made with reduced incoming inspection and no discrepancies, the part will be approved for "ship to stock" status. The importance of a good quality system should now be very obvious. Making bad parts is very costly even when they are detected during processing. Shipping defective parts not only has a direct financial effect; but, also a significant loss of both confidence and business. Future business will be affected if action is not taken.

We would be remiss if we did not discuss ISO 9000 and QS 9000 as part of this chapter on quality systems. You may already be familiar with these standards and be part of an organization which has received registration. We will not attempt to cover this subject in depth because volumes have already been written about it. Awareness of the existence of these standards and some knowledge of their requirements is important and so we will proceed with those goals in mind.

ISO (International Standards Organization) 9000 is an internationally recognized structure for establishing and maintaining rigorous quality methods and procedures. Companies of all sizes use ISO 9000 as their primary guideline for creating, refining and monitoring their quality assurance programs. There are three models under ISO 9000 - 9001, 9002 and 9003. Each applies to a different area of business; ISO 9002 is the one applicable to P/M manufacturing companies.

The concept of ISO 9000 is that an organization must say what they do - have written procedures for all areas of the business; do what they say - follow the written procedures in the daily operation of the organization; prove it - be able to show that procedures are being followed by written documentation. As you can well imagine, development and implementation of these procedures is difficult and time consuming. Most companies take 1-2 years to complete the process. Audits are required to confirm compliance. Initially these are done internally by a team from within the company. The final step before certification is a full audit carried out by a qualified ISO Registrar. The Registrar will examine each of the twenty sections (called elements) of the system. If a significant nonconformance is found in any element it can cause audit failure. Re-audit is possible but there is a time delay and additional expense as a result. Once the initial audit is passed, a company receives registration and the recognition that goes with it. This is certainly not the end of the process because re-audits are required every year in order to continue certification.

QS 9000 is kind of a big brother (maybe the one that is watching you) to ISO 9000. It was developed by the Supplier Quality Requirements Task Force in association with the Automotive Industry Action Group (AIAG). Participation by the Big Three vehicle manufacturers, and input from a wide spectrum of suppliers lead to the QS 9000 standard. It enables suppliers to meet a single standard rather than multiple standards for different manufacturers in the automotive industry. QS 9000 is much more complex than ISO 9000 and therefore more difficult to achieve.

It should be obvious at this point that significant up-front costs are involved in obtaining certification to these standards. These costs can range from \$25,000 to several hundred thousand dollars depending on the size of the organization and the length of time involved in the process. Additional audits are also costly, but necessary in order to maintain certification. The question of justification of these expenditures usually comes up and most organizations believe they gain from going through the certification process. QS 9000 is required of all first tier automotive suppliers and ISO 9001 is required of many second tier suppliers. There is not much choice for companies who want to supply automotive products. Many non-automotive manufacturing companies use ISO 9000 registration as a qualifying requirement for suppliers because they believe it ensures continuous corporate-wide attention to the quality process.

There are other companies with top reputations for quality who argue that ISO 9000 registration is not always necessary or even desirable. They believe that the whole registration process is a static, bureaucratic overlay that does little to improve product quality. Richard Buetow, Motorola's director of quality says "Our own experience has shown it (ISO 9000) to be of very limited value and it is expensive. In turn, we have informed our supply base that Motorola will not require ISO 9000 as a requirement to do business with us. Motorola makes its purchases based on excellence in products and service, not on compliance to a system of standards."

As always, there are several opinions on every issue. The important message on the quality system issue is that there needs to be some kind of system in place which is effective. Without such a system, no company can exist in today's demanding marketplace.

CHAPTER 12

Other P/M Forming Operations

Warm Pressing

As you have seen, P/M folks go to a lot of trouble to press parts that are close to full density. For instance, double pressing and double sintering, DPDS, of iron alloys, consists of pressing, presintering at 1350 to 1550°F to remove the lubricant and soften (anneal) the powders, repressing at 30 to 60 tsi to reach 7.3 to 7.5 g/cm^3 (93 to 95% of full density) and final sintering at 2050°F or higher. The reason that we use two steps is two fold: the iron powder is hard (it has a high yield strength), and the lubricant occupies 5% of the volume and keeps you from pushing the iron alloys to near full density. The same compacting pressure on something soft like copper or lead would get the part pretty close to full density on the first pressing. Is there any way to make the iron softer, so it will compact more easily? Hoeganaes Corporation developed the idea of heating the iron powder before and during the pressing step. It takes advantage of the well known fact that the heated iron has a lower yield point, which makes it more compressible. Why did it take until 1995 for this feat to be accomplished? Several obstacles had to be overcome. A special lubricant was needed that would assist metal particles in squeezing past each other at high density. The lubricant must work at very low percentage additions. The lubricant had to allow the powder to flow freely at 295°F. Someone had to prove that 295°F, a modest temperature, made any real difference in compressibility. Also, someone had to develop and build the powder heaters that would allow hot powder to be fed to the press. Finally, P/M houses needed to see some new applications with improved economics and the higher density and mechanical properties that made it worth the risk of trying warm pressing. The last reason has a lot to do with why technology changes are so slowly accepted in our industry (and probably in most industries).

The warm pressing buys you an extra .1 to .2 g/cm^3 in a single pressed part. Figure 12.1 is an example. This means that a part that

Figure 12.1. Warm formed powder metallurgy part, steel torque converter turbine hub, courtesy of Chicago Powdered Metal Products and 1997 MPIF Part-of-the-Year Competition, Grand Prize: Ferrous.

might have been capable of being pressed to 7.1 to 7.2 g/cm^3 is now being made at 7.3 to 7.4 g/cm^3. In terms of properties, the heat treated tensile strength has gone from 150 ksi to 175 ksi. Small increases in density, when we are operating in this high density region, pay big dividends in terms of tensile and impact properties. Impact measured by unnotched Charpy went up by 33%. The cost saving versus double pressing and double sintering is about 13%.

Warm Pressing of Magnetic Materials

In the earlier information on warm pressing, it was understood that the lubricant in the powder was removed during the preheating of the part, as it was heated up in the sintering furnace. No lubricant remained after the sintering. Around the same time that conventional warm pressing was being developed, it was realized that one could coat iron particles with a thin coating of plastic and then press them together. The plastic caused the electrical insulation of one particle from the other. That means that an electric current won't flow through this assembly of pressed particles. If we heat the plastic coated particles to 525°F and press them in a heated die, the plastic acts as a lubricant and allows the iron particles to squeeze past each other. This achieves the densities of 7.44 g/cm^3, getting quite close to fully dense iron which is 7.87 g/cm^3. The warm formed part is strong enough to be used without sintering having a transverse rupture strength of 18,000 psi as compacted and 36,000 psi after curing at 600°F. (Tensile strength is probably about ½ of these values.)

This dense assembly of iron particles bonded by plastic makes a very useful soft magnetic material. In a soft magnetic part, a magnetic field can easily magnetize it. When the magnetic field goes away, the particles lose their magnetism. When the magnetic field is increasing or decreasing, it tries to cause an electric current to flow inside the part. If electric current actually does flow, it causes local heating and wasted energy. An ideal soft magnetic material does not easily conduct electricity. So you can see that the P/M soft magnetic material with its insulated particles should be energy efficient. Now here is where you could reap some benefit from having waded through all the magnetic theory in Chapter 1: In a brushless motor such as that used in a cordless power tool, the rotating inner member consists of a shaft and some very small but very strong permanent magnets. (What were all those rare earth elements you thought we were kidding about in Chapter 1?) The stationary part of the motor, the stator, can be made out of this warm pressed powder. It will then have the copper windings which carry the electric current, wound about it. The P/M stator economically replaces a costly series of stacked insulated metal laminations and is energy efficient. The advantage of brushless motors is that there are none of the brushes which normally carry the electric current from stator to rotor, to wear out.

Hot Pressing

If warm pressing is so effective because it makes the iron powder soft, why not go further and really heat the powder? At present, there are some obstacles: We know of no lubricant that will work with the powders and allow them to flow. If all other obstacles were overcome, loose powder might be compacted in a die whose walls were sprayed with graphite after each cycle. Hot powder sticks together and does not flow. Hot iron powder oxidizes at 500°F in air and so would need an inert atmosphere around it at all times.

There indeed is a method of hot pressing loose powder but it is slow. Make the tools out of graphite so the powder will not stick. Put the tools in a vacuum furnace to keep the air out. Remember to weigh out the powder cold (no lube) and dump it in the tools before they go in the vacuum furnace (a.k.a. vacuum hot press, see Figure 12.2). Arrange for pressure transmitting rods to pass through the furnace walls and press on the graphite tools. Turn on the furnace or the induction heating coil that surrounds the tools, heating both tooling and the powder. Apply pressure at about 6,000 psi and wait. As the powder heats up it gets soft, the punch-

Figure 12.2. Vacuum hot press courtesy of AVS, Inc. of Ayer, MA. This press sells for $600,000 and has 350 tons of compacting force.

es squeeze out all of the porosity and give you a fully dense part. All this takes from 1 to 24 hours. The vacuum hot press in Figure 12.2 costs $600,000, exerts up to 350 tons of force, and can press materials at up to 4,000°F. The 24 hour time would apply to a thick piece 24 in. in diameter. The graphite tooling must be heated and cooled slowly. As an example, a very large molybdenum piece weighed 10,000 lbs., in the form of a die 48 by 32 inches. Hot pressing can also be done on fully dense materials, to join different parts together. This is done to join two halves of the hollow titanium blades for the engines on the Boeing 777 aircraft. The hollow blades save weight and fuel.

Presses vary from a few tons to 1,000 tons in capacity. The process is mainly used for ceramics, intermetallic compounds, and composites. People who make sputtering targets, however, actually press real metal. (Sputtering targets are used in the semi-conductor industry as a source of specialized metals such as Cr, NiAl, TiAl, Ti, Ta, Al, and Cu.) During sputtering, they are locally heated by an electron beam. Their metal evaporates and then falls onto and coats the semi-conductor product.) As noted in Chapter 1, the hot pressing process is also applied to beryllium metal powder. Other uses include diffusion bonding of metallic parts with complex shapes, and forging operations to alter grain orientation to enhance strength or to improve magnetic properties.

Powder Forging (P/F)

Perhaps you were thinking that double press double sintering is pretty efficient, and too bad that the repressing operation can't drive the part to full density. Why not just heat the presintered part and then repress it hot? The part should be fairly soft and ductile. The local blacksmith does this all the time. This remarkably good idea was first demonstrated in 1963 by the Cadillac Motor Division of General Motors Corporation, making connecting rods for their engines. The idea was not adopted then, but Toyota was the first to introduce powder forged connecting rods, followed by Ford Motor Company in the 1987 Ford Escort. Through 1996, 75 million con rods had been used in the "Big Three" automobile engines. Figure 12.3 shows a connecting rod courtesy of MascoTech Sintered Components, (now Metaldyne Sintered Components) who made the original Ford rods.

In the years after 1963, there were a few technical obstacles that had to be overcome. It was not quite as simple as doing the repressing operation hot, rather than cold. The usual way that powder forging is done is to take the 80% dense sintered part, coat it with a graphite lubricant, and heat it by induction to 1800 to 2000°F. During the heating, some nitrogen gas can flow around the part to keep it from oxidizing. Heat the M4 tool steel forging die to 700 to 1000°F. Use a robot to pick the part out of the induction heater and drop it into the forging tools on the

Figure 12.3. Powder forged connecting rod, courtesy of Mascotech Sintered Components.

press. Cycle the press at 55 to 70 tsi, driving the part to nearly full density (greater than 7.80 g/cm^3). Eject and cool the part. All this is shown in a diagram in Figure 12.4, provided by Prof. German.

There is a lot of technology in the design of the shape and density of the preform. It is designed to compress and flow out sideways, filling the die. It also can be designed to extrude forward or backward around the punches. If you don't have the powder in the right places to start with, after it flows, the part may not be uniformly fully dense. In the worst case, the preform might crack as it tries to expand out to the die, or backward and forward. However, in a properly designed powder forged part, it is possible to build in shapes by virtue of plastic flow of the preform, that cannot be done with cold powder.

The mechanical properties of the powder forged part are similar to those of ingot based steels. Properties are found in ASTM B 848 and MPIF Standard 35. As an example P/F-4640 can have a tensile strength of 230,000 psi with 11% elongation. Most of the forged parts in current production are for automotive applications. They consume about 40,000 tons of iron powder per year. Powder forged parts cost about 100% more to produce, than pressed and sintered parts of the same weight, suggesting a component value of $375 million. (By contrast, the Forging

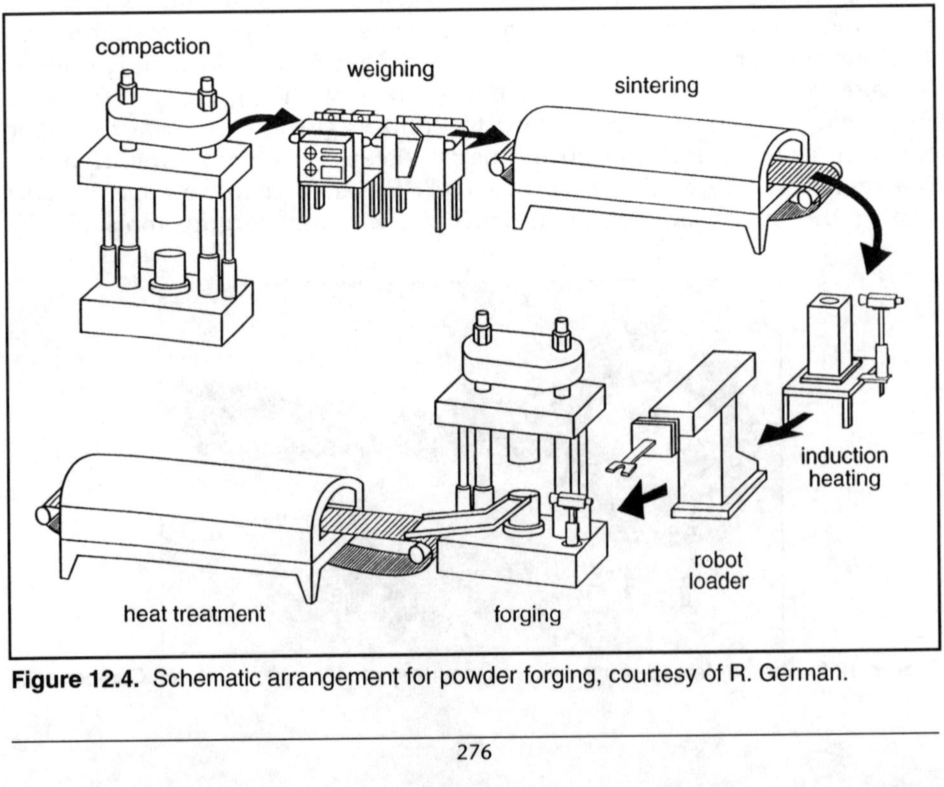

Figure 12.4. Schematic arrangement for powder forging, courtesy of R. German.

Industry Association reports that in 1999, their members closed-die forged $4,600 million worth of steel and other materials. Think of it as market potential!)

Cold Forging of P/M Preforms

Cold forging of wrought steel bar and rod is a fairly large business in North America, with annual sales around $400 to 500 million, with most of the parts made by about 25 custom producers. The process consists of shearing round rod into short pieces of relatively constant weight. (The process should be distinguished from cold forming and cold heading where elongated products like bolts are formed from coils of rod stock, or standard items like nuts are produced.) In cold forging, the pieces are fed into dies where the iron and steel is deformed by punches under high pressure, at about 5-8 times the yield strength of the steel (120 tsi). The steel spreads out to fill the die and can be extruded forward or backward around part of the punch. The steel behaves like modeling clay as it moves around to fill the die cavity. All of this happens very quickly: the cold forging presses operate at 30 to 40 strokes/min., on stock that is 1.0 to 1.75" in diameter, to make parts up to one pound. (Cold heading of wire and rod feed stock runs at 200 pieces/min.) Cold forging is a very economical process and it is hard for P/M to compete with it. But cold forging does have its limitations: there are limits to how far you can deform the round starting slug before it cracks. It works best on round or symmetrical objects. Sometimes it would make more sense to have the starting preform be some shape other than a round slug.

Would it ever make sense to make the simple round preforms as sintered metal pieces? A rotary press could turn out 60 to 100/min. A high temperature walking beam furnace could give a good sinter to about one ton of preforms, per hour. Amazingly, this was done from 1971 to 1980, producing 150,000,000 spark plug shells at the AC Electronics Div. of General Motors Corp. Figure 12.5 shows the slug preform and the final cold formed shape, which was formed in two cold forging steps. After the first step, the parts were resintered to soften them and then were cold forged a second time. What happened in 1980? The economics of the P/M process finally lost out to cold forging.

There is another problem in cold forging the P/M preforms: the ductility. The P/M preform will not flow as far as wrought steel, before it cracks. The folks at AC Electronics finessed this problem by choosing the best material for ductility, pure iron, and by giving it a tremendous sinter: 1 hour at 2400°F. Because of the porosity in the 7.0 g/cm^3 preform, the pressures on the tooling were lower than for a wrought steel slug, and the tooling lasted longer. If the P/M ductility is modest, then

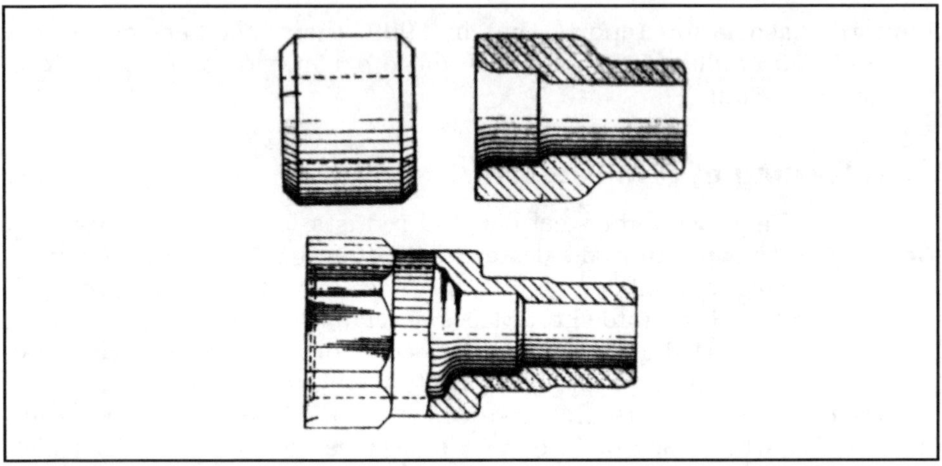

Figure 12.5. Cold forged spark plug shell. Upper left – perform. Upper right first draw, 7.6 g/cm^3. Bottom, final extrusion, 7.8 g/cm^3.

great care must be taken in the preform design so that during the cold forging, the piece is kept under compression and not given a chance to crack. Harry Antes of SPS Technologies and Howard Kuhn of Deformation Control Technologies did the early work in 1971 and formulated the beginning of a plasticity theory governing the stresses under which the P/M materials flow, and how far they flow before they crack.

With the economic problem alluded to above, you might have thought that cold forging P/M preforms would have been D.O.A. at the beginning of 1980! In the early 80's, Gleason Works owned Pennsylvania Pressed Metals, the seed from which sprouted GKN/ A-Z. They used their G-Flow process in two 75 ton horizontal forging presses, running 1800 pcs/hr. to cold forge moderate density low alloy steel preforms into parts at 7.8 g/cm^3, close to the pore-free density of steel, 7.87 g/cm^3. Figure 12.6 shows some of the parts that were made. Why did they discontinue the program? Kenneth Gerg, then President of PPM, reports that the process worked well. The materials flowed well, almost like putty, and it was not a noisy process. They had two people working full time on the project and came close to commercializing a few parts. In the end, there were not enough applications to warrant continuation. One of the drawbacks is that the development on each part required enough engineering effort that the part had to be manufactured in very large quantities, e.g. a million or more per year.

During 1999, Howard Ferguson and Steve Smith from the Metal Powder Components Div. of MPP published work on cold forming the gear shown in Figure 12.7. They made a single pressed preform out of FL-4400 prealloy (.85% Mo) and .28% graphite. The preform shape was

Figure 12.6. Several developmental cold forged parts made from P/M performs, circa 1985. courtesy of Pennsylvania Pressed Metals Div. of Gleason Works (GKN Sinter Metals, Emporium).

Figure 12.7. Powder cold forged, PCF, spur gear, .85% molybdenum + .20 to .25% carbon, 7.71 g/cm^3, courtesy of MPP Corp.

proprietary, but it looked something like a gear, and not just a doughnut. The sintering was done under commercial conditions, 20 minutes at 2080°F. This gave enough ductility for the preform to hold together and not crack, as it was cold forged out to fill a die that looked the gear in Figure 12.7. To do this, you have to have a model of how far a compressed material will expand sideways, before it cracks. You just make sure it reaches the die wall before this happens. The final gears were 7.71 g/cm^3. The cold forging closes up the pores but leaves a thin unbonded region where the pore faces came together. This has to be healed by a thermal process like heat treating. The 1600 to 2000°F heat allows the sides of the collapsed pores to sinter together. You also carburize the surface of the gear during the heat treatment. With a case depth of .025 in., the parts can be HRC 58-62, just like the competitive wrought material, AISI 8620 steel. In a fatigue test, the powder cold forged (PCF) part performed better than the 8620 because the wrought material had some stringer inclusion and did not have as smooth surface finish.

Surface Densification

Half a loaf is better than none! If it is not possible to cold press the whole gear up to full density, why not just cold form and densify the surface of the teeth? Anyway, that's where all the action is at the beginning of 2001. (The stresses imposed by the mating gear are highest just under the surface, at the flank and root of the gear.) Stackpole Limited in Mississauga, Ontario presented this idea at the 1997 MPIF/APMI International Congress, "Developing P/M Gear Tooth and Bearing Surfaces for High Stress Applications" by P.K. Jones et al. Although they do not explain their exact method, it must be something like rolling a mating machined tool steel gear form against a 7.0 g/cm^3 gear preform. This crushes the surface of the porous gear, producing local densification. A diagram of how this is done is shown in the European Patent 0 552 272 B1, 7/12/94 by Christopher Cole. The mating steel gear and the sintered gear preform have the right shape so that during the rolling, either the flank or the root can be densified. This requires a special machine to control the motion of the steel gear, toward and away from the P/M gear preform.

At Stackpole, in the example of an ID of a bearing that served as a needle bearing race, the surface was densified to a depth of .040 in., with the outer .008 in. being over 7.80 g/cm^3 or 99% of full density. The densification can be done to both the flank of the tooth and also to the base or root of the tooth. Densification at the root results in higher tooth bending fatigue endurance, 120,000 psi. When they etch this dense region, they see some of the original edges of the collapsed pores and recognize that a thermal treatment will be needed to finish the bonding. This is done by high temperature thermal treatment which effectively sinters or bonds the sides of the collapsed pores. The microstructure is fine, tempered martensite and 10 to 20% retained austenite. The races have a fatigue endurance limit for the Hertz stresses of 580,000 psi, about the same as wrought 8620 steel. Although no chemistries are given in the paper, it seems likely that the materials are low alloy steels with about .5% of elements like Cr, Mo and Mn, possibly added in master alloy form, and then high temperature sintered to bring about

Figure 12.8. Double row sprocket for automatic transmission with surfaces of teeth densified to 99% of full density, courtesy of Stackpole Ltd. Mississauga, Ontario.

alloying. It also seems likely that the initial carbon content may be low, to leave the 7.0 g/cm^3 preform in the soft condition, and relatively easy to densify. The surface strength then comes from the carburization, quenching and tempering. In 1998 and 1999, Stackpole was selectively densifying the teeth on the drive and driven sprockets for a GMC automatic transmission. This sprocket is shown in Figure 12.8.

Why is all this important? We are tempted to say "Because all the easy P/M parts have already been made!", and there is some truth in that. Stackpole believes that the 7.0 g/cm^3 core and densified surface could be used on 5th gear for the manual transmission, and for automatic transmission parking gears, and for the balancer gears in the engine. If the core were to be 7.4 to 7.8 g/cm^3, the selective densification should make it possible to make manual low speed gears and transmission gears.

Metal Injection Molding

We spent some time in Chapter 4 showing how powder is compacted in typical P/M tool sets. The basic message was that the powder flows straight ahead of the punches. It does not flow sideways to fill undercuts or protrusions off the main part. If there is more than one level to the part, we often use separate punches, each one carefully compacts the powder that is straight ahead of it. In the 1950's 60's and 70's, P/M folks would fantasize about how great P/M would be if the powder could be made to flow in every direction, like heated plastic forced into a die. We are all familiar with the complex shapes of the plastic parts we use every day. What most powder metallurgists didn't realize was that the ceramic industry was injection molding mixtures of ceramic powder and organic binders to form ceramic objects in the 1930's.

In the 1970's, Raymond Wiech, founder of Parmatech and John McPherson worked out a way to inject mixtures of fine metal powders and organic binders into a die. The mixtures consisted of 60% by volume of metal and 40% by volume of organic material. The organic material is a binder which consists of a mixture of waxes, polymers, oils, lubricants, and wetting agents. (An early example of an oil was Wesson oil!) The mixtures are made at modestly elevated temperature and then extruded into small hard pellets. This is the first step in the process shown schematically in Figure 12.9

The pellets are fed into an injection molding machine, $60,000 to

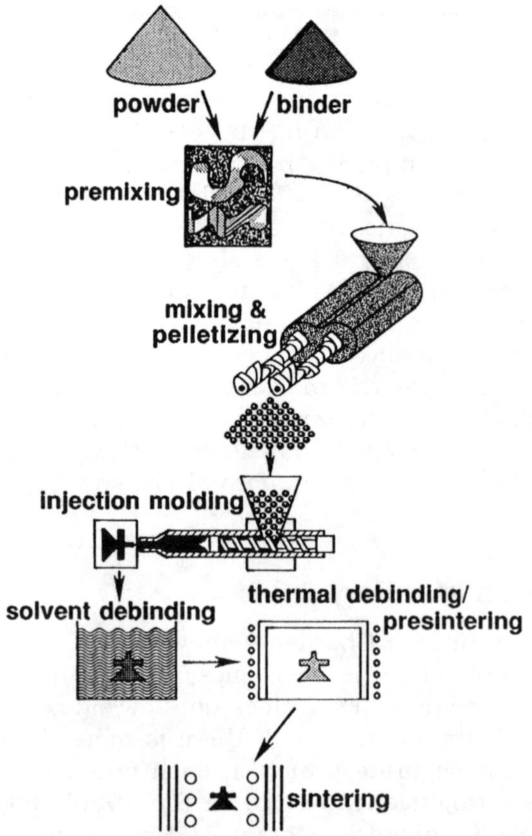

Figure 12.9. Conceptual sequence of steps involved in powder injection molding. Courtesy of R. German, Powder Metallurgy Science.

$80,000 for an 80 ton clamping force machine. There, they are heated and compacted by a screw mechanism. This brings the mixture to 300°F where it can flow in a plastic manner. Then the injection machine moves the screw toward the mold, forcing the mixture forward at up to 8000 psi. It is like squeezing toothpaste out of a tube. The mixture passes out through a nozzle and into the die, shown schematically in Figure 12.10. The die is cooler (e.g. 105°F) than the plastic mixture, which freezes as it is forced into the die. The air in the die is pushed out through vents at the far end. The part must cool for 15-20 seconds to be strong enough to be removed from the die. The overall cycle time is about 40 seconds, to inject and remove the part. The part will have attached to it the sprue, runners, and gates, shown in Figure 12.10. These are cut off and recycled.

At this point, we have a composite part which is still 60% metal and 40% organic material by volume. It has adequate strength for handling. The binders can be removed by gradual heating to 400°F in air, in up to 60 hours. This decomposes the organic material and oxidizes the metal. The oxide holds the particles together. Alternatively some of the binder can be dissolved in solvent. When this partially debindered part is slowly heated in atmosphere or vacuum the rest of the binder goes off. When

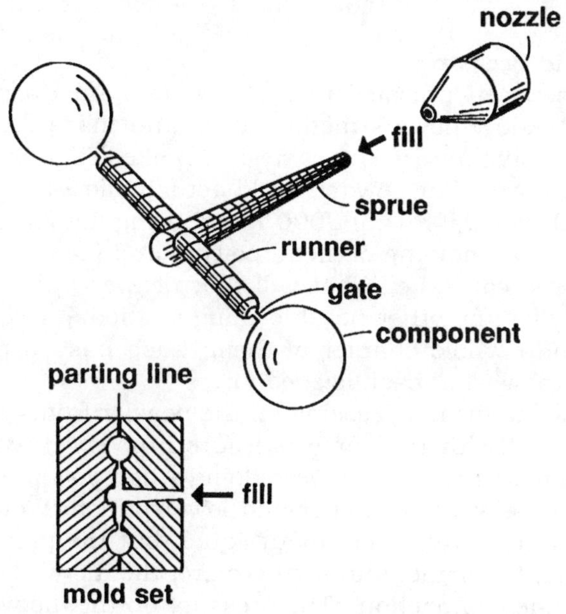

Figure 12.10. The flow path in injection molding showing the location of the nozzle, sprue, runner and gate for a two-cavity tool set. Courtesy of R. German, Powder Metallurgy Science.

things work right, the metal particles begin to sinter before all of the binder is gone, and this provides strength to hold the part together. Metal injection molding is based on the use of fine particles, 1 to 20 microns. Iron particles this small sinter very fast, resulting in significant shrinkage and densification. We need large amounts of shrinkage to get rid of the 40% volume of air space that was left when the binder was removed. When MIM parts based on carbonyl iron powder are sintered for an hour at 2400°F, they shrink about 14% in each dimension and approach full density (7.5 to 7.7 g/cm^3). Such parts have mechanical properties similar to their wrought counterparts, as shown in Table 12.I. Figure 12.11 has examples of MIM parts.

Because MIM parts shrink down and have a high density, the few pores that are left behind are unusually small, round and isolated from each other. Where the pores are isolated or not connected, the part will be pressure tight. That means you could use a MIM part in a high pressure valve or pump, and it would not leak. You could electroplate it without having to fill the pores with plastic resin. (In a porous P/M part, if you plate it without filling the pores, some plating solution will bleed out and spoil the plating). MIM parts are sometimes referred to as PIM (Powder Injection Molded) Parts, although the acronym becomes more problematic.

Cold Isostatic Pressing

When we discussed pressing in rigid tooling, you learned that compacted powder mainly flows straight ahead, and that parts that remind you of stacked cylinders are the easiest to make. Then we changed the rules by showing you that powder could actually flow plasticly, forward, backward or sideways, if it is at 2000°F and being hot forged (P/F). Then came MIM, injection molding of metal powder and plastic binders in any shape that plastic can take. What will those clever P/M folks try next? Can there really be any other novel forming methods? Yes, you are right in the middle of a whole chapter of them! Each has its place, and the real trick is to know what each is good for.

Isostatic pressing means pressing on the powder from all directions at the same time and with the same pressure in every direction. Perhaps you thought that was what you were doing when you pressed a simple cylinder in a die between an upper and lower punch. Well the pressure on the punches may have been about equal, but the pressure from the die walls was not the same, and furthermore, the die walls did not move very far during the compaction. (The pressure on the die walls was actually about 75% of the pressure on the punches, at 50 tons/in^2.)

What could we use as a source of pressure to press on all sides of a powder mass at the same time? Fluids work well in this regard. An object

Table 12.1.

Material Designation Code	Nominal Chemical Composition (%)				
	Fe	Ni	Mo	C	Si (max)
MIM-2200(1)	Bal.	1.5 – 2.5	0.5 max	0.1 max	1.0
MIM-2700	Bal.	6.5 – 8.5	0.5 max	0.1 max	1.0
MIM-4605(2)	Bal.	1.5 – 2.5	0.2 – 0.5	0.4 – 0.6	1.0

Other Elements: Total by difference may not exceed 1.0% combined.

(1) *Formerly designated as MIM-4600*
(2) *Formerly designated as MIM-4650*

Low Alloy Steels
MIM Material Properties

Material Designation Code (condition)	MINIMUM VALUES			TYPICAL VALUES							
	Tensile Properties				Tensile Properties					Hardness	
	Ultimate Strength	Yield Strength (0.2%)	Elongation (in 1 in.)	Density	Ultimate Strength	Yield Strength (0.2%)	Elongation (in 1 in.)	Young's Modulus	Unnotched Charpy Impact Energy	Macro (apparent)	Micro (converted)
	10^3 psi	10^3 psi	%	g/cm³	10^3 psi	10^3 psi	%	10^6 psi	ft•lbf	Rockwell	
MIM-2200 (as-sintered)	37	16	20.0	7.6	42	18	40.0	28.0	100	45 HRB	
MIM-2700 (as-sintered)	55	30	20.0	7.6	60	37	26.0	28.0	130	69HRB	
MIM-4605 (as-sintered)	55	25	11.0	7.5	64	30	15.0	29.0	50	62 HRB	
MIM-4605 (quenched & tempered)	215	190	<1.0	7.5	240	215	2.0	29.5	40	48HRC	55 HRC

Material Designation Code	Nominal Chemical Composition (%)								
	Fe	Ni	Cr	Mo	C (max)	Cu	Nb + Ta	Mn (max)	Si (max)
MIM-316L	Bal.	10 – 14	16 – 18	2 – 3	0.03	—	—	2.0	1.0
MIM-430L	Bal.	—	16 – 18	—	0.05	—	—	1.5	1.0
MIM-17-4 PH	Bal.	3 – 5	15.5 – 17.5	—	0.07	3 – 5	0.15 – 0.45	1.0	1.0

Other Elements: Total by difference may not exceed 1.0% combined.

Stainless Steels
MIM Material Properties

Material Designation Code (condition)	MINIMUM VALUES			TYPICAL VALUES										
	Tensile Properties				Tensile Properties					Hardness		Corrosion Resistance		
	Ultimate Strength	Yield Strength (0.2%)	Elongation (in 1 in.)	Density	Ultimate Strength	Yield Strength (0.2%)	Elongation (in 1 in.)	Young's Modulus	Unnotched Charpy Impact Energy	Macro (apparent)	Micro (converted)	H_2SO_4		Boil Test
	10^3 psi	10^3 psi	%	g/cm³	10^3 psi	10^3 psi	%	10^6 psi	ft•lbf	Rockwell		g/dm²/day	$CuSO_4$	(H_2O)
MIM-316L (as-sintered)	65	20	40.0	7.6	75	25	50.0	28.0	140	67 HRB		<0.005	Pass	Pass
MIM-430L (as-sintered)	50	30	20.0	7.5	60	35	25.0	(A)	(A)	65 HRB		(A)	(A)	(A)
MIM-17-4 PH (as-sintered)	115	94	4.0	7.5	130	106	6.0	28.0	50	27 HRC		<0.005	Pass	Pass
MIM-17-4 PH (heat treated)*	155	140	4.0	7.5	172	158	6.0	28.0	(A)	33 HRC	40 HRC	<0.005	Pass	Pass

*Heat treated parts were aged at 900°F (482°C).

NOTES:
(A) Additional data in preparation will appear in subsequent editions of this standard.

Courtesy of MPIF Standard 35.

Figure 12.11a. Metal injection molding parts, courtesy MPIF. A valve assembly in which the lower piece slides within the upper cylindrical housings, opening and closing ports to control fluid flow. Note the undercuts and cross holes that are created by virtue of the injected metal and binder mixture's ability to flow in all directions.

Figure 12.11b. Metal injection molding parts, courtesy of MPIF. Pump manifolds that are pressure tight, and have many of the undercuts and cross holes allowed by the MIM process.

immersed in water feels the same pressure on all surfaces. (This neglects the fact that the object has some height, and the effect of gravity puts a little more pressure on the deeper side than the top side. This phenomenon is very handy if you are swimming, because your body has a density less than that of water and that excess pressure on your bottom means you float rather than sink! Well people have skin, so the water does not penetrate our insides. How could we separate the iron powder from the water so we could press on it? What about filling a child's rubber toy balloon with iron powder. This would work well if you wanted to make a balloon-shaped object! So you fill the balloon with iron powder, tie a knot in the end, and drop it in the bathtub. Upon removal, the powder is still loose and you conclude that inadequate pressure was used. If you tied a string to the balloon and dropped it over the deep side of Hoover Dam and Lake Mead, where the water is 576 ft. deep, the pressure on the balloon would be .433 x 576 = 249 psi. Still not enough pressure, the powder is still pretty loose. It takes at least 10,000 psi or an ocean 23,100 ft. deep to densify the iron enough to get it to hold together. At 60,000 psi the iron powder would reach the useful density of 6.8 g/cm^3. We don't have any oceans that are 138,000 ft. deep, so we need to find a more practical way to pressurize the balloon. We also need a way to get the P/M part out of the balloon, after it has been compacted.

During the 60,000 psi compaction, the powder was compressed and the density went from 3.0 g/cm^3 up to about 6.8 g/cm^3. This means the balloon shrank about 24%. (Student exercise!). All of the starting dimensions of the inside of the balloon, with the loose powder in it, were reduced by 24%. How do we get the part out of the balloon? If it has a big mouth, maybe we stretch the balloon to get it off the part. Maybe we just use a pair of scissors; balloons are cheap: about $1. each for a standard shape, purchased by the gross!

What do real people do? They make the rubber bag out of neoprene rubber or polyurethane or silcone rubber. It should be soft enough to compress and shrink around the metal powder, but not rupture at 60,000 psi. The shape of the rubber bag can be whatever you want, but it helps if it is such that the part can be removed from the bag without destroying it after each part is made. Figure 12.12 shows rubber bag molds in which the rubber component is about $50. each in small quantities. During filling, the bag can rest in a perforated box that, with the aid of a little vacuum, will hold the bag firmly against the box's walls. This gives constant filling weight on the powder. The box and bag can be vibrated to settle the powder and produce even filling. After filling, the bags are sealed by a wire wrapped around a ring, inside the bag. The filled bag is placed in an autoclave or vessel capable of being pressurized with water or oil to 60,000 psi or more. Autoclaves come in all sizes from

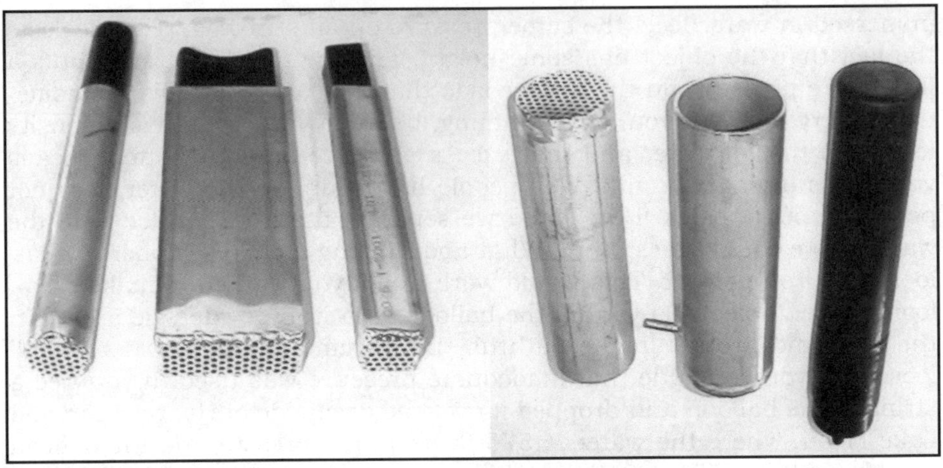

Figure 12.12. Tooling for cold isostatic pressing. The tools in this photo are 2-4" across. The rubber bag fits inside a supporting can. The space between the rubber bag and the can is evacuated, pulling the bag tightly against the can. The bag is then filled with metal powder and vibrated to settle it, and give a controlled weight in the bag. The open end of the bag is sealed with a close fitting rubber plug. Courtesy of Trexler Rubber Co., Ravenna, OH.

2 in. to 80 in. in diameter. For the laboratory, 2 in. to 12 in. in diameter is in the range of $25,000 to $130,000. For 13 in. to 24 in. in diameter at 30,000 psi for production, the cost is up to $300,000. For wire wound and prestressed vessels, 30 in. to 80 in. in diameter, the cost is $750,000 to $3 million. Figure 12.13 shows a typical autoclave. You can pack a lot of ubber bags in the larger vessels. The bags will be completely surrounded by the high pressure water, leading to the term “wet bag pressing”. It takes about 10 to 15 minutes to pump up the pressure in a 24 in. diameter unit, hold it briefly and then to de-pressurize. Compressing ordinary iron powder does not require much care at compaction, but, for fine powders like ceramics, there may be a dwell, to let some of the trapped air out. The compression rate curve is critical and it helps to de-air the compact before pressing. Then comes the fun of untying the bags and removing the metal parts. The parts can be in the range of 70 to 93% dense. They do not contain any lubricant, which makes sintering more reliable. Sintering is whatever you want to do. For iron, 30 min. at 2050°F in atmosphere is OK. To get further densification, try 1-2 hrs. at 2400°F.

The process seems well suited to making rough shapes in parts that can be much larger than the usual 1 -10 lb. pressed iron parts. If you think about making a thin walled cylinder, the CIP tooling could consist of a steel core to form the ID, and a rubber bag to form the OD. This would give a nice straight smooth ID after molding. If one uses a steel core to form the ID, we find after sintering that a 3.5 in. ID x 11 in. long

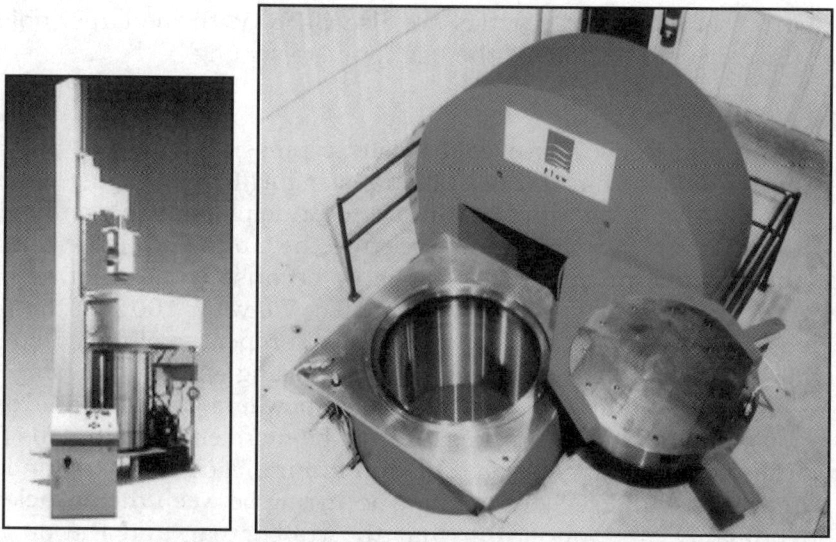

Figure 12.13. Views of two different cold isostatic presses, courtesy of Flow Autoclave.

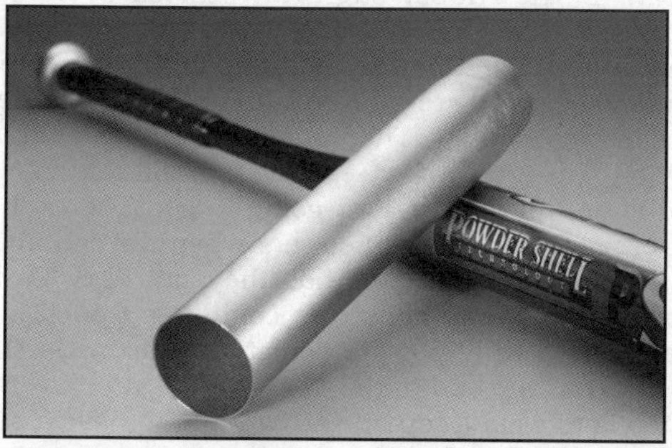

Figure 12.14. Cold isostatically pressed part (CIP'd), titanium alloy sleeve for a softball bat that prolongs the life of the wooden bat. Courtesy of Dynamet Corp. of Burlington, MA and MPIF, Part-of-the-Year Contest winner.

part can have an I.D. of .0015 to .0020 in. and O.D of ± .010 in.

Is anyone actually making anything by CIP'ing? TecSyn PMP is making the cylinder liners for small engines by CIP'ing. Also, see Figure 12.14 and the Ti sleeve developed by Dynamet Technology, to protect softball bats! Some CIP'd parts are much larger. The folks at Trexler Rubber, who make the tooling, have seen rubber enclosures 12 ft. high with a cross section of 2 ft. by 3 ft. A castable polyurethane bag is used to make rocket nozzles. Significant amounts of carbides for dies and wear parts, ceramics and plastics are also CIP'd. With the larger objects, the air is often pumped out of the bag, prior to sealing.

Hot Isostatic Pressing

Now that you know about cold isostatic pressing, you can probably guess how it differs from Hot Isostatic Pressing (HIP'ing). Its just a matter of doing the same thing, but hot. Simple in principle, but the genius is in the details! First of all, why would we go to the trouble to figure out how to do the isostatic pressing hot? Probably because we are not satisfied with making things that are only 92% dense. We want 100% of the pore-free density so we can exceed the mechanical properties of forged steel or superalloys. Having waded through warm pressing and hot pressing, you will readily believe that if we heat the metal powders, their yield strength will go down, and the applied pressure will be more effective in densifying them. The powders will creep and deform around each other, filling in all the pore spaces with excellent bonding occurring between the particles.

You probably are wondering what the rubber bag and the oil-filled autoclave thought about all the extra heat! Those two details did require some modification! The rubber bag gets replaced by a metal can. The cans are often welded sheet metal and made about 25% oversize to the final part. The unlubricated powder is dumped into the can, to fill it up. Then the can and powder are vibrated to settle the powder and get as much as possible in the can. Then some more powder is added and it may also be vibrated. After a few repetitions of fill and vibrate, you conclude the can is as full as it is ever going to get. The trick is to get the right amount of powder in the can so that when the can and powder collapse to full density, the final part has the right size. There have been numerous attempts made to construct computer models that will show you what size can and what powder density to start with, so that you get the right sized part at the end of the day.

The next step is to weld a top on the can and to have a small exit tube leading out of the top. One draws a vacuum on the can to pull out all the air and water vapor. The can may be heated to 400 to 1200°F, typically 750 to 850°F to assist in driving out the gases. The exit tube is heated to 1500°F and pinched together to seal off the can, see Figure 12.15.

You knew water would never work as a pressurizing fluid at high temperature, and so you will not be surprised to know that hot gases are a fine isostatic medium. The usual choice is argon. It has the advantage of being non-reactive or inert. At 2,000°F and 30,000 psi pressure the gas has the consistency of molasses. Special autoclave vessels are needed to contain the hot gas. The outside steel jacket is thick steel with water cooling. Inside, there is an electric furnace to heat the work. The furnace is shielded from the vessel walls by layers of insulation, so that large amounts of heat are not lost to the vessel walls. (Figure 12.16, diagram of HIP unit.) As the furnace heats the cans of powder, the pumps gradually increase the argon pressure inside the vessel. Of course the

Figure 12.15. Section from the top of HIP'd billet, showing how the evacuation tube has been heated and pinched off. Removing air from inside the can prevents oxidation of the powder and entrapped porosity air bubbles after HIP'ing. Heating to 750 to 850°F during evacuation drives off all of the volatile compounds.

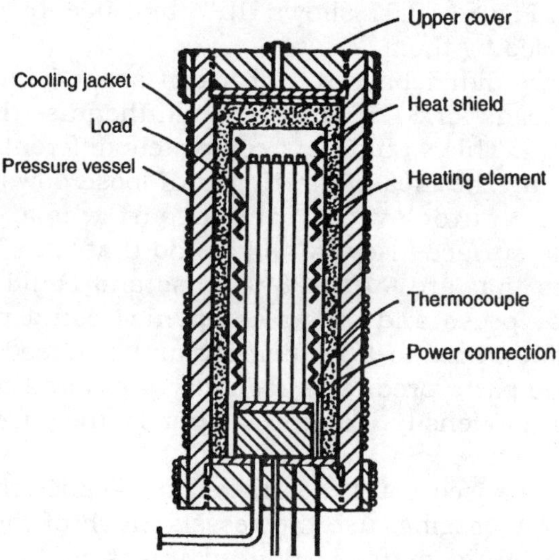

Figure 12.16. Diagram of a hot isostatic pressing vessel showing the arrangement of the load, the heating elements and the heat shields. Courtesy of ASM, Vol. 7.

argon inside the vessel reaches the same temperature as the furnace and the parts. As an example, a 2,000 lb. load of parts in a vessel 16 in. in diameter by 60 in. high could be heated to 2000°F and 30,000 psi in about 4 to 8 hrs. They could be held at pressure for 2 to 4 hours. and then cooled to room temperature in 4 to 8 hours. Such a cycle would cost about $3,000. or $1.50/lb. of parts. A smaller unit, 8 in. in diameter by 26 in. high would cost $1,000. HIP Autoclaves are expensive: the 16 in. by 60 in. unit would be $1 to $3 million depending on options such as rapid cooling capability and quick change furnaces. A large unit, 61 in. in diameter by 102 in. high, could process 10 tons of aerospace castings in 12 hrs., floor to floor, at $1.50 to $2.00/lb., or 25 tons of tool steel billets in 26 hrs. P.318 of the "ASM Handbook, Vol. 7" lists times and pressures to HIP iron alloys.

Now that we have this large, fully dense, three dimensional part, you want to know what happens to the can. Doesn't it stick to the part? Yes! You have to machine off the can or else dissolve it off, in acid. It is a good idea to make the can out of something that dissolves in acid, and make the part out of acid resisting material! About 99% of the HIP'd parts are machined and 1% are pickled in acid.

When do you choose to make a part by HIP'ing? Only when there is absolutely no other way to make the part from material of the same mechanical or chemical properties! As you can see, the whole process is expensive, from making one can for each part, to the long thermal cycle for HIP'ing, to the de-canning. Even then, you may still have to do some final machining. Figure 12.17 shows HIP'd tool steel billets, dwarfing the engineer who posed by them.

Of course, if we didn't have to put a can around the part, and then could run large loads for $1.50/lb., we could then use the parts as HIP'd i.e. no de-canning. This would be a completely different economic story. What if we gave up the idea of starting with loose powder in a can, and instead, sent the autoclave a sintered part whose pores were not connected to the surface? How would we do that? The MIM folks make millions of parts that are 94 to 98% dense and could be HIP'd to full density. Regular press and sinter companies could press, presinter, repress and final sinter to 95% density and a closed pore structure. Tungsten carbide parts producers regularly press and high temperature sinter to 98% of full density. For special grades, they use a final HIP step in a separate autoclave.

It may have surprised you to learn that companies that own HIP vessels and do toll processing, use the vessels much of the time to densify castings. When liquid metal is poured into a mold, it shrinks as it freezes. This shrinkage can result in internal flaws or pores or bubbles. These lower the strength of the casting. All critical aircraft engine or

Figure 12.17a. Hot isostatically pressed (HIP'd) billets of tool steel made by canning and hot compacting tool steel powder. Courtesy of Bodycote/IMT.

Figure 12.17b. Hot isostatically pressed (HIP'd) valve part. Although this may look like a powder injection molded part, it really is HIP'd and then machined to final shape. Actually size is several inches. This part was a winner in the MPIF Part-of-the-Year Competition, courtesy of MPIF.

airframe castings are routinely HIP'd to close up the pores and bond together the opposite sides of the internal holes. Many tons of castings are HIP'd each year in the US. The cost is about $1.5 to $2.0/lb. for iron and $3-4/lb. for aluminum. In Chapter 1 we mention that tool steel billets are made by HIP'ing metal powder. Figure 12.17 showed some HIP'd tool steel billets, courtesy of Bodycote/IMT.

Pneumatic Isostatic Forging, (PIF).

In this process a sintered or pre-sintered P/M preform is consolidated to near full density by the rapid application of high pressure nitrogen or argon gas. Pressurization to 60,000 psi takes 2-5 minutes, the work is held under full pressure for 10 sec. to 2 min. and then depressurized in 1-5 min. For iron, the work is preheated to about 2200°F prior to insertion in the pressure vessel. If the pores of the preform are non-interconnected (i.e. closed) the pressure will drive the part to full density. Because of the viscous or fluid nature of argon at this pressure, preforms of iron at 87% dense (6.85 g/cm^3) can be fully densified. In some cases, partial or full encapsulation of the preform is used, to keep the pressure from penetrating the pores. The methods are generally proprietary, but steam blackening and electroless nickel plating to .001 in. have been mentioned in "Vol. 7" of the "Metals Handbook". If the starting preform is of uniform density, as the parts densify, they shrink uniformly in all dimensions. However, most die compacted parts are not exactly uniform in density. (CIP'd parts are more uniform in density, as are MIM parts.)

This process is primarily practiced at the Ametek plant in Wallingford, CT. They have three PIF units, 5 in. diameter by 12 in. long, 8 in. diameter by 36 in. long, and 12 in. diameter by 84 in. long. In terms of productivity, you fill the chambers as full as possible with the preformed parts. The shape and density of the parts determines the number of pounds per run. Ametek sees future sales prospects for parts made this way in the 10's of millions of dollars, because the process is more economical than hot isostatic pressing. Photos of typical parts may be seen on their website at "Ametekmetals.com."

Ceracon Forging, Quasi-isostatic

This process was developed to circumvent the need to can powders, prior to hot isostatically pressing them. One starts with a green or sintered preform at any convenient fraction of full density (50 to 85% dense), as shown schematically in Figure 12.18. The preform could be made by uniaxial pressing, cold isostatic pressing in flexible tooling, or gravity sintered. The preform is made oversize, to compensate for the compression that occurs during the forging process. The preform is

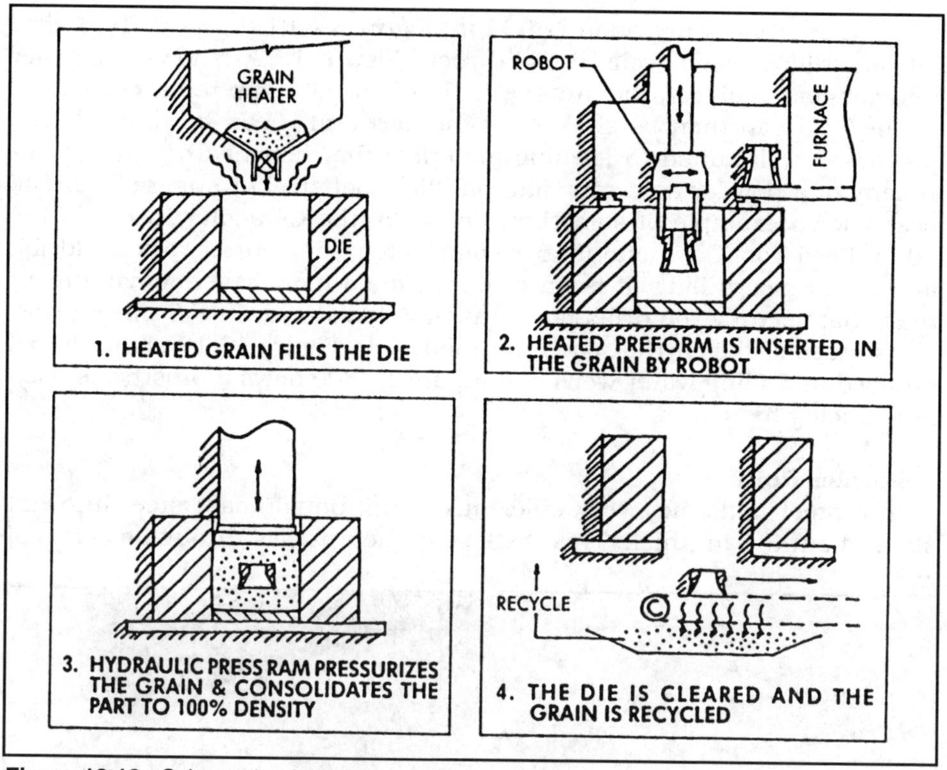

Figure 12.18. Schematic of the Ceracon Process.

heated in a protective atmosphere to the forging temperature, e.g. 2000°F for iron based material. At the same time, a pressure transmitting medium (PTM) made from a mixture of ceramic powder and a carbonaceous powder is separately heated. The PTM is poured into a simple round die on a large hydraulic press. The hot preform is quickly transferred by robot to the die, being placed into the hot PTM. (see Figure 12.18) The press is cycled. The upper punch enters the die and presses on the PTM. The PTM behaves almost like a fluid, and presses on all sides of the preform. It does not enter the pores in the preform, so its pressure causes the preform to collapse into itself. As the preform shrinks, its density approaches that of the pore-free material, during a 30 second holding period. The collapse is not perfectly uniform, because the pressure from the PTM is not exactly isostatic. The part shrinks more vertically (30%) than horizontally (10%). The preform is designed to compensate for this non-uniform shrinkage. For a 1 to 50 lb. part the process of removal from preheat furnace to ejection of forged part takes 2 to 2.5 minutes. For a 10lb. part with a 6 in. cross section and a length of 7 in., tolerances are about ±.030 in. Figure 12.19 shows a 50 lb. drill bit for drilling water wells in solid rock. Figure 12.20 shows a smaller Ceracon forged object, the prosthesis that is used in hip replacement.

The PTM can then be shaken off the part and it can be re-used. The process is well suited to forming materials that benefit from the homogeneity of a powder as a raw material, like tool steel or superalloys. The best known example of something made by the Ceracon process is the 10 lb. Reed Tool Co. metal matrix composite rolling cutter rock bit (drill), for oil and gas drilling. It is hard to imagine a more severe environment than that seen by the drills face, pushing downward, 2 or 3 miles below the surface of the earth, inside the solid rock. Figure 12.19 shows a tool bit used in drilling water wells and Figure 12.20 shows a prosthesis used in replacing hip joints.

Powder Rolling

In a steel mill, they very efficiently continuously cast steel into hot slabs about 3-6 in. thick. This thick strip then passes through a series of

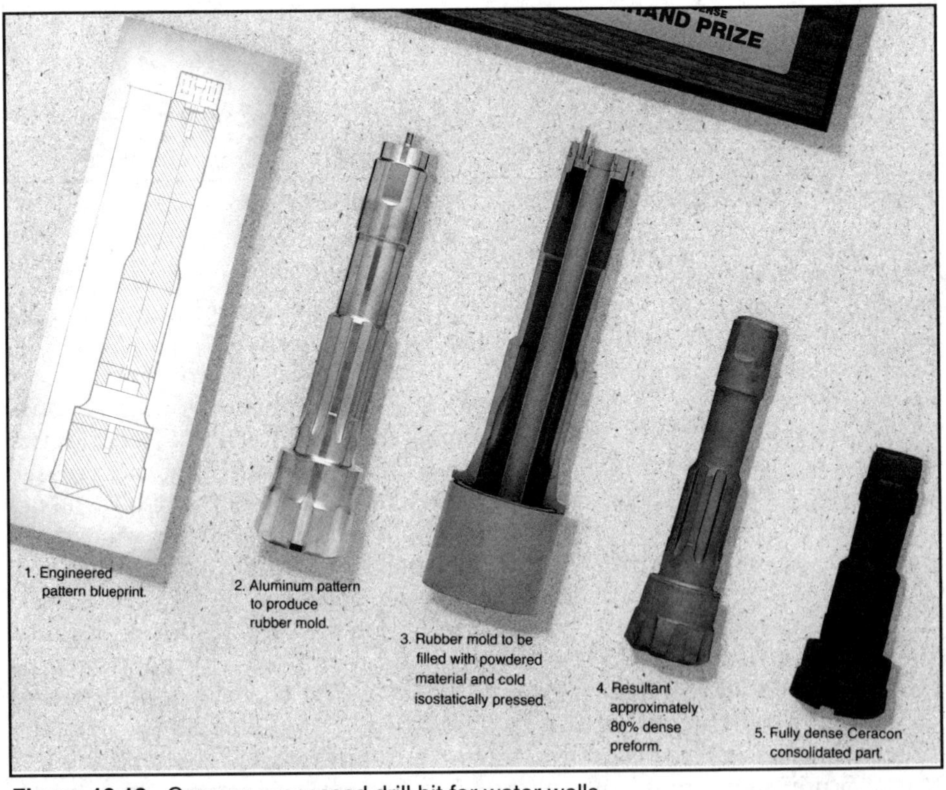

Figure 12.19. Ceracon processed drill bit for water wells.

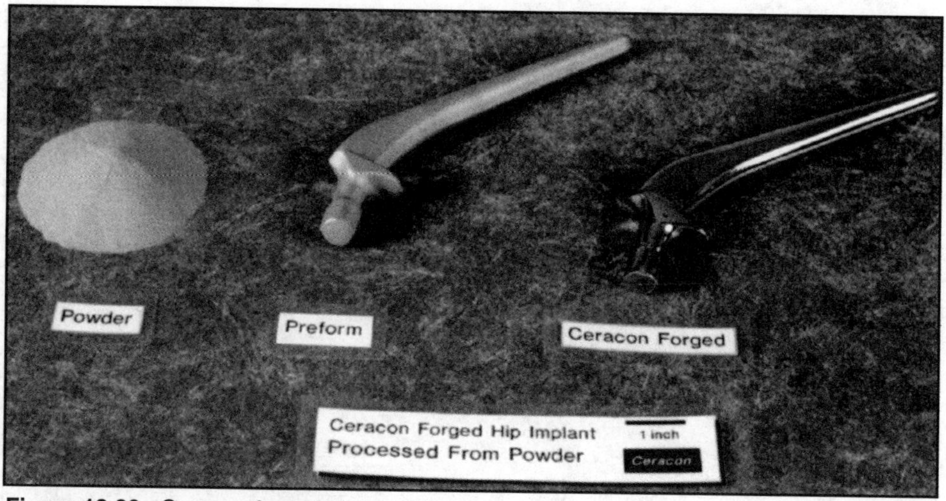

Figure 12.20. Ceracon forged HIP prosthesis.

rolling mills that roll it into thinner and thinner plates and sheets. The thinner the sheet, the more trips through the mill, and the greater the cost. It probably just occurred to you that metal powder is pretty thin; I wonder if it could be poured between two rolls and turned into thin strip? Would the rolls take a good grip on the loose powder and pull it between them, whilst compacting powder to near full density? Yes, and people earn a good living doing this! But they don't try it on iron powder because the iron powder alone sells for $.40/lb. and the finished steel sheet sells for only $.36 to .38/lb. The trick is to try it on expensive materials that are otherwise difficult to roll or which are needed only in much smaller quantities than steel sheet. Figures 12.21 and 12.22 show the arrangements of rolls to convert powder to strip, for subsequent sintering and re-rolling.

In Canada, around 1966, Sherritt Gordon (now Veridian) figured out how to make a nickel or two by powder rolling. They use a complicated hydrometallurgical process to convert nickel-containing concentrates into nickel powder. The powder is then rolled into 79% dense green strip stock 4.1 mm thick. The green strip goes right from the rolling mill into a furnace at about 2100°F. On the exiting the furnace it can be hot rolled to full density at 2.1 mm thick. This stock, in turn, can be cold rolled to 1.3 mm. This strip is about the thickness of the Canadian nickel coin. It is then punched out to form the nickel blanks which the Canadian Mint coins (embosses) with the required features, (thus justifying the initial comment about making a nickel or two).

Why doesn't the nickel sheet have the same economic problems as

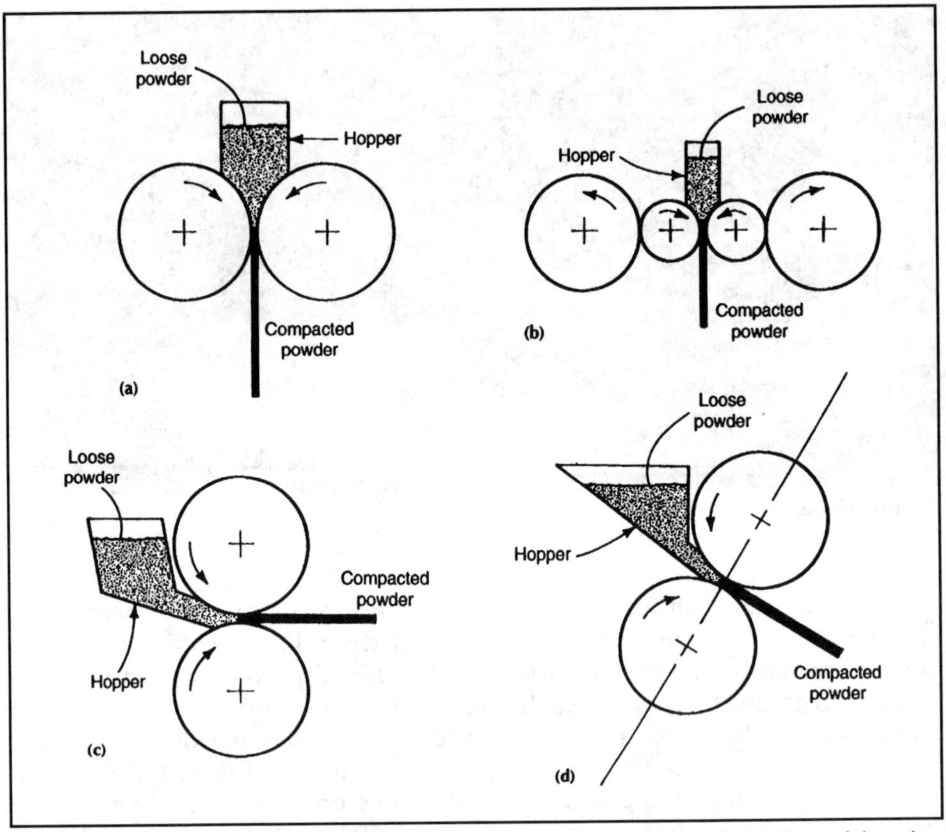

Figure 12.21. Roll compaction of metal powders showing various roll positions: (a) and (b) horizontal, (c) vertical and (d) inclined. Courtesy of ASM, Vol. 7.

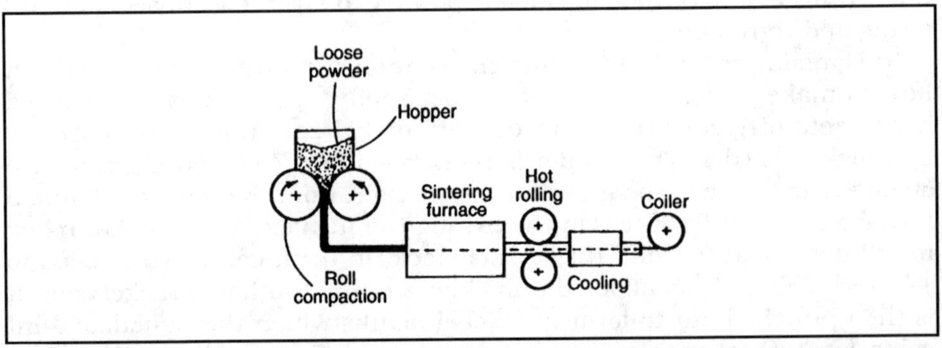

Figure 12.22. Schematic of the powder rolling process using horizontal roll arrangement, with the compacted powder feeding into a sintering furnace, flowed by hot rolling, cooling and coiling. Courtesy of ASM, Vol. 7.

sheet rolled from iron? That is, why don't they just cast the nickel and roll it down in some huge mill? The answer must lie in that the nickel is first produced in powder form, not billet form. In fact, to make the stainless steel sheet for your pots and pans, Sherritt Gordon roll compacts the nickel powder into briquettes that look like charcoal briquettes. These are sold to specialty steel mills who then add 12% nickel to 17% chromium and some iron, to be melted into ingots or continuously cast as 316L stainless steel. The slabs are then rolled in a huge series of mills. Of course, there is a lot more stainless steel sheet used every year, than there is nickel sheet. That points to one of the reasons why there is a business in rolling metal powders. If you are only going to make limited quantities of thin sheet and you are going to make it in many different alloys, you need a smaller, flexible manufacturing process. This is the route which Ametek has followed. Their powder rolled nickel and nickel alloys are used in semiconductor caps, nickel iron lead frames, and nickel iron resistor end caps. Ametek makes its powder in Eighty-four (no kidding!) Pennsylvania and ships it to Wallingford, CT, there to be converted into nickel alloy strip by rolling. In trying to visualize how big the powder rolling industry is, it should be noted that, other than the engine bearings described below, Ametek is the only U.S. supplier of roll compacted metal alloys. (These applications are kind of esoteric. They illustrate the point that it is easier to visualize a P/M process (powder rolling), than it is to really understand all the ways in which the product is used. The applications are highly engineered and extraordinarily technical in their detail!)

For a larger scale application of powder rolling, look at the manufacture of main bearings and connecting rod bearings for automobile engines. The bearing alloy is aluminum with 8.5% lead, 4% silicon 1.5% tin and 1% copper. This is one of those examples where the best way to make this alloy is by atomization to form powder, rather than casting a large slab and rolling it down. If you cast the material in massive form, the lead segregates or separates. When you atomize it, each particle has the same lead content. Figure 12.23 shows how this alloy is powder rolled as a sandwich. The bearing alloy is in the middle. One side will be pure aluminum which will later bond well to a steel backing. The other side is a disposable alloy of Al with admixed lead tin which allows the strip to exit cleanly from the rolls. That layer is later machined off. The strip is then re-oriented with the aluminum face toward a steel backing strip and the two are re-rolled, cladding them together. Sections from this assembled strip are formed into the U-shaped haves of engine bearings. The process is also carried out in such a way that the bearing alloy powder can be directly rolled onto steel backing strip. This is a significant business: there are a lot of automobile engines built every year!

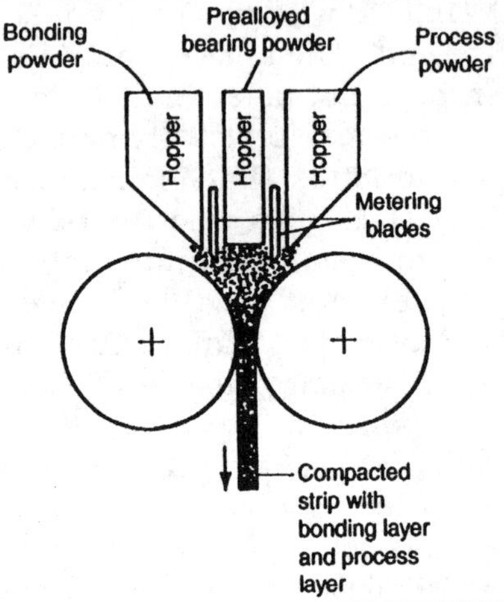

Figure 12.23. Rolling of powder into strips for bearings. The 3 layers are bonding powder, prealloyed bearing powder and process powder. Courtesy of Vol. 7, ASM.

Conform Process

Metal powder can be fed into a groove on a rotating wheel, Figure 12.24a. The wheel carries the powder into and against a tapered die shoe which gradually presses down into the groove and compacts the powder. The taper on the die shoe eventually fits so far down into the wheel groove that their is no room for the powder, even at full density. At that point, it starts to rapidly squirt or extrude forward, to get away from the die shoe. Figure 12.24b. shows the overall machine. The starting material and a piece of an extrusion, Al-Li scrap, are shown in Figure 24c. If the powder is aluminum, the severe working causes the extruded wire to be fully dense and heated to 900°F on exiting. In some instances, the wire exited at 132 mph!. John Schuldies of Industrial Ceramic Technology in Ann Arbor, MI who sells blocks of ceramic material to make dies, says that some of his materials go into the Conform dies, to resist the high temperatures and abrasion that occurs. When the starting material is powder, lower stresses are needed to form it into wire or other shapes, than when aluminum scrap or clippings are used. We have also heard that some manufactures take aluminum scrap metal turnings or chips and feed them into a Conform machine, thus forming the extrusions that are used for aluminum storm window frames. Conform is a continuous extrusion process, As long as you don't run out of feedstock, the coil or bar of extruded material can be as long as you want. In the usual extrusion process, you put a hot billet of material into a die and press on it at high pressure. The billet gets reduced 10:1 in

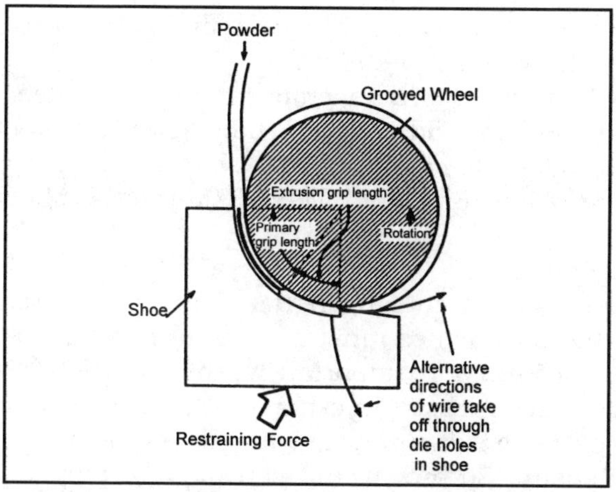

Figure 12.24a. Shows the principle of the Conform continuous extrusion, where powder may be fed in on the top left and the extruded nearly fully dense material emerges on the right.

Figure 12.24b. The conform machine.

Figure 12.24c. Starting materials and section extruded rod.

diameter and you get an extruded piece about 10 x 10=100 times longer than the billet with which you started.

Companies who use Conform machines in niche markets are very secretive about what they do, and to whom they sell their product. One such market is alloy rods made from metal powders, which are sold as alloy additions to the casting of rods or billets. (P/M process shoots self in foot!)

Powder Extrusion

Most P/M parts are rarely longer than 6 in. in the pressing direction. The reason is that die cavities must be 2 ½ times as deep as the finished pressed part, i.e. 15 in. More importantly, there is friction against the die wall, so that powders pressed into parts with great length in relation to their width, suffer from low pressure at the midpoint on the part, with resulting low density. So how would you make a part 1 in. in diameter and 10 ft. long? Cast metal billets are hot extruded everyday into long thin objects like rods and pipe. Enter hot extrusion of metal powders! If steel powders are heated to about 2200°F, they can be forced through a die. As they enter the die, the particles get forced together and approach full density. They exit the die as a long thin, fully dense object. Figure

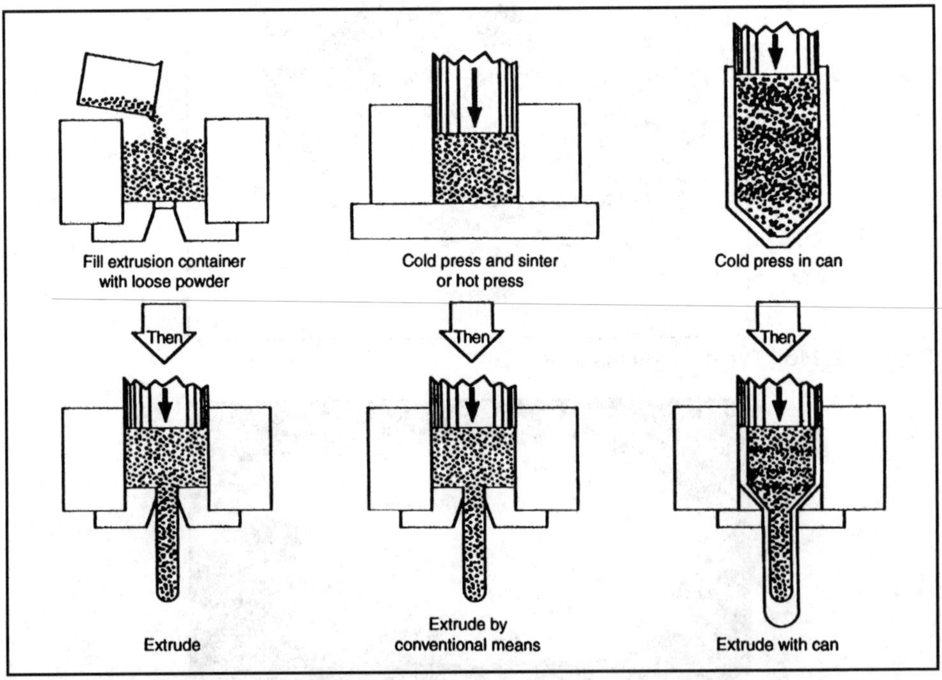

Figure 12.25. Schematic representation of 3 basic methods to hot extrude metal powder into rod shapes. Courtesy of ASM, Vol. 7.

12.25 shows schematically how this occurs. On the left, loose hot powder is forced into a heated die. There is no real protection from oxidizing the powder, and this is not a common method. In the second method, the powder is first consolidated into a billet, by cold pressing. The billet is then heated to the forging temperature and dropped into the press. With the powder billet in consolidated form, the press stroke is shorter than when loose powder is extruded. For the best protection of reactive powder, the powder can be pressed into a can. The can is then evacuated and sealed off as in Figure 12.26. The heated can with the compacted powder inside is extruded at 2200°F, as shown in the third example in Figure 12.25. The extrusion die must be lubricated with something like graphite (DAG) or molten glass (Ugine Séjournet).

Strangely enough, one of us, LP, has actually worked on extrusion of metal powders. As an undergraduate in the late 1950's, this author worked summers at M and C Nuclear. One job was to mix Zircaloy powder with boron carbide, a neutron moderator. This was handled in inert atmosphere, because the Zircaloy is pyrophoric. The mixture was poured into small cans and then compacted inside the cans. The cans were then evacuated and sealed off. The cans disappeared from the author's view and somewhere else in the building they were heated and extruded into thin rods about 4 ft. long. They then became part of the control system that is used to make a nuclear reactor run hotter or cooler. When the rods are down inside the reactor, they absorb neutrons, the chain reaction can't get going, and the reactor is cooler. No one can predict the long term future of nuclear power, but given the current climate, LP was better off spending the last 35 years pressing iron powder!

Anval Nyby, (now Dynamet Anval) in Torshalla Sweden gas atomizes thousands of tons of stainless steel and alloy powders. At the plant in France, some of these are then packed into cans at about 70% of the pore-free density of the steel. The cans are cold isostatically pressed to

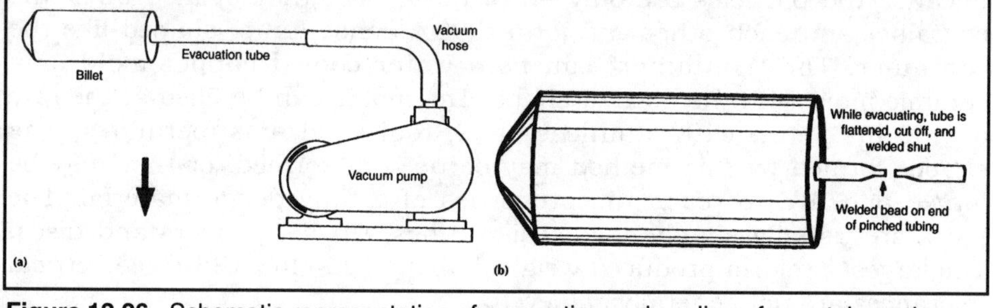

Figure 12.26. Schematic representation of evacuation and sealing of a metal powder extrusion billet. (a) Billet with evacuation tube leading to vacuum pump. (b) Billet with sealed end tube. ASM Volume 7.

85-90% density. Finally, the cans are extruded at 2200°F into special grades of stainless steel tubing Anval produces some very corrosion resistant stainless steel grades and the tubing goes into severe applications like production of urea carbamate, offshore oil rigs, chemical plants, and mechanical tubing. Current production is about 300 tons/year, but used to be 5,000 tons/year. Through the last 20 years, over 125,000 tons of tubing were extruded.

Closer to the press and sinter industry, OMG America known for copper powder, also sells a fully dense material, Glidcop . Starting with prealloyed copper and a small amount of aluminum powder, they expose the powder to an oxidizing environment and form small precipitates of aluminum oxide inside each particle. This powder is then canned and extruded into fully dense rod. The result is copper, dispersion hardened with alumina particles. It has good electrical conductivity and is ductile enough to form into welding electrode caps. These caps go onto the robotic welding arms that car manufacturers use to weld up all the sheet steel in your car. For steel that is zinc coated for corrosion resistance, the Glidcop has the happy characteristic of being able to press against the zinc, send current through it and the underlying steel, and not end up welded to or stuck to the car body. (When the welding cap or tip sticks to the car, the cap gets pulled off and the cooling water runs all over the place. The production line stops and everyone gets mad.)

Osprey Process

In Chapter 2 and 3 we described how a liquid melt could be gas atomized into a stream of powder. In making powder, we want to be sure that the collection chamber is big enough that the hot powder is pretty well solidified by the time it reaches the bottom. If not the partially liquid powder particles would tend to weld together in useless lumps. In the Osprey process, one sticks a container into the partially solidified particle stream and fills it up, as shown schematically in Figure 12.27 Because the particles are only 40 to 80% solid, they splat against the container and each other and form a 98% dense lump, shaped like the container. The "container" can be a water cooled copper mold or a ceramic mold with the desired shape. The molds can be filled at the rate of 250 lbs./min. with stainless steel, tool steel or superalloys. The shapes formed by this method may be used as formed, or they may be forged or HIP'd to reach the properties of a fully dense material. The parts are usually simply shaped like tubes, rings, cylinders and discs. The largest preform produced weighed about 1200 lbs. Close geometrical tolerances are not generally achieved.

Why not just cast the whole part by pouring liquid steel into a mold? Why bother to atomize? The atomized particles are very clean and have

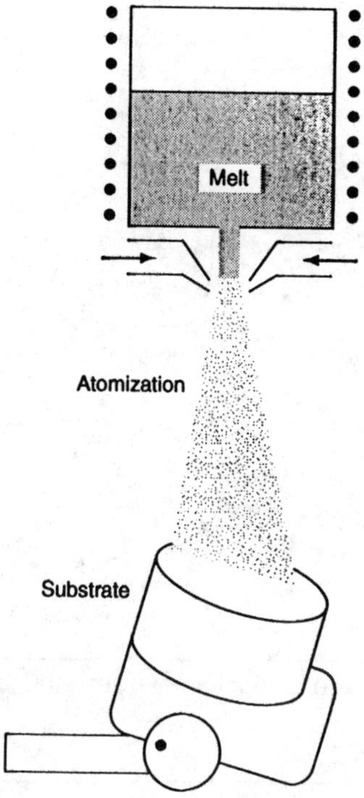

A schematic of spray forming in the Osprey mode

Figure 12.27. Schematic view of spray forming in the Osprey mode, ASM, Vol. 7.

uniform chemistry without the segregation that can occur in a casting. They also have a small and uniform grain size that gives them better mechanical properties. The process lends itself to smaller mill runs and the possibility of special alloys that are not possible by casting.

Spray Forming

Spray forming is a lot like the Osprey process except that you don't spray into a mold. Instead, you build up a shape on a starting base or substrate. There is a real cost benefit in being able to go from a liquid metal to a semifinished mill product. Figure 12.28 shows disks strips and tubes being spray formed. It is a continuous process, rather than the intermittent one of filling a number of molds. Figure 12.29 shows the continuous formation of tubing and Figure 12.30 shows the formation of a solid billet in the horizontal position. Again, the main metallurgical advantage is fine grained material free of segregation. If these materials

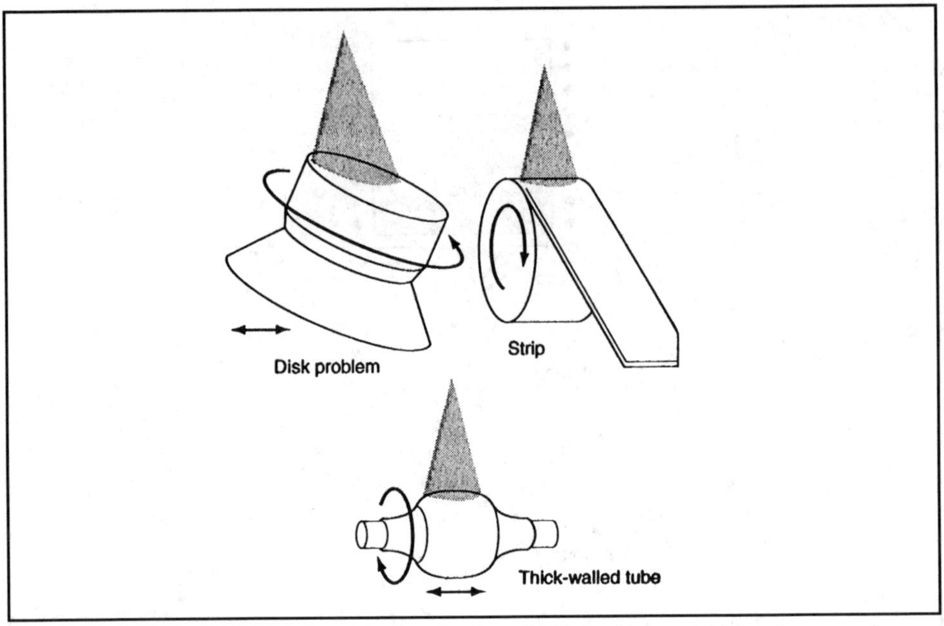

Figure 12.28. Schematic view of the production of billet, strip, and tubing by spray forming (Vol. 7).

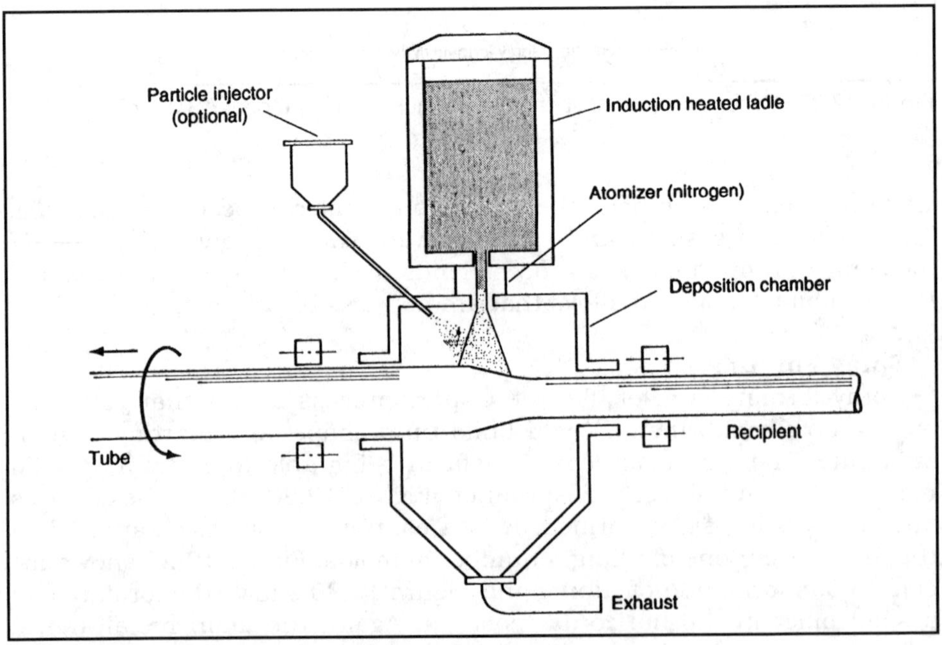

Figure 12.29. Tube fabrication by spray forming, ASM, Vol. 7.

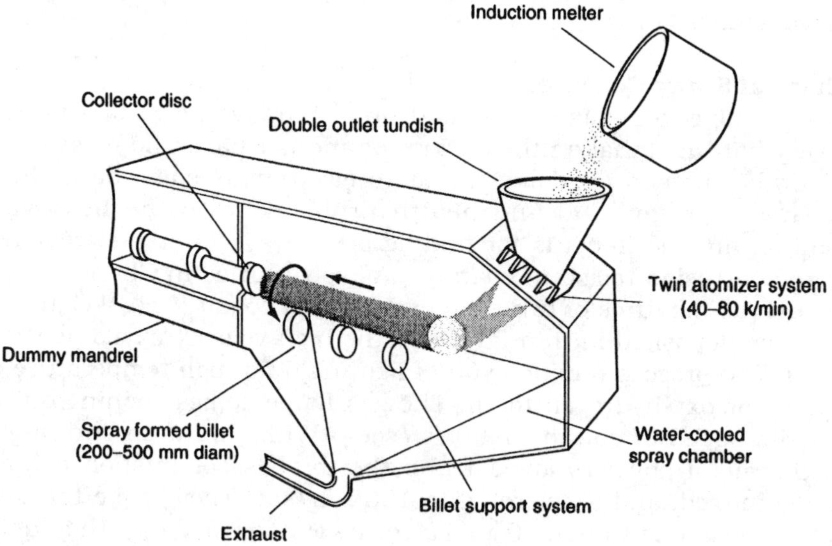

Figure 12.30. Horizontal mode billet fabrication by spray forming. Osprey Metals Ltd., ASM, Vol.7.

were melted and cast into tubes or billets, there would definitely be segregation of the metallic elements and coarser grains, both of which would lead to lower mechanical properties.

This is a real world process: Sumitomo makes very large spray formed rolls to go in steel mills for rolling of round bar, flat bar, and wire rod. They are used because of their good resistance to thermal fatigue and long life. The rolls can be 31 in. in diameter by 18 in. long. Sandvik in Sweden forms stainless steel tubes up to 26 ft. long with a diameter of 16 in. and a wall thickness up to 2 in. They have developed a special corrosion resistant alloy that can be used in waste incinerators. Howmet spray forms superalloy rings 31 in. in diameter by 19 in. long. They are used in ring or casing components in high thrust jet engines. They are generally HIP'd or ring rolled after the spray forming step.

What sort of things are in development? Aluminum alloys for conversion to automotive cylinder liners, copper alloy billets for electrical connectors, composite high alloy steel for waste incinerators and high silicon-aluminum alloys for electronic packages and heat sinks with low thermal expansion and low density. Consider also the implementation of a full sized production plant in Europe to make spray formed tool steel billets 18 in. in diameter by 8 ft. long, or the Navy Man-Tech program involving building a plant with 5 ton melt capacity to form large diameter, thick walled pipe. This is the pipe that will be used in the power plants on naval ships.

Thermal Spray Coatings

It is possible to use metal to build up a thin layer, e.g. .004 in. thick, of a very hard material on the surface of another part. The most familiar non-powder method for doing this is to electroplate some hard chromium. The environmental and health problems with the hexa-valent chromium are the impetus for finding other ways to create wear resistant and corrosion resistant coatings. As you learned in Chapter 1, powders such as mixtures of tungsten carbide and cobalt (83% WC and 17% Co) can be deposited in thin layers by the high velocity oxyflame (HVOF) method. The process reminds you of lighting up a high temperature oxy-acetylene or oxy-hydrogen torch. The gas flame comes rushing out at 5 times the speed of sound (5900 ft./sec.). With the torch running like this, the metal powders are fed into the torches combustion chamber, partially melted, and accelerated to 1500 ft./sec. They strike the target, which is positioned about 6 to 12 in. away. On striking, they deform (splat?) and stick to the target or the previous layer of sprayed material. This builds a layer that is 99% dense with less than 1% oxygen in it. In one example, it took about 20 minutes to spray a .004 in. layer on a rod 4 in. in diameter and 20 in. long.

Usually, you want only a limited region of a part to have this coating, so the rest of it must be masked off. After the spraying, it takes time to remove the masking material. Just as with hard chromium plate, the final surface needs to be finish ground. In one example (a Boeing 737 nose landing gear cylinder piston), the HVOF required 10.3 man-hours and hard chromium plated required 14.5 man-hours. The thermal spray process has also been used on propeller hubs, helicopter dynamic rotor components, and gas turbine engine components.

The flame spray gun is usually mounted on a robot or other articulating arm. It sprays molten powder out in a straight line, (line-of-sight). This means that you can not coat internal surfaces where the ID is less than 10 in. The WC/Co coatings have a hardness of about Vickers 1300 kg/mm^2, which is harder than steel's 750, and even harder than the hard chromium plate, Vickers 1,020. Further, it does not lower the fatigue resistance of the underlying heat treated steel, the way that hard chromium plating does.

Laser Forming

Much of this background has been summarized by D. Miller in 1999 at "www.optomec.com/~dmiller/technology/backgrounder.htm" as follows:

The LENS™ technology is a direct metal deposition process. "The process consists of feeding a continuous stream of fine particulate metal (metal powder) into a finely focused, high-energy laser beam, thus forming a small pool of melted metal on a substrate. By moving the substrate on which the molten pool is formed in a predetermined x and y dimensional pattern, a single layer metal part is formed by cooling the metal as it solidifies. Moving the laser and the powder feed mechanism higher above the platform with each layer allows building a 3 dimensional part. The movements in x, y and z directions are defined by the original computer aided design (CAD) design file. Because the resultant parts are made of metal, this approach has demonstrated a great deal of promise in overcoming many of the limitations imposed by the existing rapid prototyping technologies. The process provides good feature definition and exceptional material properties in the deposited structures using metal alloys. This technology has reached a sufficient level of sophistication that a wide variety of metallic shapes can be produced, encompassing a size scale ranging from tenths of millimeters to tens of centimeters. Metals ranging from solder alloys to aerospace super alloys have been employed, including stainless steels and tool steels. Both thin shell and solid objects may be constructed."

"The equipment is shown schematically in Figure 1.32 of Chapter 1. It is comprised of a powder supply unit to deliver a regulated flow of material to the deposition region, a controlled atmosphere glove box, a three axis computer controlled positioning system and a high power (several hundred Watts) cw Nd:YAG laser. Metal powder is delivered to the deposition region using a carrier gas blowing through a multi-nozzle system. Intense heat is supplied by a laser beam which is focused through the multiple nozzles to a plane below the output orifices and onto a substrate mounted on the moveable platform. The nozzle is designed to cause the powder stream to converge to the same point as the focused laser beam. As the laser beam creates a molten puddle on the metal substrate, the injected powder is consumed to build up material. Scanning the platform under the laser beam deposits the metal layer in a desired x,y pattern. Moving the vertical axis away from the substrate by the thickness of a single layer after each layer is deposited provides the third dimension of part production. Issues such as surface finish and dimensional accuracy must be addressed in detail".

The process works well in the manufacture of tooling for injection molding because, of necessity, such tools can have no overhangs. If there were overhangs, the injection molded part could not be removed from the tooling. Work is underway to aid in the production of overhanging features in laser formed parts. These include the use of a 5 axis machine to support the part or the formation of sacrificial supporting structures made in another removable material. Currently available machines have deposition rates of approximately 0.6 $in.^3/hr$. The uncompensated files allow the parts to be built .020 to .030 in. oversized or material safe so that finishing stock is available. Beam offset algorithms are currently being evaluated to provide an accuracy in the deposition plane of approximately ±.005 in. The z axis tolerance is approximately ±.015 in. After the prototype part or tool has been formed, it is necessary to cut or remove the part from the substrate to which it has been fused.

Explosive Compaction

Explosive compaction of metal powders has been practiced on a developmental basis for at least 40 years. In early work, a sheet of explosive chemical was placed on top of a bed of metal powder. On ignition, the shock wave would drive the metal particles together under very high pressure and very high velocity. It was used on materials that are difficult to compact because of high yield strength and low green strength. A more recent example would have been its use on rapidly solidified powders where it was hoped to join particles together without sintering and destroying the amorphous structure or small grain structure. In 1979, Lennon and others canned iron powder inside a steel tube with sealed ends. Explosive was ignited around the tube, forcing the tube in against the iron powder. This resulted in 98.1% dense iron. To our knowledge, there are currently no commercial products being made by this technique, but much research is still being conducted.

During the year 2000, Utron Corp. of Manassas, VA developed a device to use a controlled explosion to drive a powder metallurgy punch against a mass of powder contained in a die. They vary the composition of the combustible gases so that the speed and the maximum pressure on the punch can be controlled. Consolidation of powder is accomplished in from .001 to .015 seconds. Figure 12.31 shows the chamber in which the compaction occurs and Figure 12.32 shows a schematic view of the inside of the device. Figure 12.33 shows a 410 stainless steel part pressed with no admixed lubricant at 200,000 psi to a density of 7.44 g/cm^3. The die must be sprayed with a lubricant between each stroke. Because of the high energy contained in the combustible, an advantage of the process is that a very large force can be developed in a relatively small press. It is expected that presses capable of compacting parts of 100 to 200 $in.^2$ will be economically constructed.

Magnetic forming

In Chapter 1, we spent a lot of energy trying to explain that when the electric current flowing in a wire increases or decreases, the magnetic field around the wire changes. You must also remember that a changing

Figure 12.31. Example of device for combustion drive compaction. The ignited gases in chamber on left, drive a punch toward the right, compacting metal powders. The compaction area is then dis-assembled and the part removed. Courtesy of Utron Corp., Manassas, VA.

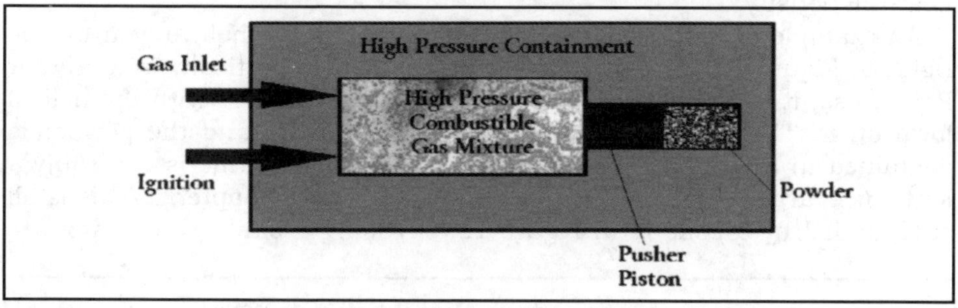

Figure 12.32. Schematic of Utron combustion drive compaction device, shows gas pressing on a piston that functions as a compacting punch.

Figure 12.33. Example of 410 stainless steel ring, compacted by the device in Figure 12.31. It was pressed at 200,000 psi to a density of 7.44 g/cm^3. Actual size about ¾ in. in outer diameter.

magnetic field can cause electric current to flow in a nearby wire. It will be opposite in direction to that in the first wire, and the two wires will try to push themselves apart. That, in turn, could make an electric motor's shaft rotate. Well the folks at IAP of Indianapolis, IN have built on that principle, to create a device where the gentle pushing action between two wires can be increased to the point where metal powders can be rapidly compacted. Figure 12.34 is a schematic drawing of the pressing of powder into a rod form. The three components of the device are the outer coil, the armature, and the powder. Discharging a large current in the outer coil causes a strong magnetic field running upward around the armature. This induces or causes a current to flow in the copper armature, and opposite in direction to that in the coil. The opposing currents push on each other, (or the magnetic fields around the currents try to exclude each other). The inner current, in the armature, loses this shoving match and the armature collapses in onto the powder, driving it to near full density.

An example of a part compacted in this way is the helical gear on the right, in Figure 12.35. Instead of being compacted vertically in a regular P/M press, it is compacted radially. Visualize a core rod with the helical form on it. The core is surrounded by iron powder and the powder is contained in a copper tube, (armature). The copper tube is the equivalent, on a large scale, of the copper wires shown in Chapter 1. This is all surrounded by a coil, as in Figure 12.34. Using a bank of capacitors to

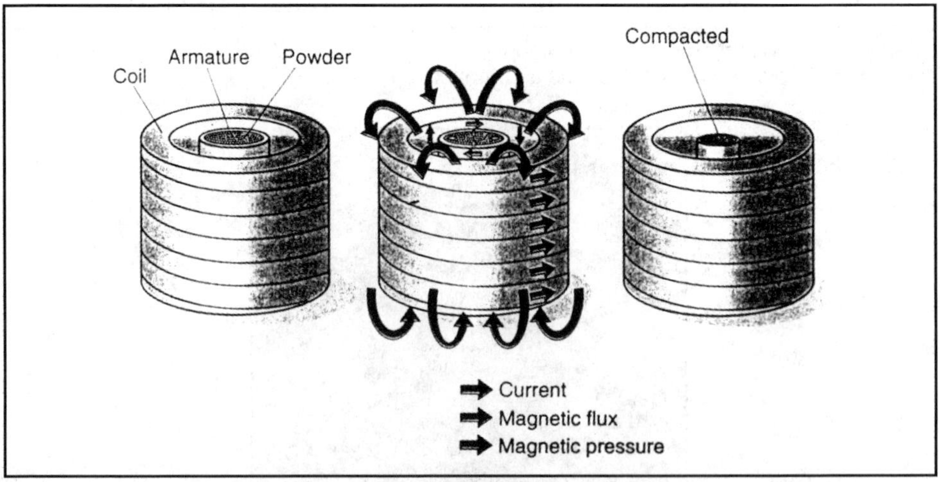

Figure 12.34. Schematic view of magnetic dynamic pressing of powder, into a rod. Current flows in the outer coil. This creates a magnetic field up and around the armature, and induces a current in it. The currents are opposed to each and this creates a force on the current in the armature, driving it in against the metal powder, and forming a nearly fully dense rod. Courtesy of MPR, February, 2000, p.22.

hold electrical charge, a huge electric current is discharged and flows through the outer copper coil. This induces a current in the inner copper tube, flowing in the opposite direction. A very large force develops between the two copper tubes, and the inner one is rapidly driven radially inward, against the iron powder, compacting it to 96 to 98% of full density. Figure 12.36 shows the IAP Magnepress in which this action

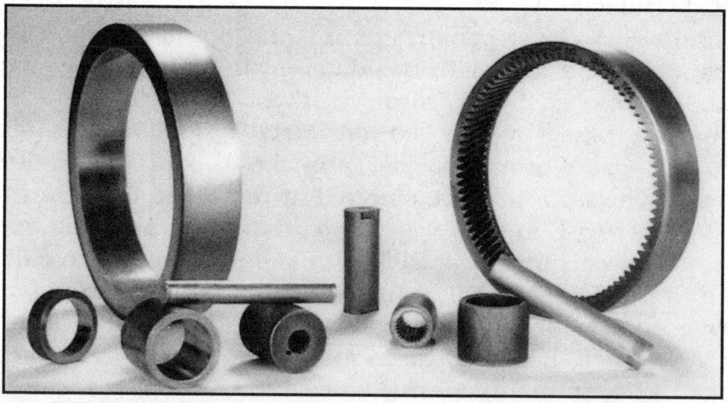

Figure 12.35. Examples of parts compacted radially, using magnetic forces, Courtesy, IAP, Indianapolis, IN.

Figure 12.36. Magnepress ™ system, used to radially compact metal powders, courtesy of IAP, Indianapolis, IN.

occurs. The press is very small compared with conventional P/M presses that would exert the same force. However, conventional uniaxial pressing of helical forms can not generate the high densities in the teeth, that the radial compaction accomplishes. In a second simple press with dual upper punches, the copper and the core rod are individually pushed down off the P/M part, to free up the gear.

Dual Materials

It is possible by clever construction of the powder feeding shoe on a P/M press to make a part with two different materials in the same part. Figure 12.37, supplied by Al Dornisch, then at Powder Metal Products, shows a bearing made by MPP, Powder Metal Products in St. Marys, PA. To save on the cost of bronze powder, only the ID is bronze, and the OD is sintered iron. The iron must be sintered at the same temperature as the bronze, about 1575°F and so will not be as strong as what would have been seen if it were sintered at 2050°F. (If the iron is low in carbon, sintering iron at 1575°F, just below the transition temperature to austenite, is a pretty good idea. Iron atoms move around (diffuse) in the BCC ferrite phase of iron faster than they do in austenite, at a higher temperature.) This design has been manufactured since 1963 at their company.

In what may have been a similar example of processing, Prof. Eudier in France worked out a way to put a layer of fine bronze powder on the ID of a bearing, and a coarser layer on the OD. The fine layer had lower permeability and allowed for better support of the rotating shaft, while the outer layer of coarser powder was less expensive and held lots of oil.

Perhaps we could also have one layer of material on top of a layer of different material. In real life we see this in pills that have two or three layers compacted together. It is not a very common procedure with metal parts but we believe that electrical contacts may have been made this way. Further, Cincinnati Inc., who builds large scale P/M presses, has built at least one press to compact two layers of powders in this fashion.

Figure 12.37. Dual compacted bearing with bronze on the ID, see arrow, and iron on the OD. Courtesy of MPP, Powder Metal Products.

Figure 12.38. Pulley that was compacted with side acting punches to form the groove, made for Motorola Corp. for use in automobile engines in the 1960's, by Powder Metal Products, MPP, courtesy of A. Dornisch.

Side Action Pressing

After reading Chapter 4 on compacting metal powders, and all the other references we have made to powder flow during compaction, you just know it is not possible to have punches or core rods acting crosswise to the direction of pressing. That said, you have also seen our many references to "those clever P/M folks". Al Dornisch and John Sterbank (now of Eastern Sintered Alloys) remember the Motorola pulley made during the 1960's at Powder Metal Products, (now part of MPP). Al Dornisch had worked in the plastics molding industry where tooling members do move in and out of the direction of plastic flow, allowing cross holes and under-cuts. The Motorola automotive pulley shown in Figure 12.38 had its V-groove formed by sliding punches that moved in and out through the die walls. This was a real cost and material saving vs machining out the groove as a secondary operation. They did not just make one of these, to show off! They went into automobile engines for many years. Eventually, V-belts in engines were replaced by the serpentine belts and toothed pulleys of today, and this model was discontinued. As you might imagine, the part probably had some significant differences in density, but, nonetheless, it was strong enough for its intended purpose. Is side acting pressing a big deal or a little deal? It is not practiced much today, if at all. We have seen one recent Japanese development on a part made this way, but it remains to be seen if it will become a significant part of P/M manufacturing.

CHAPTER 13

Case Histories

Each year as part of its spring technical conference, the Metal Powder Industries Federation sponsors the P/M Part-of-the-Year Design Competition. This competition provides an opportunity for P/M part fabricators to demonstrate powder metallurgy's advantages. Parts are judged on several criteria including cost savings, tolerance control, design uniqueness and performance reliability. Competition categories include ferrous, nonferrous, stainless steel, metal injection molding and fully dense.

We have selected award winning parts from all categories over the last several years and included them in this chapter. They are production parts and demonstrate the many varied applications of the process we have described for you in this book. Additional information about individual parts can be obtained by contacting MPIF.

Hot-forged P/M steel auto connecting rod.

The first domestically-produced, hot-forged P/M steel auto connecting rod won an award of distinction in the fully dense category. The con rod is made from a copper-steel composition, the connecting rod has an ultimate tensile strength of approximately 120,000 psi (827 N/mm^2) a hardness range of 76-89HRG, and a fatigue strength of 8,000 psi (55 N/mm^2) at 107 cycles. The part is used in the 1.9 liter engine which goes into Ford Escort and Lynx models.

P/M replaced a conventionally-forged rod, eliminating several machining operations and providing a weight uniformity that does away with the need for balancing. Secondary operations include double disc grinding, deburring, and shot peening.

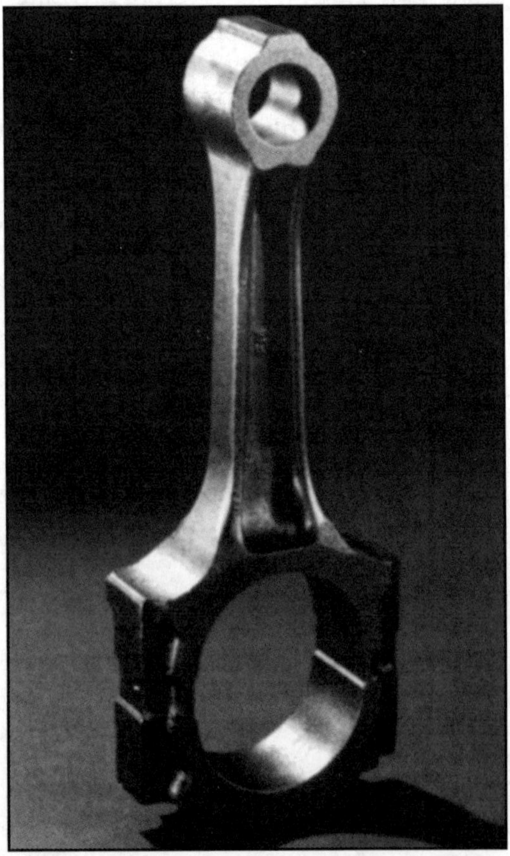

Figure 13.1. Hot-forged P/M steel auto connecting rod.

Stainless steel preload arm.

A 304 stainless steel preload arm won a stainless steel award of distinction. The part supports a rail which guides the tape head in a hard disc drive. The part has a tensile strength of 50,000 psi (344 N/mm^2), a yield strength of 40,000 psi (275 N/mm^2) and an elongation of 5%.

The part was formerly an aluminum casting. P/M provided increased strength and allowed for a more complex design including the tolerance control of the drilled holes through the five legs which must be parallel within 0.004 inches (0.1016 mm) in relationship to each other and the flange. Secondary operations include drilling of the five side holes and tapping the center hole.

Figure 13.2. Stainless steel preload arm.

Stainless steel filter

A controlled-porosity FXR filter for filtering molten thermoplastic resin won a grand prize in the stainless steel category.

Made to a density of approximately 50%, the filter removes contaminant particles in the manufacturing of nylon filaments. The material is a type 316 stainless steel modified by a small addition of molybdenum to form an austenitic-ferritic structure for increased strength. Made on a conventional compacting press, the filter must withstand 5,000 psi (34.5 N/mm^2) pressure differential without collapsing or separating from a sealing ring that is bonded to it during sintering in an H_2 atmosphere.

P/M replaced a sand filter, drastically reducing customer installation and handling costs. The patented P/M filter is more easily cleaned and has a much longer usage life than previous filters.

The inherent porosity of P/M allowed controlled pore sizes which are almost impossible to achieve by other techniques. A key design feature is the ability to make deep blind holes from both ends of the part with no secondary operations. These holes enhance filtration by increasing surface area significantly.

Figure 13.3. Stainless steel filter.

Bronze capstan bearing cap

A bronze capstan bearing cap used in a laser photo imaging system for commercial printing won a grand prize in the nonferrous category. Oil impregnated for self-lubrication, the multilevel part has a minimum tensile strength of 15,000 psi (103 N/mm^2), a yield strength of 9,600 psi (65 N/mm^2) and a hardness of 55HRH. Originally designed for powder metallurgy, the critical part supports the capstan roller assembly, that contains a mating P/M part.

Complex tooling is needed to form the part which must meet concentricity and roundness tolerances in the two radii of 0.001-0.003 inches (0.254-0.0762 mm).

Figure 13.4. Bronze capstan bearing cap.

Copper-infiltrated steel drive gear and pinion

Another award of distinction in the ferrous category was given to a copper-infiltrated steel (MPIF FX2008-90HT) drive gear and pinion used in a GM automobile electric seat.

The heat-treated part has a minimum ultimate tensile strength of 90,000 psi (620 N/mm^2). The tooth profiles on the complex part must meet minimum strength requirements and still retain good hardness properties. The pins act as bearing surfaces and must maintain a good surface finish and a 0.002 inch (0.0508 mm) tolerance on the diameter.

P/M replaced a design that consisted of a two-piece assembly. The spur gear O. D. and pins were formed on a screw machine and the tooth form was broached. The helical form was rolled. P/M combined all functions into one part and offered a projected annual savings of approximately $2 million.

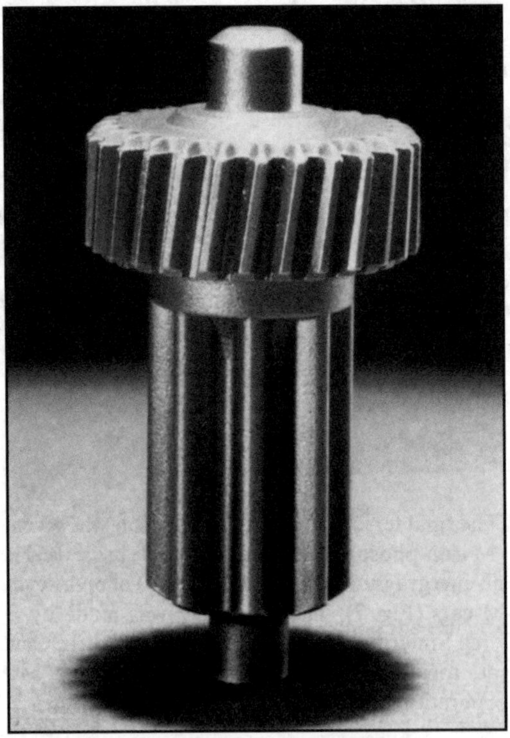

Figure 13.5. Copper-infiltrated steel drive gear and pinion.

Stainless steel index wheel

Another award of distinction in the stainless steel category was given to a 410 stainless index wheel used in a printed circuit board manufacturing machine. The multi-level part has a tensile strength of 50,000 psi (344 N/mm^2) and a yield strength of 40,000 psi (275 N/mm^2). Combining a spur gear with a flange would have made this part very difficult to machine. A critical design feature is the +0.001 inch (+0.0254 mm) I.D., and the flange thickness to the hub ratio. The keyway is pressed in and has a width tolerance of 0.002 inches (0.0508 mm). The gear has an AGMA class 6 rating.

Figure 13.6. Stainless steel index wheel.

Aluminum alloy pulley

An award of distinction in the nonferrous category went to an aluminum alloy pulley that goes into an office copying machine. The part has a tensile strength of 35,600 psi (244 N/mm^{2}), a yield strength of 30,000 psi (206 N/mm^{2}) and a 3.5% elongation.

P/M allowed the customer to rule out the need for machining the part from extruded stock and staking a flange on it. Selecting P/M also provided an estimated cost savings of about 50%. No secondary operations were performed on the P/M part.

Figure 13.7. Aluminum alloy pulley.

Brass housing

Another award of distinction in the nonferrous category was given for a brass (MPIF CZP-2000) housing used in computer tape heads for reading magnetic computer tapes, the housing has a minimum yield strength of 16,000 psi (110 N/mm^2) and an ultimate tensile strength of 30,000 psi (206 N/mm^2).

P/M provided significant cost savings along with greatly improved wear life over another design that had to be machined. P/M outlasts the wrought product by 25%. P/M also allowed the combining of three heads into one component.

The front and back of the housing must be parallel within 0.003 inches (0.0762 mm). The back flatness is especially important and must be perpendicular within 0.001 inches (0.0254 mm). The housing is coined for higher density, annealed and plastic impregnated to improve machinability.

Figure 13.8. Brass housing.

Main bearing caps for 3.8 and 3.1 liter GM engines

The main bearing caps for GM's 3.1 liter V-6 and 3.8 liter V-6 engines won the grand prize in the ferrous category. This product replaces a nodular cast iron design.

The 3.1 liter engine contains 3 caps weighing about 5 pounds (2.2 kg), and the 3.8 liter engine contains 4 caps weighing slightly more 8 pounds (3.6 kg). It is estimated that 60% of GM cars will soon be fitted with P/M main bearing caps and this application will consume more than 5,000 tons of steel powder annually.

Holding a density of $6.6 g/cm^3$ the caps have a tensile strength of 65,000psi ($448 N/mm^2$), an apparent hardness of 70HRB, an elongation of 3%, a fatigue endurance limit of 23,000psi ($159 N/mm^2$), and a Young's Modulus of 15.5×10^6psi ($107 KN/mm^2$).

The 3.8 liter engine has 2 bearing cap designs. There are 3 similar caps in positions 1, 2 and 3 (from the front of the engine), and a different design in position 4 at the rear.

The cap requires an installation profile feature for smooth fitting to the engine block. The statistically held tolerance across the "snap-width" that engages the engine block channel is 0.006 in. (0.152 mm). The flatness of the joint faces is held to 0.002 in. (0.0508 mm) individually and 0.004 in. (0.101 mm) when considered as a single plane. Since the cast iron engine block and the P/M cap are machined simultaneously during crankshaft diametral boring, the P/M and cast iron radii must match. This requires close control of the radial height and the horizontal radius to 0.010 in. (0.254 mm).

Cap number 4 is a difficult shape with a double arch and 2 different radii, causing unavoidable sintering distortion that requires machining of the snap width and feet surfaces. The complex top form including the oil return passage, required extensive tooling development to balance densities and avoid cracks on both parts and tools. Secondary operations included drilling, tapping and milling.

The conversion of cast iron automotive main bearing caps into P/M represents a major technical and commercial advancement for the industry.

Figure 13.9. Main bearing caps for 3.8 and 3.1 liter GM engines.

Auto transmission sprockets

High performance auto transmission sprockets won an award of distinction in the ferrous category. The transmission parts, a drive and driven sprocket, are compacted and sintered at 2336°F (1280°C). After sintering, the teeth and journal are selectively densified by a proprietary cold working process and then vacuum carburized. Eight parts are used in the transmission.

Vacuum carburizing produces a precisely controlled case on the critical journal tooth flank surfaces with minimum distortion. This provides high strength and exceptional bearing and tooth surface durability. Surface density of the parts is typically 7.8g/cm^3, gradually falling to lower levels towards the core. Core regions are not highly stressed and remain at a density of 7.0g/cm^3.

Parts have an ultimate tensile strength of 125,000psi (86OMPa), a yield strength of 120,000psi (826MPa), and a tooth and journal surface hardness of more than 60HRC. The endurance of the P/M sprockets exceeds induction hardened malleable iron.

Figure 13.10. Auto transmission sprockets.

Palate expander housings and plates

A stainless steel award of distinction was awarded for complex stainless steel (MPIF SS-316NI) parts which operate in the Lariat palate expander. The device expands the palates of young children with small mouths.

The two-level, very small, thin parts require uniform density throughout the cross section. The assembly consists of four P/M parts: left and right housings which are spot welded together; two guide pins are press fitted and staked into left and right P/M plates along with the assembly of the adjusting screws. The orthodontic appliance expands with the simple twist of an Allen wrench inserted into the front, a much safer procedure than with previous designs.

The parts have a minimum density of 6.7g/cm^3, an ultimate tensile strength of 50,000 psi (340Mpa) and give an estimated cost savings of 80% over other forming methods such as machining. Secondary operations are limited to tapping and burnishing.

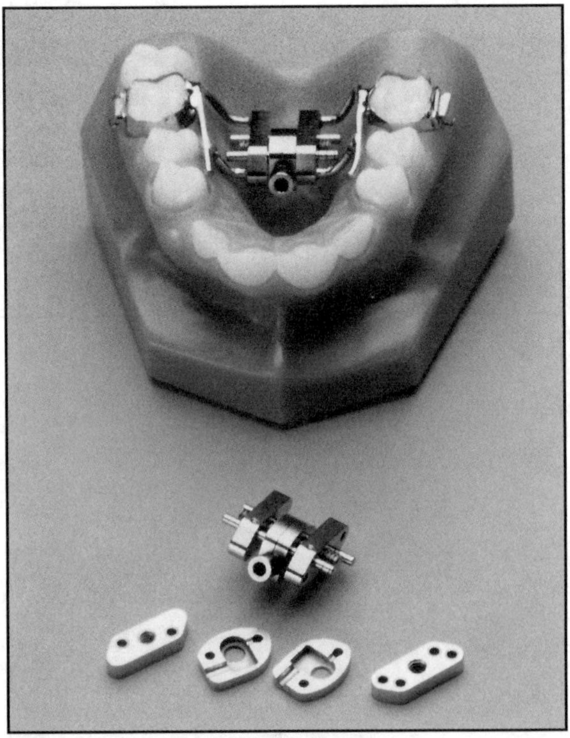

Figure 13.11. Palate expander housings and plates.

Motor coupling and adaptor

A 316 stainless steel (MPIF material SS-316N1-25) motor coupling and adaptor used in a high capacity submersible pump won an award of distinction in the stainless steel category. The 2 mating parts have a density of 6.6 g/cm^3, an ultimate tensile strength of 41,000psi (282N/mm^2) and a minimum hardness of 70HRB.

The mating fit of the 2 parts is critical for the moving action of the pump. The motor coupling hex I.D. is critical and has to be concentric to the involute spline within 0.003in (0.076mm) Total Indicator Runout (T.I.R.). The adaptor coupling has an internal and external spline, and both must be held to close tolerance.

This product was originally machined from 316 stainless bar stock and included cutting, turning, hobbing, and broaching. P/M eliminated these operations and provided a 30% cost reduction. The only secondaries are sizing and vibratory finishing.

Figure 13.12. Motor coupling and adaptor.

Golf club inserts

Copper-nickel-tin-tungsten carbide golf club face inserts won a grand prize in the nonferrous category..

The inserts are used in the new Checkmate line of golf irons (1-9 irons, plus pitching, sand and utility wedges). The high-friction inserts were first introduced in 1993 and conform to stringent U.S. Golf Association guidelines.

Made to a density range of 8.6-8.8g/cm^3, the inserts have a uniform dispersion of carbide particles. With a hardness of 85HRB, the P/M insert improves the gripping action of the club face, providing improved ball control. Carbite claims the extra friction puts more backspin on a golf ball, enhancing trajectory and controlling landing. The insert groove width, depth and radii are extremely critical.

Figure 13.13. Golf club inserts.

Planetary gear drive assembly

A compound epicyclic planetary gear drive containing eight P/M nickel-steel parts won, a grand prize in the ferrous category. Used in an automatic pedestrian commercial door.

The P/M components are: fixed ring gear; gear and insert; compound plant gear, three per unit; moving ring gear; and a planet gear cage, two per unit. The gear drive system was designed to function in a sliding door opener operating in supermarkets, hotels and stores.

P/M provided a more compact operating unit, requiring 50% less space. It also offered strength and tolerance advantages that exceed national performance codes for operation and safety, including manual operation and barrier-free entrances for the physically disabled.

The gears are made from MPIF material FN-0208-80HT to a minimum density of 6.7g/cm^3. The planet gear cage is made from MPIF material FN-0208-30.

A new P/M design, the gear tooth form, profile and tolerances are critical to avoid tolerance "stack up" and gear interference difficulties. The gears are quenched and tempered. The gear cages do not require heat treatment.

Figure 13.14. Planetary gear drive assembly. Foreground: compound planet gear (three), gear & insert and planet gear cage (two). Background: fixed ring gear and moving ring gear containing other parts.

Brass yokes and sprinkler mounting cup assembly

A brass yoke (MPIF material CZP-3502-20) used in a flush-mounted commercial sprinkler system assembly won the grand prize in the nonferrous category.

Fabricated to a minimum density of 8.0g/cm^3, the part has a minimum yield strength of 20,000psi (138N/mm^2) and a typical tensile strength of 40,000psi (276N/mm^2). The application adheres to U.S. and Canadian Underwriters Laboratories' specifications for temperature ratings and component finish. "The radius and space of the part actually contributes to the circular water spray, which is critical according to fire protection standards," reported Larry Adams, ASCO executive vice president.

The unit operates when heat from a fire reaches 135°F (57°C) which causes the solder on the cover to melt, releasing it to the floor. This allow the deflector arms guided by the P/M yoke to drop into their working position. When the temperature reaches 155°F (68°C) the liquid filled glass bulb breaks and allows up to 175psi of water to flow. The water first strikes the P/M yoke and then the deflector, creating a hemispherical water distribution pattern.

The yoke assembly is put through intensive corrosion tests such as salt spray, ammonia, carbon dioxide and sulfur dioxide. In addition, the part must withstand a 200psi water flow test for an extended period of time to rule out the possibility of failure due to metal fatigue. Yoke assemblies are also tested extensively in actual fire conditions.

Dimensions are the part are critical. The 1.313 inch (33.350mm) center line hole to hole dimension must be held to +/- 0.003 inches (+/-0.0761mm). The 0.040/0.070 inch (1.016mm/1.78mm) radii must blend very smoothly with the 0.025/0.050 inch (0.635mm/1.27mm) radii on two surfaces and create the smooth dense surface for the water to deflect properly.

Secondary operations include coining, resin impregnation and polish-deburring. The "V" knurl on the flat side of the yoke and the radii are formed during coining, eliminating costly broaching and hand polishing operations.

Figure 13.15. Brass yokes and sprinkler mounting cup assembly.

Hot forged inner race

A hot forged P/M steel inner race used in a one-way roller clutch assembly for an automatic transmission, received the award of distinction in the fully dense category.

The transmission went into the '91 model year full size front-wheel drive GM passenger cars.

This high strength P/M hot forged part (7.82 g/cm^3) has a minimum ultimate tensile strength of 290,000 psi (2000 N/mm^2) and a surface hardness range of Rockwell C 58-63. Secondary operations on this part include face grinding, I.D./O.D. turning, drilling, induction hardening and centerless O.D. grinding.

Figure 13.16. Hot forged inner race.

Fire door bolt bar

The 1990 Ferrous grand prize was awarded for a P/M copper-infiltrated steel (material FX-2008-60) bolt bar used in a high integrity, "fail-safe" fire door. The complex P/M part was designed to eliminate a stamped part which required a staking operation. Produced to meet Underwriters Laboratories' fire door specifications, the part has a nominal tensile strength of 85,000 psi (586 N/mm^2) and a minimum yield strength of 60,000 psi (413 N/mm^2). Density range is 7.2-7.6g/cm^3.

The levels on this part range from 0.070 inches (1.778 mm) thick on the tailpiece to a maximum thickness of 0.714 inches (18.136 mm); a ratio of more than 10 to 1 with the tailpiece near the middle of the part. Secondary operations are limited to deburring and nickel plating.

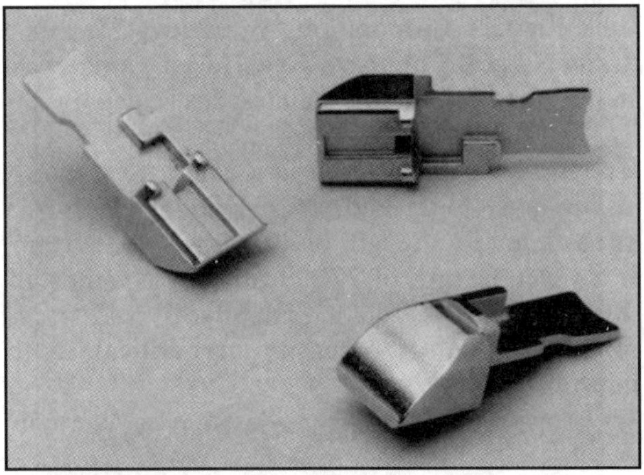

Figure 13.17. Fire door bolt bar.

Hunting knife bolsters

Nickel silver bolsters (material CZN-1818-17) used in a new West German push-button hunting knife won the grand prize in the nonferrous category.

An original P/M design, the four intricate parts are made to a minimum yield strength of 17,000 psi (117 N/mm^2). The parts have an ultimate tensile strength of 34,000 psi (234 N/mm^2) and an elongation of 11%. Density is 7.8 g/cm^3.

Post-to-post tolerances are critical along with internal tolerances. The four P/M bolsters must adhere to tolerance standards required for assembly onto the knife body. The parts are drilled and tapped for assembly.

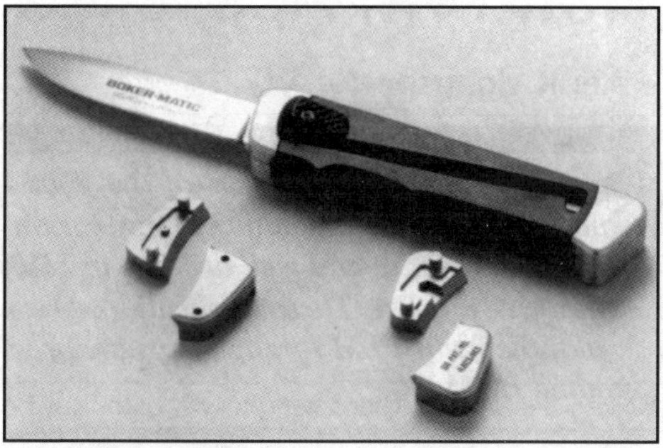

Figure 13.18. Hunting knife bolsters.

Stainless steel latch bolt head

A stainless latch bolt head (material SS-316L) won the grand prize in the stainless steel category.

The five-level part is used in a commercial mortise lock for hotel room and office building doors. The part has a tensile strength of 40,000 psi (283 N/mm^2) and a yield strength of 20,000 psi (138 N/mm^2) and a Rockwell B59 hardness. Density is 6.45 g/cm^3. Tooling and uniform density are critical in the production process.

Investment casting was considered as an alternative production process, but would have required substantial secondary machining and scrap losses. Powder metallurgy offers sufficient strength and an aesthetic finish without the need for plating.

Figure 13.19. Stainless steel latch bolt head.

P/M copper heat sinks

P/M copper heat sinks won the nonferrous award of distinction. The net-shape parts are used in a cellular phone base station antenna system. They have a minimum density of 8.2g/cm^3, a tensile strength of 193MPa (28,000psi) and a 40% elongation. The heat sinks are gold plated and assembled with a silicon chip package. No secondary machining is performed. P/M offered a 40% cost savings compared to other manufacturing techniques such as machining.

Figure 13.20. P/M copper heat sinks.

Pump body and cavity plates

A pump body and cavity plates won an award of distinction in the injection molded category. The complex 316L parts provided improved reliability and simplified assembly over previously machined pumps. Using metal injection molding minimized fluid sealing interfaces down to one O-ring seal for the basic pump. Molding flow passages of any cross sectional profile allowed maximizing the flow area within a confined space, minimizing outlet and inlet flow velocities and pressure losses in the body, reducing overall pump dimensions. MIM provided a surface finish of 16-32 RMS (microinches).

Produced to a density of 7.85g/cm^3, the parts have a yield strength of 200MPa (29,000psi), an ultimate tensile strength of 55OMPa (80,000psi), 70 HRB hardness and an elongation of 50%. Cost savings for the MIM pump body over the machined part exceeds 35%.

Figure 13.21. Pump body and cavity plates.

High Strength Helical Gears

P/M steel balancer gears with a 36.25° helix angle were awarded a ferrous grand prize. The complex parts, a balancer drive gear and driven gear, operate up to 13,000rpm in a Chrysler 2.4L engine, twice the crankshaft speed.

The gears (AGMA Class 8/9 levels) are selectively densified on the flanks to 7.8g/cm^3. The core regions do not experience high stresses near the tooth surface and remain at 7.0g/cm^3. The parts are vacuum carburized for an accurately controlled case on the critical tooth flank surfaces. Gears are hardened to 70HRA and the case is controlled to HV500 minimum at 0.200mm (0.0078in) depth.

The gears are compacted on multi-axis closed-loop hydraulic CNC presses featuring tooling that allows both the die and punches under CNC control to compact and eject a part The parts are sintered at 1280°C (2336°F).

Mechanical properties include a minimum ultimate tensile strength of 862Mpa (125,000psi) and a minimum yield strength of 827Mpa (120,000psi).

P/M replaced ductile cast iron at significant cost savings. More than 2 million gears have been made.

Figure 13.22. P/M steel balancer gears: foreground; drive gear; background, driven gear.

Aluminum cam caps

An award of distinction in the nonferrous category was given for P/M aluminum cam caps. Four thrust caps and 16 standard caps are used in the 2.7 liter dual-overhead cam engine for mounting the bearings to the cylinder head in the aluminum engine block. Chrysler mounts the caps on the engine block and then line bores them in one operation.

The caps are produced to a minimum density of 2.5g/cm^3 and have a yield strength of 26,500psi (183MPa), a tensile strength of 31,500psi (217MPa) and a 55HRE hardness. P/M parts provided a 35% cost savings over the heavily machined die cast parts they replaced. The use of aluminum allows the camshaft to rotate at higher speeds in direct contact with the cam cap bearings, eliminating the need for needle bearings.

2.7L is the base engine for the LH line of cars (Intrepid/Concorde). The P/M caps were designed for use in the 4.7L and 3.7L engines.

Figure 13.23. Aluminum cam caps.

Backhoe coupler assembly

An award of distinction was given in the ferrous category for a coupler assembly (hub and adapter) used in Caterpillar backhoes made from MPIF material FLC-4608-80HT. The mating parts transfer power from the diesel engine to a high pressure hydraulic pump which powers the backhoe loader as it digs ditches and trenches.

The adapter is formed to a minimum density of 6.9g/cm^3 and has an ultimate tensile strength of 62OMPa (90,000psi). Hardening (28HRC) takes place during the sintering cooling cycle, a technique called sinter-hardening. The critical dimension over wires on the larger external spline (48 teeth) is 107.lmm/106.8mm (4.2173/4.2073in). The dimension between wires on the internal spline (14 teeth) is 23.2mm (0.9136in).

The hub, formed in a 750st compacting press, is made from MPIF material FL-4405-100HT to a density of 6.7g/cm^3 and has an ultimate tensile strength of 758MPA (110,000 psi). It is machined and heat treated.

Figure 13.24. Backhoe coupler assembly: foreground, adaptor; background, hub.

CHAPTER 14

A Brief History of P/M, Mainly in North America

In 1994, MPIF published "Fifty Years of Service to Powder Metallurgy, 1944 to 1994". In it, their first Executive Director, Kempton "Kemp" Roll, explained how MPIF came about and how it grew into the P/M industry's trade organization. We commend it to all of our readers. In 1998, the ASM Handbook Vol. 7 lead off with an excellent historical review of P/M by MPIF Executive Director, Don White, (a.k.a. Charles White!). We also want to give you advance notice that Peter Johnson, MPIF's Director of Public Relations and Government Affairs, is hard at work on a book that will be a really authoratative history of P/M in the USA. We hope this little Chapter will whet your appetite. But this will be a different kind of History, we will not discuss the Pillar of Delphi; everybody else does that. We would rather focus on the genealogy of P/M. or, everyone came from somewhere! And especially true when it came to setting up a business to make P/M parts. But first, some old stuff.

In Chapter 5 we explained how the ancient Egyptians made tools and weapons by an old (very old) version of powder forging. Fast forward to the time period of 1755 to 1781, as recounted by Don White in Volume 7, ASM Handbook. Platinum, was very desirable for A. __ Jewelry, B__ Chemical vessels. If you chose A, you were wrong. Platinum resists oxidation and acids and doesn't melt until 3217°F; something every chemist needed to hold and heat the compounds which he wanted to react. The melting point of platinum was too high to allow it to be made as a little casting by melting it in a charcoal fire, heated by a bellows. Lewis, and then Achard, found that if the platinum were alloyed with lead or arsenic, the melting point of the alloy was much lower (1100°F). This hot mass could be heated and hammered to drive out (volatilize) the lead or arsenic. This version of getting the lead out was not a very healthful process! About 1813, Wollaston developed a process that is reminiscent of P/M today. He created Pt powder by first precipitating ammonium platinum chloride from a water solution. The precipitate was then heated to drive off the ammonium ion and the chlorine, This left a sponge of pure platinum which was carefully ground into platinum powder. He separated out the fine powder by wet sedimentation. The wet particles were pressed into a shape, using a horizontal toggle press of his own invention. The wet cakes were dried slowly and then heated to about 1800°F, at which temperature they were hot hammered to densify them. The platinum was next rolled into thin sheets, to be bent and shaped to make the chemical crucibles. All of this technology came to an end in 1859 when Sainte Claire Deville and Debray made a hot enough gas flame to directly melt and cast platinum. (sic transit omnia gloria mundi, or Peter Lynch never did like technology stocks!)

In 1830, Osann made copper powder by first making a powdery precipitate of copper carbonate. That was then reduced to copper powder. The finest of the copper powder was formed into coins by compression in a die, under the action of hammer blows or in a knuckle press. The green coins were sealed in copper sheet and clay, and sintered at about 1900°F. We do not know about the furnace, but it probably was a charcoal fire. The coins shrank 20% from die size and were harder than the pure copper of that day. Osnann noted how the shrinkage was affected by temperature and was considering using the compacted copper as sort of a temperature gage: after the sintering, measure the shrinkage and go back to the graph of shrinkage vs temperature and pick off the actual temperature in the furnace. (Students are welcome to offer their opinions on this idea!) By 1870 Gwynn had a patent on making bearings from 99% tin powder. The tin powder was made by filing. It was mixed with 1% of petroleum still residue and stirred while hot. To make a solid form, the mixture was pressed at high pressure. The object so formed was used *without further sintering,* and was supposed to be self-lubricating.

Once Edison had worked out the methods to generate electricity, it was necessary to extract the electric current from the rotating dynamos, and at the other end of the line, it was necessary to get the electric current back into a rotating electric motor. The contact point in each device is a stationary brush which contacts and rubs against the rotating commutator. Prior to 1902, there were two kinds of brushes: ones made from carbon and ones made from metal. The carbon brushes were made as mixtures of graphite and a coked adhesive. They did not wear out the commutator, and they generally did not spark. But at high electrical load, they did spark, and more importantly, they threw off carbon that had to be removed on a daily basis. The sparking was thought to be caused by air gaps (pores) and non-conducting coke. On July 31st, 1902, Victor Lowendahl of Stockholm Sweden applied for British patent 17,002. By galvanic action, he coated graphite powder with a thin layer of copper . This was done by immersing the graphite powder in aqueous copper sulphate solution, and then dropping in a little (5%) zinc powder. On the electromotive scale, the zinc displaces the copper and the copper lands all over the graphite, as a thin layer. The copper coated graphite is compacted at room temperature, preferably in vacuum, into the brush shape. The brush is used as pressed, *no sintering* is needed. The copper forms a continuous electrically conducting network in the brush. There are no sparks and no graphite dust being generated. He also claims that the brushes are pore-free. You can also use the galvanic method to put a layer of tin on the copper. After, or during, the pressing of this powder, heat can be applied to fuse or solder the tin and copper, resulting in even greater strength. As an aside, Victor mentions that other folks have been making brushes by mixing graphite and metal powder, or by making mixtures of metal powders and then pressing the mixtures. These other brushes are not as strong as his, and they wear out commutators about as fast as metallic brushes. So much for the competition!

Well, we haven't heard the last of Victor L. In 1909 he filed application for US Patent 1,051, 814, finally issued January 28, 1913. It begins I, Victor Lowendahl, a subject of the King of Sweden ... He then refers back to a 1902 patent and says, in effect, and by the way, graphite wears out after a while and the surface of the brush is then largely metallic! What is needed is a lubricant that won't disappear the way the graphite does. He does this by taking his tin coated copper and mixes it with 10% ammonium nitrate. The mixture is pressed in a hydraulic press into the desired shape or block. He seems to believe that the pressure is high enough that there are no pores in the compact. The compact is then heated to 1562°F so that "the tin shall solder the copper without melting the whole". (Not a bad guess. Nowadays, we sinter bronze at about 1550°F) On the way up in temperature, the ammonium nitrate volatilizes, leaving a network of pores behind. He does not describe his furnace, but the work must somehow be protected from air. Perhaps he keeps it immersed in a bed of burning charcoal. The sintered blocks are next placed in a bath of melted paraffin for two hours, to fill the pores by capillary action. The patent also mentions mineral oil or oil plus added fine material, for impregnation. Lowendahl wanders away from the idea of just using the material as an electrical brush and talks about using his blocks in shaft boxes, pressing them against the shaft. That sounds an awful lot like using them as a self lubricating bearing! Finally he foresees modern P/M materials that are nearly pore-free: "These properties may, moreover, by pressing, hammering and so on be increased to nearly the same amount that they have in the compact metal, while the last-mentioned property (porosity) by the same means, and also by the mechanical finishing of the surface, may be reduced at will." We assume that he uses the term "compact metal" to mean the metal in its pore-free state.

On July 3, 1914, Emery Gilson of Schenectady, NY filed an application for Bearing Material Suitable for Internal-Combustion Engines, finally issued March 28, 1916, as 1,177,407. He was patenting a self lubricating material to be used as the bearing to surround and guide a valve in an internal combustion engine. His words say it best of all: "I preferably use an alloy of copper, tin and lead, and to this end I may prepare the material from black copper oxide, tin oxide and lead oxide. Suitable proportions of these materials, together with the graphite , are first finely pulverized and thoroughly mixed. For this purpose they can conveniently be pulverized in a ball mill. This mixture is placed in a suitable retort consisting for example, of copper, and is heated for about three hours to a temperature of about 600°C (1128°F), or somewhat higher. At the end of this time, a part of the metals will have been reduced by the graphite to a metallic state, but the mixture should contain small amounts of unreduced oxides.

Upon cooling, the mixture is put through a one hundred mesh sieve and is compressed into the desired form, The compressed articles are then packed in charcoal so as to exclude air, and are again heated to a temperature of about 600°C, (1128°F), or somewhat higher, until the reduction of the metals is substantially complete and alloyage between the copper, tin, and lead has taken place. As the metal articles when fired, preserve their desired form and are porous, the metal may be said to be sintered, that is, the higher melting components may be cemented by the components of lower melting point, as distinguished from complete fusion at a higher temperature. To obtain these results, it is advisable to regulate the heat and the duration of heating so as to reach the above temperature at the end of eighteen hours, the heating being continued for sixty hours or longer. At the present time I prefer to use such a mixture that the final product will consist of approximately 73% Cu, 13% Sn, 10% Pb, and 4% graphite.

Material made by the above process is more or less porous and is capable of being impregnated with some other material, and I may take advantage of this fact in forming my bearing and impregnate it with a suitable substance for improving the lubricating qualities of the material. I find that the material will retain as high as 5% by weight of lubricating oil ..."

We have no way of knowing if this is related to Mr. Gilson's work, but in 1941, H.E. "Bud" Hall of Metals Disintegrating related that in 1921, General Electric Co. worked out a way to make "Genelite" bearings. These were made by pressing mixtures of copper oxide, tin oxide, and graphite (a reducing agent). During sintering, the graphite reduced the oxides to Cu and Sn metal. Tin diffused into copper, and created a bronze bearing with some excess flakes of graphite, for extra lubrication. There evidently were production problems, and this material was never commercialized on the scale needed by the automotive industry.

We are grateful to ACuPowder International for its advertisement in the 1994 MPIF commemorative history noted above. They discuss their founder Prof. Joel Hall and the early history of their original company, Metals Disintegrating, located in New Jersey. During World War I (1917-1921) Professor Hall helped set up an aluminum powder plant for Army ordnance as well as a plant to make zinc powder. After the war, in 1921, the people at Metals Disintegrating worked with Harry Williams of the General Motors Research Center. MD supplied copper and tin powders and Mr. Williams developed and patented self lubricated bronze bearings. To us, this seems like a logical extension of Victor Lowendahl's 1913 patent, Mr. Gilson's 1916 patent, and the "Genelite" described above. The new idea in 1921 was to make mixtures of elemental powders, rather than to galvanically coat the powders or create your own metal powders by reduction of oxides. Harry Williams filed his patent application on August 17, 1922, and it was issued September 27, 1927 as number 1,642,347, with the title "Bearing". The author claims," The process which comprises intimately mixing finely divided copper, tin, and salicylic acid; compressing the mixture into the desired form under a pressure of about 80,000 psi and heating the resultant formed article to about 675°C, (1247°F) in a non-oxidizing atmosphere." Satisfactory bushings have absorbed up to 17% by volume of oil. Oildag was also used for impregnation. We believe the salicylic acid (aspirin!) was added with the idea that it would volatilize on heating, and not hurt the metal. It was a pore-forming agent. The heating time was 3 hrs. to reach 675°C and then the parts spent 3-6 hrs. at temperature. He does not say how one achieves a non-oxidizing atmosphere, but this is covered in his next patent. In his third claim of the present patent, he covered all of powder metallurgy, "The process of producing a porous body which comprises, compressing a mass of finely divided, metallic, alloyable ingredients mingled with a finely divided, completely volatile organic substance, then heating the mass sufficiently to cause the metallic ingredients to alloy with each other and also to effect removal of said organic volatile substance." Had that not eventually expired, we would not be writing this history!

Williams applied for his next patent on January 19, 1923 and it was issued September 13, 1927 as patent number 1,642,348, "Alloy Structure". He advocates the use of ammonium chloride as a volatile flux in the amount of .5-1.0 % by weight, added to the metal powders. We also learn that the closed container used for sintering was packed with charcoal. The charcoal would react with any air sneaking into the container and keep the copper from oxidizing. (You may remember that ancient Egyptians, Chapter 5, also used charcoal fires, but they were reducing iron ore to sponge iron.) Williams goes on to say that the ammonium chloride need not actually be mixed into the powder. If you have a false perforated bottom under your sintering can, the ammonium chloride can be placed below. It then breaks down on heating and provides its fluxing action. Part of the fluxing mechanism is the liberation of hydrogen and possibly HCl gas. The hydrogen would certainly help the sintering! This sintering atmosphere was used well before the advent of exothermic generators .

You may be wondering if we intend to review each and every P/M patent through the beginning of the Third Millennium. The answer is "No.", but how about if we just give one more example. Carl Claus of Plainfield New Jersey applied for a patent on May 28, 1926 and received his patent on November 8, 1927, number 1,648, 722, "Bearing Material and Method of Making the Same". His mix was 88% Cu, 7-14% Sn 2-3% Zn and 2-6% graphite. He also uses a mix without graphite. The initial compaction is double ended. The compressed bearings are placed in a closed container and sintered for 10-20 min. at 1400°F to 1500°F, depending on the size of parts and the character of the container. The bearings are next oiled, and finally are sized for dimensions. The key advantage of this patented method is the short sintering time vs Williams 3-6 hours. How did Mr. Claus avoid oxidizing the bearings? If the container is close fitting, perhaps with sand sealed edges, then maybe the zinc vapors protect the work. Maybe he threw in a handful of ammonium chloride? Maybe the container was a graphite box? Does anybody, today, really know?

Fast forward to 1940-1942. In that year, Prof. John Wulff of M.I.T. wrote several papers on pressed and sintered iron, and edited the American Society for Metals book: "Powder Metallurgy". It was a terrific collection of easily understood papers that document P/M at that point in time, covering all materials, as well as methods of manufacturing powders, and methods and equipment for compaction and sintering. It fills in the gap from the beginnings in the '20's to the then state of the art, based on the 51 papers presented at the M.I.T. Conferences on Powder Metallurgy, 1940 and 1941. We often read that iron based P/M in the U.S. did not begin until after World war II, in 1946. It is suggested that iron based P/M began after Prof. Comstock returned from Germany about 1946 and explained how the Germans had made sintered iron rotating bands during the war. Well, on p. 297 of Prof. Wulff's book, Harlan Webber of General Electric shows a picture of a continuous roller hearth, atmosphere furnace. Through it were passing wire trays loaded with iron bearings! That's a long way from the General Motors covered cans!

In our enthusiasm to make the point about the iron bearings, we left the "fast forward" button on too long. From 1920 to 1940, much progress was made in powders, presses, sintering furnaces, and atmospheres. Starting with presses, Lawrence Bailey from Stokes Machine Co. explained that in 1920, people were secretly using high speed single action presses to compact metal powders. Prior to 1920, metallic Cu-Pb brushes for automotive starting motors were pressed in large blocks on hydraulic presses. After sintering, they were machined into the small brushes and the pigtails were inserted. (The pigtails are the electrical wires that carry current to the brushes.) In 1920, Stokes was asked to produce a press that could compact the brush powders to size, with the pigtails being later inserted. The first automatic press could make 40-50 brushes per minute. Rotary presses had long been used to make pharmaceutical tablets, explosive pellets, moth balls and laundry tablets. Special presses could create a hole through the tablet, e.g. in making Life-Savers. These presses were used to make brushes at 200 pieces/min. A later version called a double rotary press would make 500 small cup shaped parts/min. for Ford Motor Co. shock absorbers. Each shock absorber contained 40 of these small cups.

Subsequent presses for compacting bearings one at a time had powder fill up to 6 in., movable core rods (Claus patent in 1926) and double ended compaction. These features were needed to mold bearings with uniform density, as opposed to just pressing them from one end. In 1928 E. J. Hall of Metals Disintegrating obtained patent 1,728,273 for a process to mold the wire pigtail into the brush, at the compaction stage. It then stayed in place during the sintering. By 1940, Richard Seelig of Powder Metallurgy Inc. was making many complicated structural part shapes. He would use two lower punches so as to adjust the fill for two different levels, to get uniform density. He despaired of ever learning the rules that governed the powder flow, off at an angle, from one level to another. (A problem that is still with us!)

Roland Koehring of General Motors Moraine Products Div. explained that in 1921, bearings molded from Cu, Sn, and graphite were packed in lampblack and heated in air overnight at 1400°F in a box furnace. They were free of oxide but the process was not commercially viable. Other possible atmospheres in 1921 were pure hydrogen, (produced electrolytically), raw natural gas, ammonium chloride vapor, and carbon dioxide made by decomposition of limestone. None of these were commercially acceptable. In 1922, they began to use alloy metal boxes with tight fitting covers. Inside the boxes, the work was covered with a carburizing compound. They were sintered overnight at 1400°F. This method was used for about six years. In 1928, they began to send their covered pans through a continuous gas fired, air atmosphere furnace. The parts formed their own atmosphere in the cans, in part because they contained salicylic acid as a pore former.

In 1933 Koehring began work on a furnace whose atmosphere was based on flue gas. The flue gas was created by burning natural gas in a closed container with so little air that the flame was green. This gas could then be admitted into a furnace with a cast muffle and served to protect exposed parts that were not inside alloy cans or boxes. The parts were on trays that could be pushed through the furnace. By 1940, this burning of natural gas had evolved into our exothermic generator. The gas from it could be dried to 40°F dewpoint and used to sinter pure iron. (It would not have effectively protected any carbon in the iron.) This exothermic gas had the following composition: 4% CO_2, 11% CO, 15% H_2, 2% CH_4, 63% N_2, and 3% H_2O. In 1940 dollars, the costs for atmosphere gases were :

Exothermic gas	$.05 to $.40/1000cf
Dissociated ammonia	$1.50 to $4.00/1000cf
Hydrogen	$10.00/1000cf

In current dollars, noting that in 1941 the consumer price index was 13.9 and that in 2001 it is 175.1. Using a straight ratio factor of 12.6X , this would be:

Exothermic gas	$.62 to $5.03/1000cf
Dissociated ammonia	$18.89 to $50.30/1000cf
Hydrogen	$125.97/1000cf

That assumes that these commercial goods went up in price at the same rate as consumer goods. Actually, we can purchase these gases in 2001 for:

Exothermic gas	$1.40/1000cf
Dissociated ammonia	$5.00 to $7.00/1000cf
Hydrogen	$5.00 to $30.00/1000cf
(On site generation vs cylinders)	
Nitrogen	$2.50 to $5.00/1000cf
(On site generation vs liquid)	
Endothermic gas	$1.80 to $2.00/1000cf
(large generator)	

Sintering furnaces use about 1000 to 2000 cu. ft./hr. for protective atmosphere.

(If you are wondering if the 12.6X increase in the CPI from 1941 to 2001 is erroneous, this author can testify that as a lad in 1944, the ice cream cones at Jaeger's corner store were $.05 each. Can you buy ice cream cones where you live for less than $.05 x 12.6 = $.63? Not!)

We began this 1940 update with Harlan Webber's roller hearth furnace. Let's conclude the furnace discussion with the little jewel shown in Figure 14.1. P/M parts entered at one end and were continuously pushed up hill. (We wonder if it was taking advantage of gravity segregation of hydrogen where it floats to the high point in the furnace, and the heavier impurity gases sink down and back toward the entrance door.) At the top of the hill the hot parts exited through a flap door. They were then grabbed with tongs and transferred while hot to a sizing press! Did he build just one of these furnaces? Who, in 1941, was doing hot repressing of P/M parts?

In 1940, H. E. Hall reported that ferrous parts had been made for the last year or two. This included substituting iron for copper in bearings. The growth momentum of sintered iron parts came to halt in 1940 when the blockade in Europe cut off the supply of the cheap sponge iron powder from Sweden (Höganäs). Folks in the U.S. were scrambling to find substitutes. One of these may have been hydrogen reduced mill scale from Pyron. (In our eternal search for the meaning of the phrase "What goes around, comes around", we think that Höganäs' 2000 purchase of Pyron may be a good example!) Prof. Fritz Lenel, then an employee of Moraine Products Div. of GMC, reported on the manufacture of the pressed and sintered iron oil pump gears. At that point in 1940, all of the GM oil pump gears were being made by P/M. He indicated that other

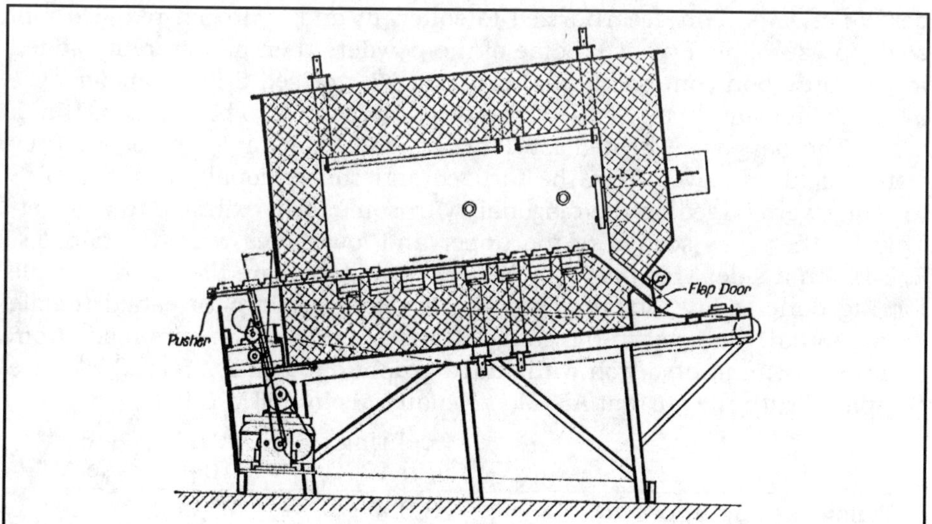

Figure 14.1. View of special sintering furnace circa 1941. Parts enter at one end and move up hill. They exit hot and moved to a sizing press for hot repressing. This seems like a very early version of hot powder metal forging and its actual purpose is unknown today. The photo is from the Wulff text published by ASM in 1942.

P/M iron parts had been made, in quantities up to one million. These parts were pressed at only 20 to 30 tsi, to cut down on die wear. Die life for tool steel dies was often about 10,000 pieces. They were sintered in continuous furnaces, belts or roller hearths, with combusted gas atmosphere. Shrinkage was about .5%, resulting in a piece to piece dimensional variation of 20% of that, or .1%. Parts were sintered for 20 to 40 minutes at 2000°F. At the end of the furnace they were oil impregnated. The tooth form on the oil pump gear was not touched, but they burnished the ID, turned the OD and finished the end surfaces. This part was previously made from a simple cast iron round slug with only 36% material utilization. The P/M part had a tensile strength of 20 to 25,000 psi and a hardness of HRB 30. Prof. Lenel did a second paper at the conference, and it was on steam blackening of P/M parts. He showed what we still know: steam blackening helps hardness, wear resistance and compressive yield strength. Corrosion resistance of steam blackened parts is not wonderful, in a salt spray test, but it is about twice as effective as no coating at all on the iron. As examples of wear resistance he cited iron P/M guides for a metal cutting saw, rollers for supporting drawers in ovens and refrigerators, and the vanes in a diesel engine fuel pump.

Finally, a bit of early history of P/M forging of sintered preforms. By 1942, Roland Koehring of Moraine Products, GMC, tested the forged properties of three forms of iron powder. These included 1/8 in. steel chips made from 1112 steel that had been ground to -100 mesh, Fe-C steel powder which was then decarburized to soften it, and hydrogen reduced mill scale (presumably Pyrons'). Some of the powders then had graphite added to give a carbon containing final product. He pressed 5 in. diameter by 1 in. thick slugs at 30 tsi. These green discs were cut into bars 1 in. x 1 in. x 5 in.. The bars were heated for 1 hr. at 2000°F in tank hydrogen. They were quickly removed from the furnace and conventionally forged to 7/8 in. They were forged on the diagonal, with some flash exiting into a gutter. That is, the cross section of the upper and lower dies were 45 triangles, 7/8 in. on a side. This made for easy removal, compared to the closed die forging done today. These pieces were machined into threaded tensile bars, .450 in. diameter in the gauge. The properties of the bars made from their own comminuted iron with added graphite to equal .25 to .35%C are compared with the current ASTM P/F materials in ASTM B 848:

	Koehring AISI 1030	ASTM B 848 1040, grade A
Tensile strength, psi	73,000	75,000
.2% Yield Strength, psi	51,200	45,000
% elongation in 2 or 1 in	23	27
Hardness HRB	74	80
Density, g/cm^3	7.81	7.80

This powder forging was technical triumph that did not see commercial reality until the powder forging work of the early 1980's.

One last example of P/M pressed and sintered iron parts from the early 40's is shown in Figure 14.2. It is a bullet given to the author by Bob Marshall, retired from Chrysler Amplex. The story was that he and Andy Langhammer, a manager at Chrysler, worked on this during World War II. The metallurgist part of the author wants to cut it up and check structure and density uniformity. The historian part says "Stifle yourself!"

With regard to sintering bronze, Arnold Kelley, now retired from The Wakefield Corporation, remembers that in 1946, the Neveroil Bearing Co. of Wakefield, MA was sintering bronze bar stock in oil-fired bee hive ovens. The bronze was placed in a steel box and packed in charcoal. The oven was heated by what amounts to a home style oil burner, with flames and combustion products entering the oven and surrounding the boxes.

Norbert Arnold of St. Marys, PA was an employee of Keystone Carbon Co. in St Marys, PA from 1939 until his retirement in 1985. In 1989, he published a brief history of Powder Metallurgy in Elk County, PA which includes St. Marys and other nearby towns. This is the heartland of P/M in the U.S. with 45 P/M parts companies located within a radius of 50 miles. What is really fascinating is the way that the suppliers to the P/M parts producing industry sprang up in this lovely rural area. When someone wants to start a new P/M business in that area, he or she has close at hand businesses that sell: powder from warehouses, P/M compacting tooling, secondary machining, heat treating, steam blackening, plating, crack detection, presses and furnaces. And of course there is a skilled P/M labor pool like no other! With Norbert's permission and words, we reprint P/M history as he remembers it. (His chronology of who started what and when has also been incorporated into our own list at the end of this chapter.)

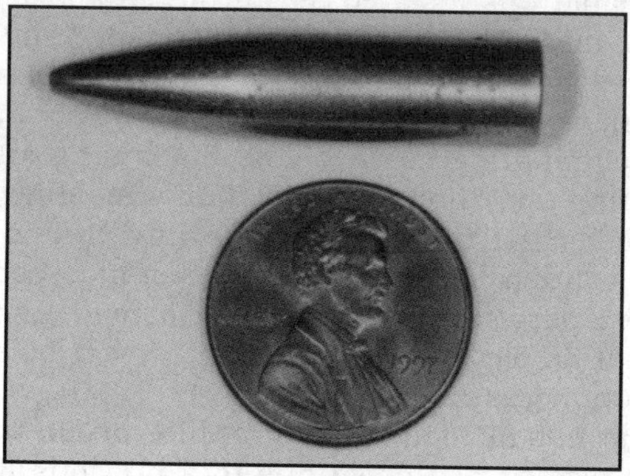

Figure 14.2. World War II P/M bullet. Courtesy of Bob Marshall, made at Chrysler Amplex Corp.

AS I REMEMBER IT
BY NORB ARNOLD

THE HISTORY OF POWDER METAL IN ST. MARYS

St. Marys has long enjoyed an involvement with powder. As early as 1900 lampblack and carbon powders were blended with pitch to make carbon rods for the Battery Industry.

Electrodes were next on the agenda and were made for the electric arc steel furnaces. The carbon brush, that indispensable part of the electric motor, either followed the production of the electrode or developed simultaneously.

The growth of the steel industry requiring more electrodes for their melting furnaces and the advent of the automotive industry requiring carbon brushes for automotive generators became the big growth factors for the two carbon industries in St. Marys: Speer Carbon and Stackpole Carbon Company.

Brushes with higher current carrying capacities were needed for rotary converters and plating generators. These brushes were primarily imported from Germany and England. Speer Carbon Company started a development to duplicate these brushes. In the early years the brushes were crudely made in as much as they could not be baked in the same manner as the plates for the carbon brushes. The first metal graphite brushes were wrapped in copper foil, and even then the outer surface oxidized during the baking operation. The oxidation was machined off in the finishing of the brush. The metal graphite brush was in reality the start of powder metallurgy in St. Marys.

Metal graphite brushes were machined from a block or a plate. After the brush form was made a hole was drilled into the brush into which abraded copper cable was inserted and firmly attached to the brush by tamping a copper mercury mix which amalgamated the metals to make the electrical connection between the brush and copper shunt or cable.

Two young men, working at Speer Carbon Company in the late 1920's, had the idea of eliminating much of the machining and especially the drilling and tamping operation, by individually molding each brush and simultaneously inserting the copper cable into the powder mix which would result in a much lower electrical resistance between the shunt and brush. This operation was to be performed under a protective atmosphere and at an elevated temperature so that no subsequent baking or sintering was necessary.

The idea of the hot molded metal graphite brush was presented to management but was turned down due to a previous financial decision to expand the electrode business by forming a new subsidiary, The

International Graphite and Electrode Corporation at Niagara Falls. This facility was completed in 1930.

The two young men, Ben Reuscher and Adolph Schmidthammer, felt so strongly that their idea was sound that, in 1927, they left Speer Carbon and formed a new company in Emporium, PA. The new industry, Keystone Carbon Company, moved to St. Marys in 1931.

Powders available at that time had poor compressibility and were very limited as was the equipment necessary for briquetting and sintering. Alfred Showie Gleixner had the task to design and build the necessary equipment which made the ideas of Ben and Adolph a production reality. As a point of interest Mr. Gleixner also is the father of the floating die table which is so common today on powder metal presses. At this time Cliff Sorg was in charge of manufacturing, while Fred Krellner had sales responsibility.

In 1932 a group from Keystone Carbon attended the World's Fair in Chicago and were exposed to the oilless bearing - a development by the Amplex Division of the Chrysler Corporation.

Since Keystone had conquered many of the techniques necessary to form the available powders of that time, it was a natural evolution that powdered metal bearings be investigated.

Keystone had captured a good share of the automotive metal graphite heater motor brush business in the middle of the 30's. Realizing that on every heater motor two bearings were also needed, gave some urgency to the development of the P/M bearing.

After many years of testing, late 1938 marks the advent of Keystone Carbon as a serious contender into the oilless powder metal bearing business and the beginning of the Powder Metal Industry as we know it today. In the early 40's Ray Hoffman came on the scene followed by William Gies. These two gentlemen augmented the early technical and production aspects which were laid down by the founders. In recognition of his exceptional abilities, Joe Krellner was made supervisor of the electrical department. Today Keystone's manufacturing and technical facilities are second to none.

The progress of powder metallurgy in St. Marys grew rapidly during the next few years, especially because of the needs of our nation during World War II.

The development of the nickel cup by Keystone Carbon Company, as a safety device for the proximity fuze for artillery shells did much to turn the tide of the war.

(In reviewing Norbert Arnold's information on the fuze, Kemp Roll recalls that Gib Hoehn who had worked at Keystone at that time said he was the one who perfected the porous nickel cup. Kemp continues: "Since I was the Bomb Disposal Officer on the USS Saratoga during the war, I was very much aware of the importance of this new weapon. It literally saved the U.S. Navy from being destroyed by Kamikazes. Me along with it!)

Keystone continued its growth, but like any household, sons leave the nest to make their mark in the world, so some of the family of Keystone left to start their own companies. And from these companies, more powder metal facilities have evolved.

Keystone Carbon Company is considered as the largest independent Powder Metal Fabricator in the World. The North Central Area of PA accounts for approximately 40% of all domestic powder metal parts production.

The following is a chronological listing of current powder metal industries in Elk County as compiled by Joan Smith.

POWDER METAL INDUSTRIES IN ELK COUNTY SEPTEMBER 1989

In 1927 Keystone Carbon Company was founded in Emporium by Ben Reuscher and Adolf Schmidthammer. In 1931 they moved to St. Marys at 1935 State Street. In 1938 first powder metal shipment was made. The Reuscher Family are the current owners.

(National Molded Metals 1941 or 1942)

In 1947 Powder Metal Products was begun at 879 Washington Road, St. Marys by James M. Schloder and Gerald Schloder. Current owners are the same.

In 1948 IPM was begun at Montmorenci Road in Ridgway by Gib Hoehn and Mario Victor. In 1968 IPM was sold to Allegheny International. The name was changed and in 1984 Clarion Sintered Metals was bought by James Delhunty, Rudy Ferragine, Howard Peterson, Ben Marzella, Dallas Kelly and private investors.

In 1956 Carbon City Products was set up at 150 Ford Road, St. Marys by Larry Geitner, Dan DeLullo, Maurice Haines, and Vincent Heindl. The current owners are Metal Powder Products, Inc., Indianapolis. An interim name change in 1978 Prometech.

1958 Elco Sintered Alloys Co. Inc., Kersey was formed. Current owner Peter Herzing.

In 1961 Imperial Sintered Metals, Inc. at 850 Washington Road, St. Marys was founded by Arthur VanSlander, Kenny Renwick, Paul Garner, George Yound, Herb Schreiber, Allie Wegemer. The current owner is Thomas VanSlander.

In October, 1967, Alpha Sintered Metals, Montmorenci Road in Ridgway was founded by Frank Cartwright and Chester Viola. The current owners are Chester Viola and Roger Hasselman.

In 1968 St. Marys Pressed Metals Inc., Main Street, Kersey, was founded by Walter Schwabenbauer and a group of St. Marys people. Presently it is Corporation owned with President Art Aiello.

In July, 1968, Exotic Metals, Inc., Boot Jack Road, Ridgway, was founded by Bob Howard and family. It was sold to Matco Corp., New York.

In 1972 Windfall Products began operations on Kallas Road, St. Marys. In 1982 the operation was moved to Industrial Park, West Creek Road, St. Marys. Windfall was founded by and is currently owned by Richard Wolfe, Arnold Wolfe, Jim Ryan and Joe Holjencin.

In 1978 Ridgway Powdered Metals, Inc., Route 120 Ridgway, began operations. The current owners are Jim Olay and Joe Schaefer.

In 1978 Fairview Sintered Metals, Inc., Fairview Road, St. Marys was founded and is currently owned by Luke Wehler, Tim Wehler, Jerry Wehler, Paul Shaffer and Tom Holjencin. It began as Fairview Machine Co., a tool and die company. In 1980 it became a Powder Metal Industry.

1978 saw Elite Metal Products, Inc. begin operations in Ridgway. Following a fire it relocated at 400 S. St. Marys St., St. Marys. Its founder and current owner: Richard A. Wendel, and others. At this site, a former powder metal industry, National Molded Metals, was founded in 1940 and owned by Hans Kerner, Walter Rigard, Ed Meier, Ray Harrigan, and Ray's father-in-law. It was sold to McCrosky Tool in 1972. Following a strike, the company moved its facility to Franklin, Tennessee, in 1982.

In 1981 Metco Industries, Inc., began on Washington Street, St. Marys, founded by and currently owner by Walter Spencer. In 1984 Metco moved to 1241 Brussells Street, St. Marys.

In 1981 Alpine Pressed Metals, Inc. began at their location on Route 120, Ridgway. Founded by Tony Vigilone, John Urmann and Joe Urmann.

In 1985 Intech Metals, Inc., PO Box 506, Route 120 East, Ridgway, was founded by David Hanlin, the current owner.

In 1987 Comtec began operations on Windfall Road, St. Marys.

In 1989 Bucktail Pressed Metals, Inc. began operations at 1283 Brussells Street, St. Marys.

Note: St. Marys Carbon Powder Metal Facility is located in Galeton. The Elk County plant is carbon and metal graphite.

1975 saw R.C. Metals on Montmorenci Road, Ridgway founded and still owned by Richard Lewis. Its purpose: Recycling powder metal.

In 1987 ARC Metals, Inc. River Road, Ridgway began operations in providing and mixing powder metal for industry.

There are many secondary operation plants in Elk County, though not powder metal producers, that are vital to the primary facilities. They are not listed because of space limitations. The above information was provided by the industries, St. Marys Chamber of Commerce, Norb Arnold, Archie DeLoge, and Larry Geitner.

Well this is now the 3rd Millennium, and P/M has moved far beyond the confines of Pennsylvania and the U.S.A. Earlier in the book we talked about the test methods and material specifications that have been developed to assist buyer and seller here in the U.S.A. But P/M sales by companies in the rest of the world add about 200% to what is produced in North America. International P/M standards are becoming important. The person who has lead the participation of the U.S.A. in international standardization of P/M materials is William Buerkel, Chairman of ASTM B09.09, the Technical Advisory Group (TAG) to the International Standards Organization (ISO) of Geneva, Switzerland. The powder metallurgy committee in ISO is TC-119. In a recent article published in the International Journal of Powder Metallurgy, Bill did a wonderful job recounting the history and progress of this participation. With his permission, we have reprinted the article for you.

INTRODUCTION

"Jacques Brel is alive and well and living in Paris", but are North American interests in P/M standards in Brussels, Geneva, New York and Philadelphia?

Why Brussels? Because this is where the headquarters of the European Union (EU) is located. In relation to the P/M industry in the USA, it is also the headquarters of the European Committee for Standardization (CEN). CEN is responsible for approving and issuing European Standards (EN) for the entire P/M industry throughout the EU. These standards are mandatory for all 18 EU member nations when issued and will replace all national standards throughout the EU.

Why is this important to the P/M industry in the USA? It is because the industry does not have any direct representation or vote in the CEN for the development and approval of EN Standards. This could lead to the creation of standards that might be unfavorable to our industry and which could possibly hamper or exclude our business opportunities and competition in the European and, possibly, even global P/M markets.

WHY GENEVA? Because this is the location of the headquarters of the International Organization for Standardization (ISO). The United States does have direct representation in ISO and can therefore influence its specifications as a Participating Member (P-member)*. ISO Technical Committee #119, Powder Metallurgy, (ISO/TC 119) is responsible for all P/M specifications developed and published by ISO.

WHY NEW YORK AND PHILADELPHIA?

Because these cities are where the American National Standards Institute (ANSI) and the American Society for Testing and Materials (ASTM), respectively are located. Both organizations are important because they are responsible for representing United States P/M interests in ISO's Technical Committee #119, Powder Metallurgy. This committee produces standards that have been accepted favorably by CEN and issued as EN's. ASTM subcommittee B-09.09, Technical Advisory Group (TAG) for ISO/TC 119, is responsible directly (through ANSI) for technical input to ISO standards from the USA. All USA and European locations are related and all must be "alive and healthy' for the P/M industry in the USA to prosper.

HISTORY AND RELATIONSHIPS:
IEC, ISO, ANSI, ASTM AND ISO/TC 119

To understand the participation of the USA P/M industry in International standards, one must have a familiarity with the history of the two primary organizations involved in International Standards - the International Electrochemical Commission (IEC) and the International Organization of Standardization (ISO)" The IEC was founded in 1906 and represents currently about 95% of the world's electrical generating capacity. It develops standards related to safety, performance, construction and installation of electrical equipment and services in the electrical and electronics fields and some areas of the telecommunications industry. The American National Standards Institute (ANSI) is the member body to IEC for the USA.

The ISO was founded in 1947 after an earlier meeting of 25 countries in London. This organization has grown to 110 members, one for each member country. It has over 30,000 experts in their fields and has published over 9,000 ISO standards, about three times the numbers published by IEC. ISO standards are similar in purpose to IEC standards but are different in scope. Not by coincidence, the scope of ISO covers all fields except those covered by the IEC. Again, there is only one member per country and ANSI is recognized as the one and only representative of the interests of the USA.

ANSI goes back to the American Standards Association (ASA) and the United States National Committee (for IEC) which was incorporated into ANSI as a standing committee within the Institute. While not a monumental item, it may be of interest to the readers and especially camera enthusiasts to know, that the film speed shown on all packages of photographic film uses an ASA designation, i.e., 100 ASA, 200 ASA, etc. ANSI membership includes approximately 300 companies, 35 government agencies and more than 260 technical, trade, labor and consumer groups. It promotes voluntary consensus standards and does not write standards per se. ANSI is a non-government supported organization that promotes voluntary consensus standards through it's members. Each organization in ANSI does, however, pay membership dues to support ANSI. In turn, ANSI pays the dues for membership of the USA in ISO and IEC.

Prior to 1967, there was no Technical Committee within ISO for the P/M industry, and standards were written by a number of different organizations in the USA. The American Society for Testing and Materials (ASTM) and the Metal Powder Industries Federation (MPIF) wrote standards for the P/M industry in general. ASTM and the Cemented Carbide Producers Association (CCPA) wrote standards for the carbide industry. ANSI worked with ASTM as an organization producing consensus standards for the P/M industry and the standards were issued under both ANSI and ASTM logos. ASTM issued, and still continues to issue, their own standards and is known widely throughout the world.

Early in 1967, ISO called a meeting of its members to form a new Technical Committee specifically for P/M Standards. It contacted ANSI, its USA member, and requested it to send representatives/technical experts in the field to a meeting (June, 1967) in Moscow, USSR, for the purpose of founding this committee. ANSI, in turn, contacted ASTM. ASTM, organized in 1898, is one of the largest standards development organizations in the world. It is a not-for-profit organization of producers, users, ultimate consumers and general interest groups (government, academia and others) who meet to produce and publish standards for products, systems and services. It was selected because of its prior relationships with ANSI and, in particular, because of its mode of operation involving a voluntary, full consensus approach to standards. Full consensus is where full consideration is given to the votes of each committee member throughout the entire standards development process. Each committee must also have balanced interest in that the number of producers on a committee cannot exceed the number of non-producers.

ASTM Committee B-09 on Metal Powders and Metal Powder Products was selected by ANSI to provide representation to the Moscow meeting and, eventually, to form a committee of technical experts. Further, it met twice a year, with administrative help from ASTM Headquarters. It includes most of the technical experts covering the entire P/M field and members of MPIF, SAE, CCPA, etc. on its committees. Athan Stousy, (Hoeganaes Corp.) was able to attend the meeting in Moscow to represent the USA by means of a company sponsored business trip to Sweden and with further monetary assistance from MPIF, to Moscow. The meeting was a success and resulted in the formation of ISO/TC 119, Powder Metallurgy (present name).

At the time it was founded, Committee ISO/TC 119 was composed of four Working Groups, which in 1970 became subcommittees (SC):

* Subcommittee 1 - Terminology Secretariat: UK
* Subcommittee 2 - Sampling and Testing Methods for, Powders (including Hard Metals) - Secretariat: USA
* Subcommittee 3 - Sampling and Testing Methods for Sintered Metal Products (Excluding Hard Metals) - Secretariat: France
* Subcommittee 4 - Sampling and Testing Methods for Hard Metals - Secretariat: USSR

Since 1967, the composition of ISO/TC 119 has changed somewhat but relatively little, as compared to overall international conditions. A new subcommittee (SC 5) was proposed in 1967 to cover material specifications but was delayed until 1974 when appropriate testing methods were available through development of the necessary standards in SC 3. France became Secretariat of SC 5.

The current subcommittee structure of ISO/TC 119 is:
Subcommittee 1 - Terminology - Secretariat: UK

Subcommittee 2 - Sampling and Testing Methods for Powders (including Hard Metals) - Secretariat: Sweden

Subcommittee 3 - Sampling and Testing Methods for Sintered Metal Products (excluding Hard Metals) - Secretariat: Germany

Subcommittee 4 - Sampling and Testing Methods for Hard Metals - Secretariat: Open

Subcommittee 5 - Specifications for Powder Metallurgical Materials (excluding hardmetals) - Secretariat - UK

The Secretariat for four of the subcommittees changed primarily due to monetary reasons.

EU AND THE EUROPEAN COMMITTEE FOR STANDARDIZATION (CEN)

Since approval of the Lisbon agreement in 1989, and the further development and approval of the Vienna agreement in 1991, a good working relationship seems to have developed between ISO/TC 119 and CEN. One reason for this is that in late 1991, Per Lindskog, President of the European Powder Metallurgy Association (EPMA) and Chair of ISO/TC 119, suggested that an ISO/TC 119 Standards be submitted to CEN for approval as EN's, (standards). In this manner, and in accord with the Vienna Agreement, standards could be sent out for acceptance via the Primary Questionnaire (PQ) procedure, thereby avoiding a duplication of technical effort by CEN.

In June of 1993, the CEN Central Secretariat in Brussels reported that of the 49 standards submitted by ISO/TC 11 9, 35 had been accepted as European Standards (EN). The standards that were not accepted by CEN were either under revision or were to be revised under the ISO "Five Year Review" procedure. The eight standards (of the 57 total ISO/TC 119 standards in place at the time) that were not submitted to CEN for approval were held back only because German language versions were not available at the time. In the latest report from CEN (1995), 37 standards are now approved, five items were added from New Work, two other standards are in the process of being voted on and one old work item was being withdrawn.

Standards are similar to living items and need to be created, reviewed, and revised to keep abreast with current needs and technology, and even cancelled if the need has disappeared. Thus, the number of completed and approved standards in place is changing continually.

As of July 15, 1997 application to CEN for the formation of a Technical Committee in CEN to "introduce completely new standards through CEN" was circulated to members by the Technical Board for voting and comments. The Technical Board considered all the votes and comments by its member nations and decided not to create a CEN/TC on Powder Metallurgy to develop new standards. Two of the comments are especially worth noting:

1. "... premise for a CEN Powder Metallurgy committee is based wholly on receiving CEN funding ..."
2. (No need) "... as ISO/TC 119, Powder Metallurgy, is a well-functioning committee, covering, to a very large degree, the working program, proposed ..."

Are United States interests in P/M standards still "alive and well in Brussels and Geneva? You bet they are!

US PARTICIPATION IN ISO/TC 119, POWDER METALLURGY

After the formation of ISO/TC 119 in Moscow in 1967, the representative from the USA reported back to MPIF and ASTM Committee B-09 and recommended that a special subcommittee be set up within the ASTM B-09 Committee. This subcommittee should work through ANSI to ISO/TC 119. Subcommittee B-09.09, Technical Advisory Group (TAG) for ISO/TC 119, Powder Metallurgy was established with Athan Stousy its Chair. The Chairmen of the B-09 Subcommittees and others were to act as delegates, advisors and assistants to handle the technical documents for ISO/TC 119 work. This group was under- staffed, over worked and, grossly under funded. The representatives from the USA were unable to hold or attend ISO subcommittee meetings in order to participate directly in the standards work. They had to rely on mail for communication which, in those days, was relatively slow and proved to be ineffective for international standards work. The optimum situation is to be present at meetings and to work and negotiate in person.

At the second plenary meeting in New York in 1970 (sponsored by MPIF during the, International Powder Metallurgy Conference), the United States declared that it wished to withdraw from being Secretariat of SC 2 on Metal Powders. The reality was that SC 2 had only one meeting to discuss SC 2 work in the three years since its formation - held the day before this second plenary meeting! The resignation was accepted and, two years later Sweden was appointed as Secretariat of SC2.

In 1972, the USA TAG was then organized by a new chairman and with delegates that were experts in their fields and who could attend ISO/TC 119 subcommittee meetings (through support of their companies, or otherwise) were selected. At the same time, a solicitation was made to fund the USA TAG with funds sufficient to pay for delegate travel costs, if necessary, for ISO meetings. The initial solicitation was to request donations from various technical organizations. While the results were disappointing, it turned out to be beneficial in the longer term. The only donor in this initial solicitation was MPIF with an annual donation of $2,000. Over the years this annual donation has been increased twice to a total of $5,000 annually in 1996. This donation, along with other monies secured through a successful ASTM subcommittee B-09.03 Round-Robin on metal powders in 1974, was sufficient to provide money for transportation only for delegates to ISO meetings.

ASTM Subcommittee B-09.03 on Refractory Metal Powders conducted this world-wide Round-Robin incorporating all known methods of measuring and characterizing tungsten metal powder for size, surface area, density, etc. The B09.03 Subcommittee solicited donations of tungsten powder from suppliers in the refractory metals industry. These powders were blended and packaged under argon gas for storage so the participants in the Round-Robin could test, store and retest the powder as desired. Participants were charged a "slight handling fee".

After the economic problem of the 1970's was solved, the TAG worked at building a roster of experts to act as Liaison/Consultants and Technical Advisors to the, now, five ISO/TC 119 Subcommittees. The ISO subcommittee meetings had also changed for the better. Instead of meetings of only one or two subcommittees at a time in a given country, annual meetings of all subcommittees were now generally held at one rotation. Plenary meetings were also scheduled every three years and held at the same time. This allowed the US TAG to have Liaison/Consultants experts in two or more subcommittees available for these annual meetings. Meetings were also held in the USA at ASTM headquarters in the Philadelphia area; these meetings were less costly than the counterpart European meetings.

During the late 70's and through the 80's, the TAG was able to cover their financial needs for travel to meetings through a combination of (1) the support of the companies for whom they worked, (2) the annual MPIF donation and (3) the Round-Robin portion of the Travel Fund built up previously (and added to on occasion with additional donations from ASTM subcommittee B-09.03). However, during the 90's more of the Liaison/Consultants were retiring from active work and fewer active technical people with support from their companies (time and travel money) were available for ISO work. Down-sizing of the industry in the USA caused many companies to become conservative with respect to employees time and travel for the "extracurricular" standards work required.

The current structure of ASTM B-09.09 TAG for ISO/TC 119, Powder Metallurgy, and the Liaison Consultants and Officers are:

LIAISON/CONSULTANT

SC I	Terminology*	William A. Buerkel, (and Vernon Potter in 1999)
SC 2	Metal Powders	Harry D. Ambs
SC 3	Sintered Metal Powders	Erhard Klar
SC4	Hardmetals	James W. Kratofil
SC 5	Specifications	William Q. Judge

* Names shortened for convenience.

Chairman - William A. Buerkel
Secretary - Leander F. Pease III

During the late 90's, a survey of the cost needed to attend ISO meetings was made and it was projected that in less than two years there would not be sufficient money available to continue the P/M standards work. In addition, as a further burden, ANSI demanded an annual assessment of $3,000 from our TAG to cover their headquarters costs in order to have our technical experts represent the USA through their organization. Without payment, the national membership of the USA in ISO/TC 119 would be changed from a participating (P-member) status to an observer (0-member) status. The powder industry in the USA would then have no direct representation and vote in ISO and, as a result, no part in the formation of ISO standards presented to CEN for adoption.

Since an increase in MPIF's annual donation was not available in the near future to cover the extra travel expenses and the ANSI assessment, another attempt was made to solicit funds. This time, however, individual companies that should be the most concerned, and stand to benefit the most from friendly ISO and CEN EN's, were contacted instead of technical organizations. The solicitation is still in progress in that not all companies have given their final response. However, of the eight companies contacted, five have already committed to pledges and two are still considering the request. While the amount pledged to-date will be adequate to cover the needs of the TAG for increased travel expenses and the ANSI assessment, it is hoped that more pledges will be made so that a reduction can be made in the future in the amount requested from the companies that have already pledged. The five companies already making pledges are:

Federal-Mogul Sintered Products
GKN Sinter Metals
Hoeganaes Corp.
Kennametal Inc.
Stackpole Limited

These companies have helped to resolve the end of the century economic problems of ASTM B-09.09 Technical Advisory Group for ISO/TC 119, Powder Metallurgy

Are the P/M interests of the USA in standards still "alive and well" in New York and Philadelphia? You bet - for now!

ACKNOWLEDGEMENT

The author is indebted to Ms. Henrietta Scully, ANSI; Mr. Pat Picariello, ASTM; Dr. Leander F. Pease III, Powder-Tech Associates Inc.; and Mr. Owe Claesson, Secretariat ISO/TC 119 for providing material used in this article. Appreciation is also extended to the Liaison Consultants, ASTM B-09.09 members, the Technical Advisors to the Liaison Consultants and the Secretary of ASTM B-09.09 Subcommittee; these individuals are a valuable resource and indispensable to our work in P/M standards.

The origin of the Metal Powder Industries Federation may be dated from the formation and first meeting of the Metal Powder Association (MPA) on May 4-5, 1944 in New York City. It was initially a trade organization of companies that made metal powders. Kempton Roll's article noted above gives an excellent detailed description of the growth of MPA into MPIF. On May 5th, 1944, there was a portion of the meeting open to people other than powder producers. At 11:00 AM on that day, one Professor John Wulff, an MIT Professor and consultant to S.K. Wellman, gave a talk entitled "Powder Metallurgy Merry-Go-Round" (On a personal note, 14 years, later in September of 1958, Prof. Wulff asked aspiring young physicist, LFP, III, if he would like to switch out of Physics and go to graduate school in Metallurgy. Answer: "Yes!")

For the year 1944, Kempton Roll gives us a quite complete list of companies who were making powder metallurgy bearings. That is where our genealogy began. Our idea of genealogy is as follows:

Company name
When and where it started
Who started it
Where did they come from, i.e. where did they get their P/M experience
What did they make
What became of the company

This a funny kind of list. The names listed alphabetically are generally the P/M company names at their founding. Many names were changed, often several times, over the years. If you are interested in a particular company, today, you may well not find it listed alphabetically in the list, because it is a successor name. The best example is GKN, which purchased many P/M operations in the U.S., (including one it purchased twice!), but never started one in the U.S. One penultimate thought, if you or anyone you know ever decides to start a P/M parts business, we hope that you will peruse this list before you choose a name! Also, if you are part of a large corporation thinking about setting up a captive P/M parts operation, you should carefully note births and deaths of such operations! And the last thought: there is no way that we have been able to list every P/M operation that ever existed in North America, or every entrepreneur. Think of it as a work in progress and send us the corrections and additions. We could have a career just keeping the list up to date! And if you know of any software into which we could input these data and show all the connections among the houses, sort of like a line of descent, please pass the reference along to us.

If reading this genealogy is not your idea of a good time, you could always move ahead and read the index to our little book!

The Genealogy of All the P/M Houses that Ever Existed in North America

It includes when they were started, who started them, where the entrepreneurs came from (almost everyone came from somewhere!), what became of the operation, what did they produce for parts? When you see Produce: regular P/M, "regular" is in the eye of the beholder, and we see it as meaning conventionally pressed and sintered P/M parts! If you do not see a ceased operations date, it is likely that the company was actively in business in 2002. The information came from many sources. The folks who helped us with this list knew the entrepreneurs on a first name or nickname basis, and passed on the name in that fashion. We pass them on to you in the same way, not making any attempt to determine what their proper first names might be!

Acknowledgements

We acknowledge considerable help from Kempton Roll, George Gasbarre, Gene Andreotti, Dave Connelly, Maurice Bridgman, Jess Helsel, Robert Burns, Don Gustafson, Phil Booker, Ken Gleszer, Louis Baum, Bob Owens, and Athan Stosuy, Earl Johnson and John Gerzina.

CHAPTER 14 - APPENDIX

P/M Genealogy

A

Accraform
Palatine, IL

Started prior to 1970. Purchased by Sealed Power in 1971 and moved to Zeeland, MI.

See also Sealed Power.

Produce: regular P/M

Accumet
Ridgway, PA

Started in 1995 by Paul Stillwell who had worked at MascoTech

Produce: regular P/M

A.C. Gilbert
New Haven, CT

A famous toy producer. About 1950, started to make P/M parts with technical help from consultant Robert Talmadge. Made P/M nuts and gears for Erector Sets. Ceased operations between 1955 and 1960.

Produce: captive, regular P/M

Acushnet Co.
New Bedford, MA

A golf ball manufacturer active in P/M in 1969 with employee Wilfred Bowler. Were they making golf ball molds, similar to what Karsten was doing?

Produce:

Adma Products, Inc.
Solon, OH

Started in 1991 by Dr. Vladimir Moxson who had worked at PCC Air Foils and Pure Industries, Brownsville Bearing, in Brownsville, TN

Produce: Titanium, Zr, Nb

Advance Products Corp.
Beaver Dam, WI

Started in 1969 by Peter Seippel

Sold to Burgess Norton in 1999 and renamed Burgess Norton Plant #4

Produce: regular P/M

Advanced Forming Technology
Longmont, Colorado

Started by the Bruce brothers, David and Robert, and John Popken in 1987.

The founders were from Cad-Cam, an engineering firm with experience in design and building of tooling for plastic injection molding. Sold to PCC Airfoils in 1991

Produce: metal injection molding

Advanced Materials Technologies
Roseville, MN

Started in 1969 by Dale Stein PhD, Louis Toth, Martin Gimpi, Arthur Kydd and two others. None had worked in powder metallurgy but they had strong technical backgrounds. Operations discontinued in two years. The excellent equipment was purchased by a dealer and resold to Hennepin Technical Institute, becoming the basis for their Powder Metallurgy course.

Produce: Al P/M with Alcoa's technical help.

Advanced Metallurgy Inc.
McKeesport, PA

Started by Norman Hoyer in 1964. He had worked at Gibson Electric from 1946 to 1964, making electrical contacts. He sold Advanced Metallurgy to Technitrol in 1977. Name change to AMI Dudoco in 1997.

Produce: Cu-W, electrical contacts

Advantage Sintered Metals
Battle Creek, MI

Started in 1990 by Getulio Perelli, Jr. who had worked at Bear Metals. His father had managed Major Gauge and Tool in Livonia, MI.

Produce: regular P/M

Aeromet, Inc.
Eden Prairie, MN

Started in 1998 as a part of testing machine builder MTI. Technical direction from Dr. Frank Arcella who had been a manger of several programs at Johns Hopkins, APL.

Produce: laser formed P/M parts

Alanco, Ltd.
Montreal, Quebec, Canada

Started in 1968 by the owner of Aluminum Anodizing and Coating in Montreal. He was encouraged to do so by the President of Fina Metal, which company was just beginning to produce sponge iron powder. Technical help from Clem Sutinen from Bussman Simetag and Arthur Adler from Easton Metal Powders, Easton, PA.

Ceased operations in 1979.

Produce: regular P/M

Allegheny Powder Metal
Falls Creek, PA 15840

Started in 1982 by Francis "Bud" Grieneisen, Steve Miller, Gerd Genevro, Ron Young, Bob Trunzo and Al Curry, all of whom had worked at Brockway Pressed Metals.

Sold to Hawk Group in 1999

Produce: regular P/M

Allied Sinterings Corp.
Danbury, CT

Started in 1959 by Gifford Foster and Axel Madsen, both of whom had worked at Norwalk Powdered Metals

Alex Madsen had worked earlier at American Sintered Alloys. Gifford Foster had earlier worked at Watertown Manufacturing.

Produce: regular P/M

Allread Products
Terryville, CT

Started in 1974 by William O. Allread who had worked at

General Motors Research Center, GMC Saginaw, MI, RB&W, Michigan Spline (Master Powdered Metals Div.), Chase Copper and Brass, and Sintered Alloys.

Produce: regular P/M, filters

Allsinter Inc.
Compton, CA

Sanford W. Brown, active in 1981

Alpha Sintered Metals
Ridgway, PA

Started in 1967 by Roger Hasselman, who had worked at International Powder Metallurgy, Norman Bankovich who had worked at Keystone Carbon Co., and Frank Cartwright and Chester Viola. The latter two were from the West Penn Power Co.

Produce: regular P/M

Alloy Technology
West Nyack, NY

Started as a laboratory project in 1952-54, and then as a company, Sintercast, producing steel bonded carbides, (FerroTic) in 1955. Founded by Stu Tarken and others. Sold to Chromalloy. A Chromalloy management buy out in 1978 changed name to Alloy Technology. Sold to Pacific Sintered Metals in early 2002. In 2001, Pacific had purchased a technically related small company, Superwear, from Jay Ponchal, who had worked at Alloy Technology.

Produce: Ferro-Tic, a steel bonded carbide.

Alpine Pressed Metals
Ridgway, PA

Started in 1981 by Anthony Vigilione, who had worked at Exotic Metals and Alpha Sintered Metals, and by John Urmann and Joe Urmann who were in the coal business and were investors.

They purchased the assets of Compacted Powder Metals in Ridgway.

Sold to Metal Powder Products in 1999

Produce: regular P/M

Alpine Specialty Products
Ridgway, PA

Started in 1992 by Anthony Vigilione, who owned Alpine Pressed Metals and Richard Phillips, who owned Engineered Pressed Materials.

Sold to Metal Powder Products in 1999

Produce: Stainless steel and alloys

AMAX Metal Injection Molding, Inc.
Escondido, CA

Started in Sept. 1989 as a joint venture between AMAX and RISI Industries. Technical and managerial help from Yancey Smith and Kenneth Johnson. RISI produced special industrial equipment related to munitions manufacture. It was intended that the metal injection molding plant would make use of some RISI technology to make a missile fin base. In 1993, Cypress bought AMAX and had no interest in the injection molding operation in Escondido. It sold its interest and the company was renamed Injectamax.

Produce: metal injection molded

American Brake Shoe
Mahwah, NJ

In the 1980's much activity at their lab in Mahwah, NJ, with friction materials being made at other plants.

Produce: friction materials

American Can Co.
Barrington, IL

In the early 1980's under William H. L. Hooper, Albert Klein and Timothy Volin, considerable development was done on strip rolling metal powders to make tin can stock for drawn an ironed cans . Operation discontinued in the mid 1980's.

Produce: strip rolled iron powder

American Electro Metals
Yonkers, NY

Started by the Paul Schwarzkopf whose family was famous for carbides and heavy metals in Austria, (Metallwerk Plansee). They left Austria during the early 1940's. General Electric Co. helped to bring the family to the U.S. for their knowledge of carbide manufacture. He later started a business making pressed and sintered iron parts, and other more advanced materials, such as copper infiltrated turbine blades for jet engines. At Yonkers, he had assistance from Henry Hausner, Claus Goetzel, Richard Seelig and Robert Pettibone. Still active in 1946, but its existence after that time is not known.

American Meter
Nebraska City, NE

Started in about 1954 as a captive P/M operation for the company to make parts (couplings) for their gas meters. Early parts were simple bronze bearings. No one knows the source of technical expertise in P/M that got the operation up and running.

Produce: regular P/M, captive

American Orthodontics
Sheboygan, WI

Started in 1985

Paul Riddle used the Brunswick Technetics technology

Produce: metal injection molded parts for orthodontics.

American Powdered Metals
North Haven, CT

Founded in 1953 by Gerry Fellows and Gerald Struble with James Struble and Walter Krieg, all from Keystone Carbon Co. Backing by Bob Leeson from Leesona. In 1958 they bought the P/M operation of Eaton Yale and Town, (Yale Lock and Hardware) in Franklin Park, IL and moved it to North Haven.

Sold to Barden Corp. in 1970, and then sold to Alcoa in 1972. In 1972, Alcoa introduced P/M aluminum parts at APM. In 1983 the Conover, NC plant was started and operations were gradually moved there. Alcoa sold APM to George

Haymaker in 1988 and it was next sold to Pennsylvania Pressed Metals in 1992. See also Pennsylvania Pressed Metals for the transition to GKN ownership.) GKN are the initials for Guest Keen Nettlefolds, an English company.

Produce: regular P/M and aluminum

American Sintered Alloys
Bethel, CT

Started in 1945 by Kenneth Gleszer and Raymond Engstrand and John Polinetz who bought the P/M operations of Micro Ferrocart Products of Stamford, CT. Sold to Yale Lock and Hardware in Franklin Park, IL 1950. See also Micro Ferrocart. Kenneth Gleszer had worked at Moraine Products Div. of GMC. John Polinetz later worked at RB & W.

American Sintered Technologies, Inc.
Emporium, PA

Started in 1988 by William Werner, Jay Edel, Wes Fye and Bill Fox, all of whom had worked at Pennsylvania Pressed Metals. Sold to Fansteel Corp. in 1996

Produce: regular P/M

American Yard Products
P/M Department
Orangeburg, SC

Started by Roper Outdoor Products in 1979 as a captive P/M operation. Sold to American Powder Metals in 1989.

Produce: regular P/M, captive

Ames Reese, Inc.
Was Reese Metal Products
Bird-In-Hand, PA

Purchased by Caesar Molins of Ames in Spain, in 1997.

Produce: regular P/M

Ametek, Inc.
Specialty Metal Products
Wallingford, CT

Started strip rolling in 1963. The original operation was started by Brian Haworth, and later sold to Pfizer and then sold to Ametek.

Produce: powder rolled metal strip and pneumatically iso-forged material

AMKAD Metal Components, Inc.
New Hamburg, Ontario, Canada

Started in 1995 by Ken Kropf who had worked 26 years at Perth Metal Industries.

Produce: regular P/M

AMP Incorporated
Harrisburg, PA

Had been purchasing MIM parts since 1975 when they took a Parmatech license in 1982 to make parts on a captive basis for the crimping tools for use on electrical connectors. Engineering support from George Lurie.

Ceased operations at a date now unknown.

Produce: Metal injection molded, captive

Amplex Div. of Chrysler Corp.
Detroit, MI

In 1926, Walter Chrysler reviewed complaints about P/M bearings purchased from a supplier of their clutch pilot shaft bearing. It screeched and wore out. He began research in Dec. 1927 on P/M bearings at Engineering in Highland Park, MI. In December 1929 production started in the Hamtramck section of Detroit MI, in the Dodge Brothers building. Starting in 1929, it was managed by Anthony (Andy) J. Langhammer, until 1956. Later expanded to the Harper St. plant, (finally 194,000 sq.ft., but started in 1937). Other later plants at Trenton, MI, (42,000 sq. ft., 1957) and Van Wert, OH (1965-6) (179,000 sq. ft.) and Perth Metals in Stratford, Ontario, Canada, 1958, (13,000 sq. ft.).

It was said to be the first U.S. P/M parts producer, however we wonder who supplied the screeching bearings in 1926!. (Bound Brook Bearing Co. or GM?) It was also started for the purpose of making the bearings that go into windshield wiper motors. During World War II, in the early 40's, the plant made millions of bullets, (of which we have shown and example above), shell bands, gun mounts, sight components and sirens.

The Canadian plant was sold to Stackpole Ltd. of Toronto, in 1991.

Balance sold to ICM then GKN.

Van Wert plant closed in 2000

Produce: regular P/M, filters

Amstead Research Labs
Bensenville, IL

During the early 1980's under Earl Carlson and others, they developed ways to make high density alloys by liquid phase sintering. Some of these were sold as commercial parts. The technology was later sold to Sintered Specialties.

Produce: Near fully dense materials

Anchor Technologies, Inc.
Dubois, PA

Started in 1997 by Rich Smochek, who had worked at Clarion Sintered Metals

Produce: regular P/M

Applied Porous Technologies
Tariffville, CT

Started in 1999 by Ed Swiniarski who had worked at Chand Associates.

Produce: filters

Arrow Sintered Products Co.
Forest Park, IL

Started in 1946 by Mr. Brown, who developed his own process to make bronze filters

Produce: filters

ASAP Sintered Metals Co.
Falls Creek, PA 15840

Started in 1992 by Perry Phillips, who had previously worked at his father Paul Phillips' company, Capital Sintered Products. ASAP was sold to Matthew Nerone in 1997 after the death of Perry Phillips.

Produce: regular P/M

ASCO Sintering Company
Commerce, CA

Started by two people, one of whom was Richard Sinclair who represented Chrysler Amplex for the sale of bearings on the West Coast. He bought the assets of a P/M company in Connecticut called American Sintered Company and moved it to California. His partner was Richard Kuhn who had worked both for Bound Brook and Chrysler Amplex. Some think that the name ASCO is the acronym for the earlier CT company, although Robert Burgess says that they never really learned the origin of the acronym ASCO. Burgess Norton Co. bought ASCO in 1964 and then sold it to Robert C. Burgess in 1971. Amstead Industries had purchased Burgess Norton Mfg. in 1968. In the 1980's ASCO Sintering Co. was sold to the employees as part of an ESOP.

Produce: regular P/M

Atlas Pressed Metals
DuBois, PA

Started in 1979- 1980 by Dennis Heindl who had worked at St Marys Carbon Co. and Paul Harris. In 1983, it was sold to Maurice M. Hanes, who had founded Carbon City Products, and Tom and Dick Pfingstler.

Produce: regular P/M

Avenger Inc.
Ipswich, MA

Started in May, 1983 by John and Russell Morris, Patricia Kavanaugh and John Ed Enos, all from Sintered Metals Inc., the then owner of Sinterbond, of Gloucester, MA

Produce: filters

B

Bassick
Bridgeport, Connecticut

They were the largest producer of casters in the U.S. About 1955 they set up a department to make roller casters from sintered iron, self-taught. (What do you suppose they used as a lubricant to keep their product from squeaking? (See answer at end of listing of P/M houses) They had 125 ton presses and a roller hearth furnace that was more than adequate for their needs. Around 1970, they were bought by Stewart Warner and the P/M plant was moved to the midwest. No information on what happened next.

Produce: regular P/M, captive

Bearing Sales Corp.
Chicago, IL

Started in 1965-67 by David Hilton who had worked at Chrysler Amplex. They both distribute bearings and produce by CIP. See also Dymet.

Produce: sintered bearings

Bear Metals
Battle Creek, MI

Started in 1968 by Earl Burlingame who had worked at Cleveland Powdered Metal in 1964 and later at Southern Sintered Metals.

Sold in 1998 to Anjan Sabbani, who had worked as an engineer at General Motors Corp. Changed name to Paradigm Sintered Products in 1998.

Produce: regular P/M

Beemer Precision Inc.
Philadelphia, PA and later moved to Fort Washington, PA

Started in 1930 by Frank Beemer to whom Walter Chrysler offered the Eastern bearing distribution for the newly formed Chrysler P/M bearing operation, Chrysler Amplex. Later added bearing finishing services and bought the Oilite name.

Beemer Precision
West Chester, PA

Started in 1998 by Hal Meyer of Beemer at Ft. Washington, PA

Produce: bearings and structurals

Bendix Corp.
Friction Materials Department
Troy, NY

Started in 1949 to make copper powder based aircraft brakes. In 1957, they were approved as a supplier of iron powder based riveted brake shoes for the Ford Thunderbird. There were 3 square pads, 2.5 in. on a side, that were riveted onto each shoe. Operation later transferred to Troy, MI. Became part of Allied Signal and then Honeywell.

Bestmetal Corp.
Woodstock, IL

Started in 1989 by James Eddins and Janet Beaucher who had both worked at Sinterloy. Janet Beaucher's father, Bill Randolph owned Sinterloy.

Produce: regular P/M

BIC Pen Corp.
Milford, CT

During the 1970's and 1980's, under Arthur Fuhrmann, they pressed the tungsten carbide balls that are used as the tip in ball point pens. Not sure of present status.

Produce: WC pen balls

Birmingham Bolt Co.
Ripley, TN

Frank Thompson was active there until the 1990's making fasteners that are used in mining roof support.

Produce: regular P/M

Black and Decker Company
P/M Parts Department
626 Hanover Pike
Hampstead, MD

Department started in 1972. By 2001, producing about 90,000,000 P/M parts per year.

Produce: regular P/M captive

Blount, Inc.
Wilsonville, OR

Purchased Omark Industries Div. that made MIM parts Later sold to Walter Weyler and renamed Kinetics. *See also* Kinetics.

Produce: metal injection molded

Bodycote, IMT
Andover, MA

See also Industrial Materials Technology.

Borg Warner Automotive Inc.
Powdered Metal Products
Livonia, MI

See also Michigan Powdered Metals.

Borg Warner Morse-TEC Inc.
Ithaca, NY

Started as a captive operation in 1987 using equipment obtained from FMC, in Indianapolis. Technical help in 1996 with Rocco Petrilli who had worked at Keystone Carbon and Zenith Sintered Metals. Plant being relocated to Cortland, NY, and enlarged, 2001, with technical help from Sandy Jaffee.

Produce: regular P/M , captive

Borg-Warner Automotive
Romulus, MI

Purchased this powder forging plant from Federal Mogul in 1995. Sold it in 1998 to GKN Sinter Metals. Sold the Gallapolis, OH plant to GKN 1999.

See also Haller Powdered Metals.

Produce: powder forged parts

Bound Brook Oilless Bearing Co.
Bound Brook, NJ

Started in 1925 by Dr. Claus, and believed to be the second P/M parts manufacture in the U.S. During the 1930's. there was a joint venture that resulted in starting Bound Brook Oilless Bearing in Litchfield, England. Bill Toeplitz was the metallurgist at Bound Brook new Jersey during the 1930's and gradually purchased the ownership of the company. After 1945, an English company, bought the Bound Brook Litchfield operation. They then bought into the US Bound Brook operation and eventually owned all of it. The company that owned the Bound Brook operations was in turn bought by GKN in the early 1960's. Bound Brook bought land out in Auburn, PA and built a 60,000 sq. ft plant to make atomized copper powder and bearings. A decision was made to build another bearing plant as a satellite, in Brownsville, TN, occupying 30,000 sg. ft., to make high volume bearings. GKN also wanted to have a U.S. operation to make P/M structural parts, and not just bearings at Bound Brook and Brownsville, TN. In 1968, they bought 30% of PresMet from Morris Boorky, its founder. In 1973, they purchased the remainder. In 1975, Mac Bumphrey came from GKN in England to manage PresMet and Bound Brook in Auburn and Brownsville. He then built the PresMet Holden, MA plant with the intent of making high volume ferrous P/M parts for automotive. In 1974 there was an energy shortage and the Auburn, PA plant could not get an uninterrupted natural gas supply. GKN shut the Auburn plant and moved its bearing making operations to the PresMet Holden, MA plant. The Auburn plant had a tool room, and continued making bearing tools for Brownsville and Holden.

Marvin Reed had been at the New Jersey plant, and was then running the Brownsville plant. He assembled a group and bought the Brownsville, TN operation from GKN in 1978. He and he and his partners sold to Pure Industries in 1979.

Pure Industries was a wholly owned subsidiary of Stackpole Corp. and operated this plant as Brownsville Bearing until 1987 when the management was transferred to Stackpole Ltd. Stackpole Ltd. made an IPO on the Toronto Stock Exchange in 1997. The operation at Brownsville was then purchased by the local management and became Dynametal Technologies in 1997.

Produce: regular P/M

Brazed Parts, Lt.
St. Thomas, Ontario, Canada

Started in 1973 by Roy Goodwin and Herbert Johnston. Roy Goodwin had P/M experience in the U.K.

Produce: regular P/M

Breed Technologies
Lakeland, FL

Started in 1995 by Alan Breed who owned a company in New York state that made the seat ejector system for military aircraft. Similar technology was used in early automotive air bags that had five small parts in an electromechanical assembly. Technical help from John Gaspervich, who had worked in MIM at Advanced Forming Technologies. Successive technology for air bags was based on a solid state device. Operations ceased about 1998. Produce: MIM, captive

Brico
Dayton, OH

See also Supermet.

Brockway Pressed Metals
Brockway, PA 15824

Started in 1953 by George Gasbarre, who had P/M experience at Symmco and IPM, and investors: Marshall Jetty, John Vern, Bill Josephson, and Hyman Kantor. In 1954, Marshall Jetty left Pure Carbon to become General Manager at Brockway Pressed Metals. In 1963, George Gasbarre sold his stock back to Brockway Pressed Metals.

Managed for many years by Robert Reitz, the company was sold to Inland Engineered Materials in 1998, who in turn sold it to Engineered Materials Corp. 1998

Produce: regular P/M and magnetic materials

BrownCo Powder Products
Emporium, PA 15834

Started in 1994 by Steve Brown who had also started Pennsylvania Sintered Metals, in 1990. No prior industrial P/M experience but formal education included P/M.

Produce: regular P/M

Brownsville Bearing Co.
Brownsville, TN

Part of Pure Industries and Stackpole Corp. *See also* Bound Brook Oilless Bearing Co.

Brunswick Technetics Weitec,
Deland, FL.

Started in 1982 with technology and license from Raymond Wiech and the purchase of Weitec California, of San Diego CA. The latter was a provider of MIM feedstock and sintering furnaces (reactors). Weitec California was a subsidiary of Raymond Wiech's company, Weitec Cayman Islands. Brunswick did sell technology, feedstock, and reactors for a short time, and then concentrated on MIM parts.

Sold to Metal Powder Products in 1990, changed name to FloMet. Retained by Arlan Clayton, after Metal Powder Products was sold in 1998.

Produce: metal injection molded

Brush Wellman
Tucson, AZ

Started making Cu-W parts in 1997.Reorganized in 2001, name changed to Zentrix

Bucktail Pressed Metals, Inc
St. Marys, PA

Started around 1989-92 by Melvin Luhr who had worked at Keystone Carbon and Tony Pfeufer who had experience at Exotic Metals and Windfall.

Ceased operation 1996

Produce: regular P/M

Bundy of Canada
Guelph, Ontario, Canada

Sheepbridge Engineering from England started a casting factory at Guelph to support the proposed Avro aircraft. The plane was never built and the factory was converted to P/M in the 1960's. Bundy tubing bought it in 1968. It was sold to ISM (International Sintered Metals) in 1983. In 1987 it was sold to Maag. Maag was a partner with Krebsoge. In 1990 the plant was sold to ICM/Krebsoge which was sold to Sinter Metals and then to GKN/ Sinter Metals.

Bunting Bearings Corp.
Holland, OH

See also Detroit Sintered Metals, its predecessor.

Produce: regular P/M

Burgess Norton Mfg. Company
Geneva, IL

Started in the 1920's by the Burgess family making wrought piston pins for automobile engines. In World War II, they made tank treads. It was a very large screw machine operation. About 1957-58, Robert Burgess established a P/M operation to make oil pump gears. Technical help from consultants Robert Talmadge and Gerald Pearson.

Sold to Amstead Industries in 1968. Bob Burgess was very active in promoting and fostering MPIF. *See also* ASCO Sintering Co.

Produce: regular P/M

Burroughs Corp.
Detroit, MI

Member in-plant P/M group in 1963. Technical help from employee Archie DeLoge.

Produce: regular P/M, captive

C

Cameron Diversified Products, Inc.
Wilcox, PA

Started by Edward Anderson in Emporium, PA, 1968 as a tooling business. Pioneered the use of the Crucible CPM 10V grade in compacting tooling such as punches. Began to make P/M parts in 1987 and moved to Wilcox in 1989.

Produce: regular P/M

Capital Sintered Metals
Falls Creek, PA

Started in about 1974-78 by Paul Phillips who has started ASAP and worked at IPM. Sold to Winston Corp., plant moved from Johnstown to Evansberg. Assets sold to Ridgway Powdered Metals in 1995.

Produce: regular P/M

Capstan/Atlantic

Wrentham, MA

See also Sintered Metals.

Produce: regular P/M

Capstan/Inland
Hazen, AR

See also Remington Arms.

Capstan/Mexico
Jalisco C. P. 45685
Mexico

Started by Capstan Industries as a greenfield operation in 1997.

Produce: regular P/M

Capstan/PermaFlow
Gardena, CA

Started as a separate division and location in 1991 from Capstan's own filter making operation. Renamed Capstan/California

Produce Filters

Carbon City Products, Inc.
St. Marys, PA

Started in 1956-57 by Larry Geitner, who worked for King Chevrolet, Dan DeLullo, a shoe salesman, Maurice Hanes, Harry Gerg, Art McQuown and Vincent Heidl. Maurice Hanes was the only one with P/M experience. Sold in 1978 to Howard Bates and then to National Forge, in 1981, with name change to Prometech in 1983. Then sold to Metal Powder Products, in 1986, with return of original name.

Produce: regular P/M

Carpenter Technology
Reading, PA

Started P/M operations in 1968 to make prealloyed gas atomized powder to be hot pressed and HIP'd, e.g superalloy Rene 95 for G.E.. Purchased Dynamet, Inc. in 1997. Dynamet had purchased Universal Cyclops Cytemp process in 1990. Strip rolled N-42 from 1976 to 1982.

Produce: fully dense high alloy billets

Caterpillar Inc.
ACT Group
Rockwood, TN

The P/M pilot plant began in Illinois in 1971 with production at Peoria, IL in 1973. Moved to Rockwood, TN in 1994. Specialize in very large and sometimes low volume parts for the large scale Caterpillar earth-moving equipment.

Produce: regular P/M, mainly captive

Ceracon Technologies Co.
Fair Oaks, CA

Sprang from Wheeling Pittsburgh, which was alive in 1971

Produce: pseudo-isostatically forged P/M parts ("Ceracon forged")

Ceromet Inc.
Anaheim, CA

Started in 1953 by Al Williams and his father-in-law, and his brother-in-law Ray Anderson. Ray Anderson had worked for the Stokes Press Company, which made P/M presses. Sold to Metal Powder Products in 1990

Produce: regular P/M

Cer-Plas
Huntington Beach, CA

Started by Raymond Wiech and Raymond Millett, prior to 1971. Ceased operations by 1973 when they became partners in Parmatech.

Produce: metal injection molded

Certified Metal Products, Inc.
Little Rock, AR

Started in 1971-72 by Steve Tieman who had worked at Remington. Sold in 1977-78 to Southern Sintered Metals.

Chase Copper and Brass
Div. of Kennecott Copper

Waterbury, CT

Started about 1966 by Dr. Malcom Burr, a Chase employee, to make copper powder by water atomization and to convert the powder to cold formed plumbing fittings. Copper was pressed into parts of switch gear, requiring good electrical conductivity. Also making copper sheet by rolling powder. Kemp Roll reports that Malcom Burr also attempted to make copper coin blanks for the U.S. Mint but that it was not economically feasible. (LP can attest to the fact that Malcom went on business trips with his bagpipes and practiced in hotel room!) Operations ended in 1973. Powder-Tech Associates purchased some of the assets.

Chicago Powdered Metal Products
Schiller Park, IL 60167

Founded in March, 1948 by John E. Miller, Sr. who had been Manufacturing Manager at Powder Metal Products Co. of America, Franklin Park, IL., which company had been started in the 1940's by his uncle and his grandfather, Edward Bagner. (The Bagner family was well known in plastics molding, e.g. the cases for Motorola radios in the 1930's, and later, television sets.)

Produce: regular P/M and warm formed parts, mainly large parts

Clarion Sintered Metals
Ridgway, PA

See also International Powder Metallurgy.

Clearfield Powdered Metals Inc.
Clearfield, PA

Started in 1984 by Dennis L. Merrey, a key person at Clearfield Machine Co.; Don Wilson and Dave Wilson, both of whom had worked at Symmco.

Sold in 1998 to Hawk Group of Cleveland OH.

Produce: regular P/M

Cleveland Graphite Bronze
Cleveland, OH

Made steel backed bronze bearings from an early date. Established a P/M division in North Carolina to make sintered aluminum bearings in the 1960's. This division only produced bearings for a very few years. Cleveland Graphite Bronze became part of a holding company, Clevite, at an early date.

Produce: steel backed bronze bearings

Cleveland Powder Metal Co.
Kent, OH

Started for IPM in 1962 by Ken Gerg who had worked at IPM, Bob Capello who had P/M experience, Bill Seacrist who also had P/M experience, Joe Schaefer who had worked at IPM, and Walter Goens. In 1970, company active with Walter Goens and Joseph Schaefer, who later founded Ridgway Powder Metals. Shut down by Allegheny Industries after they bought IPM sometime after 1968.

Clevite Corporation
Cleveland, OH

Had a manufacturing plant in St. Thomas Ontario to make automotive sleeve bearings (steel backed) and rubber vibration insulating bearings. In 1952, it did a joint venture with Manganese Bronze from the UK to make pressed and sintered bronze bearings in St. Thomas. This J.V. was called Paxol. In 1954, Clevite bought out the Manganese Bronze share of Paxol and it became Clevite's St. Thomas plant for making P/M parts and bearings, and later for powder forged parts. About 1970, as Clevite was getting ready to buy Waltham Industries and Johnson Bronzes's P/M operation in Salem, IN (c.f. Ferraloy), Gould Inc. purchased Clevite. After that, Pullman bought the Salem and St. Thomas P/M plants from Gould. ICM, which had bought the Chrysler Amplex P/M business, then bought the P/M plants from Pullman. ICM was then purchased by Sinter Metals, a publicly held company from Cleveland, which in turn was purchased by GKN. Thus, in 2002, the plant in Salem, IN is known as GKN Sinter Metals, Salem

Clevite Corp.
Mechanical Research Div.
Cleveland, OH

Developed sintered aluminum bearings, sold through Cleveland Graphite Bronze, sintered titanium, and powder forging. Member in-plant P/M group in 1963

Cloyes Gear & Products, Inc.
Powdered Metal Division
Subiaco, AR

Started in 1989. Cloyes had owned Rush Metals since 1979 and was familiar with P/M.

Produce: regular P/M

CMW Incorporated
Indianapolis, IN 46206

Contact Metal Welding

In 1918 Philip Rogers Mallory developed a way to make tungsten wire filaments for the small electric lights used on Christmas trees. This business was then sold to General Electric.

About 1928 a part of General Electric in New Jersey, was making all forms of P/M tungsten, trade name Elkon. In 1934, P.R. Mallory left GE and founded Mallory Metallurgical in Indianapolis, IN, to make W and Mo P/M, with infiltration with Cu and Ag. In 1978, a management buyout of the Mallory estate resulted in CMW, Inc.

Produce: contacts

Colorado Sintered Metals Inc.
Colorado Springs, CO

Started in 1967 by John Gerzina and Mr. Salscheider, with John Gerzina as majority owner. John Gerzina had a long history in P/M and most recently worked at Rocky Mountain Powdered Metals. Sold to Precision Powdered Metals, 1996

Produce: regular P/M

Colt Industries
Crucible Compaction Metals Operation
Oakdale, PA

Under Ed Dulis, in late 1960's, they began to produce atomized tool steel powder and consolidate it into tool steel billets by HIP'ing. In 1973, Compaction Operation began making superalloy materials and parts for the aerospace industry, using a ceramic can. Metal cans have also been used. The HIP'ing concept came from Battelle Memorial Institute who had used it on nuclear fuel.

Produce: HIP

Colt Industries
Gore, Oklahoma

Between 1980-85, they had a 60 ton press and were making 3 in. diameter P/M parts for use in emission control systems. Claimed tolerance of .0005 in. on 3 in. OD.

Produce: regular P/M, captive

Compacted Powdered Metals, Inc.
Ridgway, PA

Started about 1968-69 or 1975 by Dennis Heindl who had worked at National Molded Products, and by Paul Phillips who had worked at IPM in 1949, and others. Assets sold to Tony Vigilione, Joe Buehler and others, in starting Alpine Pressed Metals in Ridgway, PA. in 1981.

Produce; regular P/M

Compacted Metals Corp.
Waukegan, IL

Started in 1940's by Joseph and Rudy Cepon. Sold to Joe Nava circa 1985. Merged with Timing Gears Corp.. Moved to St. Charles, IL in 1987. Sold to Cloyes Gear, 1989. Operations discontinued in 1996.

Produce: regular P/M

Compax, Inc.
Anaheim, CA

Started in 1962 by Orville "Bob" Bowen who had worked at National Cash register in Dayton, OH. Sold to Earl Johnson in 1976. Earl Johnson had P/M experience from early work designing tooling for ceramic molding and then worked at Ceromet, sold presses with AC Compacting and Dorst, and then worked at Kwikset.

Produce: regular P/M

Components General Corp. Inc.
Kokomo, IN

Started in 1980's by Dr. Marina S. Levin, who had worked at Cabot Corporation. Ceased operation by 2000.

Produce: metal injection molded

Comtec Manufacturing, Inc.
St Marys, PA

Started in 1988 by David L. DeLullo, Jr. who had prior P/M experience at Carbon City Products, Fairview Sintered Products and St. Mary's Pressed Metals, and Kenneth E. Leithner, former Head of the Sintering Department at Windfall Products. Mr. Leithner later sold his interest to Mr. DeLullo.

Produce: regular P/M

Comtech
Waupun, WI

Started in 1986 by Peter Seippel and Stephen Clinch from Advance Products and by Wells Mfg. of Skokie, IL. Wells later assumed full control of the company and sold it in 1988 to MPR (Michigan Piston Ring) Goetze. It was later sold to T and N, then to Brico, then to Federal Mogul.

Produce: regular P/M, primarily valve seat inserts.

Consolidated Metallurgical Corp.
Dexter, MI

Started in 1976 by Richard Hervey; John Anderson and John Mollison to make liquid phase sintered tool steel parts based on the Robert Matt, Aerojet patent of 1967. Glidden-SCM supplied M-2 powder and gradually assumed control until 1979-80. Operations were shut down and assets sold to the Joseph Krupp family in late 1980. Renamed Krupp P/M, then Krupp Engineering and in 1997 changed to P/M Krupp Technology, all under the Krupp family. Sold to Pacific Sintered Metals in 2000.

Continuous Metal Technology
Ridgway, PA

Started in 1995 by Roger Smith who had worked at Keystone Carbon Co. and Tim Smith an engineering graduate of Penn State.

Produce: regular P/M

Coors Porcelain
Golden, CO

Compacted substrate and heat sinks for semi-conductors. Also, roll compacted ceramic strip.

Copress, Inc.
Wilton, CT

Started in 1947 by Earl Patch who had worked as a Vice President at Moraine Products Div. GMC. Robert Talmadge had also worked at Moraine Products and was an employee at Copress. In January 1948, John Gerzina and John Caputo had started Standard Bearing Company in Norwalk, CT. Standard Bearing Co. was merged into Copress in 1949. In 1951, Earl Patch sold the business to Prescott Hotchkins of Chicago IL. Hotchkins changed the name to Pressco, and John Gerzina returned to work there. In 1952 Prescott Hotchkins sold the business to Andre Blumenthal who renamed the company as Norwalk Powdered Metals.

Produce: regular P/M

CTI
Owens Sound, Ontario, Canada

George and Peter Aller(sp)

Produce: refractory metals

Cytemp Powder Products

Bridgeville, PA

Sold in 1991 to Dynamet, Inc.

D

Deloro Stellite

Belleville, Ontario, Canada

Started by Cabot Corp. in Speedway, IN in 1967-68 to make pressed and sintered Co based parts, from materials such as Stellite 6, 12, 3, Star J and Tribaloy 400. Early technical help from Herb Rogers and Ed Foley. Moved to Kokomo, IN in early 1970's. Set up in Canada in 1985.

Produce: press and sinter Co base

Detroit Sintered Metals
Kalamazoo, MI

Founded in 1947. Milton F. Smith sold it to Bunting Bearings in 1955. Was located at the Kilgore Rd. site in Kalamazoo, MI later occupied by Bunting Bearings Corp.

Produce: regular P/M

Deutsch Fastener Corp.
ElSegundo, CA

Around 1981-84, active in making sintered aluminum fasteners.

Technical help from employees T.W. Allsworth, Vern Sanders, Farrokh Farzin-Nia and possibly Peter Schwarzkopf (if my memory serves me right!). Uncertain about when it ceased operations.

Produce: sintered aluminum fasteners.

Diamet Corporation
Columbus, IN

Started in 1987 by Mitsubishi

Produce: regular P/M

Dixon Sintaloy
Stamford, CT

Started in 1955 by Kenneth Gleszer for the Joseph Dixon Company. Mr. Gleszer had been a principal of American Sintered Alloys.

Some assets sold to Engineered Sinterings and Plastics in 1985.

Produce: regular P/M

D. M. Steward Mfg. Co.
Chattanooga, TN

Started in the mid 1950's. Jimmy Jones was Technical Director.

Produce: ferrite components

Dougherty Pressed Metals, Inc.
St. Petersburgh, FL

Started in 1978 by Tom Dougherty who had worked for press builders: Kux Machine, Mannesmann, and Stokes. This is like the "Unbegun Symphony"! Tom Dougherty started his P/M business by building a press to make P/M parts. A visitor bought his press, and so he decided to make specialized presses rather than P/M parts! To reflect this, the company was renamed Dougherty Precision Systems in 1983. The presses are often used to press rare earth magnets. The presses incorporate a very strong magnetizing coil. The powder is preweighed and enters the die. The upper punch closes the die and the magnet aligns all the north poles of the particles. The powder is pressed, and then, prior to ejection, the powder is demagnetized, but maintains it correct physical orientation. Without demagnetization, you could not remove the green pieces from the tooling! After sintering and grinding, the material is remagnetized.

Produce: no P/M parts, then specialized tools and presses

DuBois Powdered Metals
DuBois, PA

See also Rockwell Meter.

Dymet Corporation
Muskegon, MI

Started in 1973 by Carl Hilton who had experience at Chrysler Amplex. The intent was to make bearings to be sold through his other company, Bearings Incorporated, in Chicago, which had been started about 1965-67. Dymet was sold to John Holman in 1990. John Holman had worked at Pure Carbon, IPM, and Advance Products.

Produce: CIP and regular P/M

Dynamet, Inc.
Bridgeville, PA

Sold to Carpenter Technology in 1997

Dynamet Technology,Inc
Burlington, MA

Started in 1972 by Stanley Abkowitz, who had done some preliminary work with Ti P/M in the 1960's. Joined by son Miles Abkowitz and daughter Susan Abkowitz in 1989.

Produce: CIP and R and D of Ti and its alloys, and production for military, sporting goods and automotive.

Dynametal Technologies
Brownsville, TN

See also Bound Brook Oilless Bearing.

E

Eastern Sintered Alloys Inc.
St Marys, PA 15857

Started in 1979 by John Sterbank and Art Hillebrand both of whom had worked at Powder Metal Products.

Produce: regular P/M

Eastman Kodak Company
Rochester, NY

Started P/M department in 1963 at State Street by Jim Carey, a Kodak trained engineer, and Ed Walker, their metallurgist. Technical help from Jim Ross who had worked at Symmco. They started with very small camera gears used to advance film, because there was no custom P/M parts producer who could make the parts. They were made with a Stokes 575 press, and a combination of a robot arm and vacuum pick-off to handle a very delicate part. Later made magnets used in shutters on instant cameras.

Ceased operations in 1999.

Produce: regular P/M, captive

East West Industries

Started in 1963 in St Louis, MO by George Gasbarre and others. Moved to Piggot, AR. Sold to Martin Sprocket in 1970

Produce: regular P/M

Eaton Lock and Hardware Div.
Eaton Yale and Towne
Franklin Park, IL

Founded a P/M business in 1950 to make lock hardware.It was based on the 1950 purchase of American Sintered Alloys of Bethel CT from Kenneth Gleszer et al. Technical help from Bill Dorney, who had worked at Delco Moraine Div. of GMC. Sold in 1958 to American Powdered Metals and moved back to CT!

Produce regular P/M, captive

Elco Sintered Alloys Co., Inc.
Started in 1952 in St. Marys, PA , moved to Kersey, PA in 1973

According to founder Peter "Bud" Herzing, started in 1952 with Norman Bankovic, a graduate of the U. Missouri School of Mines at Rolla, MO and other Bankovic family members. No one but Norman Bankovic (Keystone Carbon Co.) had previously worked in a P/M house.

Produce: regular P/M

Electro-Motive Div., GMC
P/M Parts Department
La Grange, IL

Started in 1976 to makes P/M parts in limited quantities for locomotives. Closed in 1995

Produce: regular P/M, captive

Elite Metal Products Inc.
Ridgway, PA

Started in 1979-80 by Richard A. Wendel and Walter Spencer, both of whom having worked at Powder Metal Products. Moved to St Marys, PA in 1984. Sold in 1999 to GKN-Sinter Metals

Produce: regular P/M

Elk Metals
St, Marys, PA

Started in 1998 by Melvin Luhr who had worked at Keystone Carbon and had been a principal at Bucktail.

Produce: regular P/M

Embassy Powdered Metals, Inc.
Emporium, PA

Started in 1996 by Jim Buerk, David Hampton, John Heverly and Jim Belin, all of whom had worked at Pennsylvania Pressed Metals in Emporium.

Produce: regular P/M

Emporium Specialties Co.,Inc.
Emporium, PA, and shortly moved to Austin, PA

Started in 1947 by George Erskine, son of the founder of Sylvania Electric Products. Initially the company made toys, then hand tools, and finally wire forms with some cold heading. To make copper wire forms with larger heads, they developed a way to form copper powder around the wire, and create the larger head. In 1968, Gib Hoehn, of press-builder Mohawk Industries, helped set up a P/M compaction department. Sold in 1978 to Marvin Deupree.

Produce: regular P/M

Engineered Pressed Materials

St. Marys, PA Started in 1988 by Richard R. Phillips who had worked at Pitney Bowes, Carbon City Products, and Intech Metals. .

Produce: titanium, tool steel and SS parts

Engineered Sintered Components
Troutman, NC

Started in 1989 as joint venture between Sumitomo and Eaton Corp. with technical help from Heron Rodriques who had worked at Supermet and Yasu Saegusa who was a Sumitomo employee.

Produce: regular P/M

Engineered Sinterings & Plastics, Inc.
Watertown, CT

In 1944, its founder Alexander L. Alves attended a Society of Mechanical Engineers meeting and was much impressed by powder metallurgy as presented by a speaker from Metals Disintegrating of Elizabeth, NJ. With great forethought, in 1944, he bought a house with a small separate building out in back. The building was then the incubator for four successive businesses. Starting in 1944, as a hobby, using the little building, a small press and a furnace, Alves made P/M parts under the name Molded Metals. From 1944 to 1951 he worked as General Mgr. at Watertown Mfg., a maker of thermoset molded products. A fellow employee Bill Quinn, a toolmaker, used to make the P/M tools during the evening. About 1949, Watertown Mfg. bought Molded Metals and put Bill Quinn in charge of running it. By 1951 there was a change in ownership at Watertown Mfg. and Alves decided to make a change. One Clayton Gengras owned American Tool Works. In 1951, at American Tool Works, the owner and Alves started American Sintering. Bill Quinn joined them in March 1951. In 1953, at age 50, Alex Alves and Bill Quinn, and some other smaller investors purchased American Sintering from American Tool Works for $100,000. The business moved into a new 6,000 sq. ft. building which they outgrew in 8 years. They added a line of molding equipment and renamed the company Engineered Plastics and American Sintering. The P/M part of the business grew the fastest and so in 1960 the name was changed to Engineered Sinterings and Plastics. In later years, ESP purchased the assets of Dixon Sintaloy, New Industrial Techniques, and Pitney Bowes captive P/M department.

Produce: regular P/M, P/F, and metal injection molding

Enhanced Sintered Products, Inc.
St Marys, PA

Started in 2001 by Joseph Lecker and others who had worked at Fairview Sintered Metals.

Produce: regular P/M

Exotic Metals
Ridgway, PA

Started in 1967-68 by Robert Howard who had worked at IPM (International Powder Metallurgy)

Sold to Masco Corp. in 1990's and renamed MascoTech Sintered Components. Renamed Metaldyne Sintered Components in 2001.

Do regular P/M and P/F

F

Fafnir Bearing Corp.
P/M parts department

started in 1975 in Newington, CT, to make bearing housings;self taught. Later moved the department to Pulaski, TN, and closed it in 1997.

Produce: regular P/M, captive

Fairview Sintered Metals Inc.
St Marys, PA

Started in 1978 by Luke Wehler who had worked at Keystone Carbon Co., Tim Wehler, Jerry Wehler, Paul Shaffer, and Tom Holjencin who worked at Powdered Metal Products

Sold in 1999 to GKN/Sinter Metals, St. Marys

Produce: regular P/M

Falls Creek Powdered Metals

Wilcox, PA

Started in 1999 by Hamid Attas who had worked at Cameron Diversified Products.

Produce: Regular P/M

Fansteel Corp.

Chicago, IL

During World War II, U.S. government set up the Tantalum Defense Corp. managed by Fansteel to make tantalum powder and to form it into wire, following a procedure similar to what we earlier describe for making tungsten wire. The tantalum wire was woven into a mesh which could be implanted in a persons body who had a serious war wound. The human body does not reject tantalum. By 1949 they were trying to press iron powder into parts to make gun sights. This effort was assisted by Gene Andreotti in 1950. By 1953, Fansteel recognized that the business was not as profitable as their other lines like the contact points for cars, and the iron P/M parts were discontinued. As an example of what goes around comes around, in 1996, they bought American Sintered Technologies of Emporium, PA, a producer of regular P/M parts!

Federal Mogul Division

Federal Mogul Bower Bearings

Greenville, MI

Forging technology purchased from John Haller, between 1960 and 1965,

See also Haller Powdered Metals. Member in-plant P/M group in 1963

Produce: P/F

Federal Mogul Sintered Products

Dayton, OH and Marshall, IL

Marshall plant sold to GKN in 2000

See also Supermet.

Produce: regular P/M

Federal Mogul Sintered Products

Waupun, WI

See also Comtech.

Produce; regular P/M

Federal Mogul, Precision Forged Products Div.

Romulus, MI, also a plant in Gallapolis, OH. Sold to Borg-Warner in 1995 and then to GKN Sinter Metals

See also Haller Powdered Metals.

Produce: powder forged materials

Federal Mogul Assembled Camshafts

Grand Haven, MI

Started as Weyburn Assembled Camshafts, a joint venture between Nippon Piston Ring and Weyburn. They produce pressed and sintered lobes for cam shafts. The green or presintered lobes are placed on the shafts, and the whole assembly is vacuum sintered with the cams going through a liquid phase. They shrink and bond to the shaft. Federal Mogul bought this operation about 1998-99.

Produce sintered lobes for cam shafts

Ferraloy, Inc.
Salem, IN

See also Ferro Powder Metals and Wel-Met.

Produce: regular P/M

Ferro Powder Metals
Kent, OH

In 1954, Jess Helsel and Ferro Corp. bought Wel-Met and changed the name to Ferro. In 1957, Ferro was then sold it share to Jess Helsel who changed the name to Ferraloy and moved it to Salem, Indiana *See also* also Wel-Met.

Produce: regular P/M

Flomet, LLC.
DeLand, FL

See also Brunswick Technetics Weitec.

Fine Particle Technology
Camarillo, CA

Holding company for Multi Material Molding of San Diego, CA

Produce: metal injection molding

FMC Chain Division of Link Belt
Indianapolis, IN

Link Belt was founded in 1875 and by 1915 was making cast or wrought sprockets and timing chains for automobiles. P/M department started about 1960-61 with technical support from Duane Gambrell at Cincinnati, Inc., a press manufacturer. Produced P/M rollers and crank and cam shaft sprockets for the company's line of timing chain. Chains were sold for commercial industrial applications and also for automotive use. Prior to 1967, using sponge iron, sprockets were compacted with no carbon, sintered, carburized, and then induction hardened. After 1967 with the advent of atomized iron powder, sprockets were compacted with Fe + .9% graphite added, sintered and then heat treated. At peak, making about 3,000,000 sprockets per year. To settle litigation in 1987, the P/M equipment and related business was sold to Borg-Warner, and set up in Ithaca, NY. The commercial and industrial chain business went to Rexnord.

Produce: regular P/M captive

FMS Corporation
Minneapolis, MN

Founded in 1946 by Fred Sweet and originally manufactured parts for the outdoor advertising electric sign industry. Initially, the powder metal manufacturing process was limited to a small department within FMS, manufacturing cores, nuts, and other structural components for FMS's sign controllers. Gerry and Doug Sweet took over ownership in 1971. From the 1970's to the 1980's, the P/M department grew from a captive to a contract manufacturer of P/M components. Greg, Mike, and John Sweet took over ownership in 1998, and FMS formally split into two divisions, The Powder Metal division and the Sign Products Division.

Produce: Regular P/M parts

Ford Aerospace and Communications Corp.,
Aeronutronics Div.
Newport Beach, CA

In 1981, started to develop a way to make a missile fin base. Not clear if they ever went into mass production.

Produce: metal injection molded, captive

Ford Motor Company
P/M Department
Rawsonville, MI

Started in the Rouge River Plant about 1945 ± 3. John Gerzina remembers that when he worked at Keystone Carbon in 1940 that there were only three other P/M houses in the U.S., namely Chrysler Amplex, GMC's Moraine Products and Bound Brook Bearing Co. This sets the early date at 1940. Without many cars being made during WWII from 1941 to 1945, there was probably no reason to start a P/M parts department.) The P/M operation was managed at an early date by Ed Ensign who in turn hired Pete Schneider. (Later, IBM hired Pete Schneider to run their captive P/M operation). About 1952, John Gerzina visited/worked at the Rouge River plant for about two weeks. They were making oil pump gears: they had thirteen presses of which only two were running. He suggested that a P/M operation did not belong in the Rouge River plant, with its foundry, forge, and steel making. The P/M operation was subsequently moved to Rawsonville, MI around 1953. At that time they had 15 P/M presses. Joe Farmer was instrumental in the move to Rawsonville and the plant was managed by "Jet" Perrelli. They were trying to make shock absorber bushing parts. Between 1964 and 1974, they added 26 Cincinnati presses. John Cenko was very active at Ford Rawsonville.

Ceased P/M operations about 1994.

Produce: regular P/M

Forged Performance Products
Oak ridge, TN

Started in 1993 by Robert Taverner and Ed Hodge, both of whom had HIP'ing experience at Battelle. In 1995, added investment from others, renamed Format Industries and moved to Knoxville, TN. Sold to Ametek in 1997, moved to Wallingford, CT and absorbed into Ametek. In pneumatic isostatic forging, a preform at 85% density or higher is heated in a protective atmosphere, with or without an impermeable coating. It is transferred to an isostatic press and rapidly pressurized in high pressure, high viscosity gas, (30 to 60 sec. cycle). The pores collapse under the onslaught of the viscous gas, without significant gas penetration, and the part is driven to full density or near full density.

Produce: pneumatic isostatic forged parts

Form Physics, Inc
San Diego, CA

Started in 1984 by John MacPherson, Jim Adee, and Mark Eidson. Since 1972 MacPherson had owned a prior MIM company called Powmet and had worked earlier at Rohr and then Parmatech. Later managed by Stan Zalkind who had considerable experience operating other companies. Liquidated in 1988.

Produce: MIM, metal injection molding

Form Physics LP.
San Diego, CA

Started in 1996 by Ian Clark and Stan Zalkind in the same building as the prior Form Physics.

Produce: MIM, metal injection molding

Forsman Sintered Products
Caribou, ME

Started in 1991 by Ron Chand of Chand Associates to make sintered filters, primarily for sale to Chand Associates.

Produce: filters, captive

Friction Products Co.
Medina, OH

Started by Fran Loewy in 1961, after he left Sintermet.

See also S. K. Wellman.

Sold to Hawk Group

Produce: friction materials

G

General Electric Co.
Plainville, CT

P/M operations managed by Charles E. Ricker, active in 1981. Ceased operation in 1995?

Produce: brushes for electric motors

G.M. Root, Inc.
Lackawanna, NY

Started in 1989 by Richard Root. Technical help from Joe Drzewiecki who had worked at the captive Trico P/M operation in Buffalo, NY. They make a stainless steel insert that goes in tire vents (you know, that funny little rubber thing through which you re-pressurize your tire.)

Produce: regular P/M, SS

General Metals Powder Co.
Akron, OH

In 1929-30, John S. Fisher developed a way to make electrolytic copper powder. In 1931, he worked out a way to consolidate the copper into a friction material that was sold to the Twin Coach Company of New York, as a clutch lining. 1932 saw the formal incorporation of General Metal Powders. In the 1930's Cu-based clutch materials were sold for use in machine tools, aircraft, and farm implements. In World War II, the company made friction discs for auto-pilots on airplanes and gun turret clutches. After the war, they were heavily involved in friction materials for automatic transmissions in cars.

Produce: friction materials

General Motors Corp.
Buick Motor Div.
Flint, MI

In 1969 actively powder forging side pinion gears and trying to to do the ring gear for the differential. (LP has one of the Buick side pinion gears which the metallurgist him urged him to cut up and mount for metallography.) Neither Buick nor anyone else ever developed an economical P/F side pinion gear.

Produce: P/F gears

General Motors Corp.
Cadillac Motor Div.
Detroit, MI

In 1969 George Sukes was actively interested in P/M, possibly because Cadillac had successfully made and tested powder forged connecting rods in 1964. Not adopted for engines until the 1990's. (As they say on PEI, "Live horse, 'til you get hay!")

Produce: never in production by Cadillac Div.

General Motors Corp.
Delco Moraine Div.
Dayton, OH

See also Moraine Products Div.

General Motors Corp.
Delco Remy Div.
Meridian, MS

Started in 1977, making the 700 g inner frame for a G.M. 5MT Cranking Motor, using 40,000 lbs/day of iron powder. Prior to P/M manufacturing, the plant had been making starting motors and cranking motor frames by cold forming or casting. Developed technology to make the powder forged connecting rods. Name change to Delco Remy America. Sold in 1996 to Sinter Metals then to GKN

Produce: regular P/M and P/F

General Motors Corp.
Delco Remy, Magnaquench Div.

Started in 1985. Spun off as Magnaquench, International.

Produce: Hot and Cold pressed NdFeB magnets

General Motors Corp.
New Departure Hyatt Div.
Sandusky, OH

Started a P/M department in 1978. Technical help from John Cenko.

Ceased operations at a time now unknown.

Produce: P/F bearing races and in-house grinding wheels

General Motors Corp.
Saginaw Steering Gear Div.
Saginaw, MI

Starting around 1964 with technical help from William Allread, who had worked at GM Tech. Center, they used a double press and double sinter process to make a near net shape preform. The preform was then cold forged to make a pump housing. At least two different parts were tooled. A 3000 ton press was available to do the cold forming. This may have been the incubator for later work on cold forging done at AC Electronics Div.

General Motors Corp.
AC Electronics Div.
Flint, MI

Between 1971 and 1980, 150,000,000 spark plug shells were cold formed from iron powder. The powder was pressed to 7 g/cm^3, sintered at 2400 to 2475°F in a walking beam furnace, cold extruded to 7.6 g/cm^3, resintered as above, and extrude cold to 7.8 g/cm^3. Discontinued after 1980 when cold forming of wire became more economical.

Produce: cold forge P/M, captive

Gibson Electric
Delmont, PA

Started prior to 1946

Sold to GTE in 1988.

Produce: electrical contacts

H

HazenTec Llc.
Hazen, AR

See also Remington Arms.

Haller Powdered Metals
Northville, MI

Uncertain about the exact name, and uncertain about the timing and sequence of John Haller's business ventures.

At the earliest time of which we have knowledge, John Haller had a a tool and die business that somehow resulted in production of (stamped?) toys like metal shovels and rakes for children. He then made P/M tooling for the Ford Motor Company captive P/M operation at the Rouge River Plant. From that experience, he began making P/M parts and some of his own presses, probably about 1946. Location of this original business is uncertain but may have been closer to Detroit. We do not know its name. He sold the operation in 1950 to Allied Products, but continued his interest in P/M and powder forging. Under Allied, the P/M plant was renamed Michigan Powdered Metals and was managed by John Gerzina from 1951 to 1954. This plant was in Livonia, MI.

Haller started another P/M operation at Northville, MI, some time after 1950. It had a separate research building where he pursued development of powder forging around 1958. The press and P/M parts business was using about 500 ton/month of iron powder. Between 1960 and 1965, Federal Mogul bought this business and the related forging technology. It was then called the Haller Div. of Federal-Mogul-Bower Bearing, Inc. They closed the Northville plant and set up for powder forging in Gallipolis, OH. Later they built a powder forging plant at Romulus, MI. The whole powder forging business was later sold to Borg Warner, and then to GKN Sinter Metals.

Produce: regular P/M and P/F

Hamilton Standard Div. UTC
Windsor Locks, CT

Operated a small captive P/M department in the 1970' and 1980's to make smaller quantities of P/M parts.

Produce: regular P/M, captive

Helsel Inc.
Campbellsburg, IN

In 1974, Jess Helsel bought a heat treating company in Indianapolis, Qualimet, and moved it to Campbellsburg, IN, re-naming it Helsel Metallurgical and then Helsel, Inc., again starting in the P/M parts business. Jess Helsel had previously operated Ferraloy for many years.

Helsel Inc. was sold to Hawk Group in 1994.

Produce: regular P/M

Hi-Tech Sintered Metals, Inc.
Naugatuck, CT

Started in 1991 by John Delos Reyes who had worked at Engineered Sinterings and Plastics

Produce: regular P/M

Hitachi Powdered Metals
Edmore, MI

Started by General Electric in 1952 as a P/M plant to make magnets. Purchased by Hitachi in 1973.

Produce: magnets

HJE Company, Inc.
Glen Falls, NY

Started in 1991 by Joseph Tunick Strauss who had P/M experience at RPI, Valeron Corp. on tool steels, and at National Diamond Lab in Los Angeles CA, a producer of hot pressed diamond and metal powder cutting tools.

Produce: metal injection molding

Hogen Industries
Mentor, OH, and Willoughby, OH

Started in 1959 by a Thomas Hogen. Sold to Mi-Tech Metals in 1995.

Produce: Cu-W alloys

Holley Carburetor
Warren, MI

Department started in 1974 and ceased operations in 1987

Produce: regular P/M, parts for pollution control, air pump, in cars, captive

Homogeneous Metals, Inc.
Herkimer, NY and then Clayville NY

Started by Joseph Wentzell in 1965 to make superalloy powders. Sold to United Technology Corp. in 1975. Prior to 1975, they began canning and HIP'ing or extruding superalloy billets for aircraft engine applications.

Produce: superalloy mill forms

Honeywell, Inc.
Golden Valley, MN

Started in 1959 under Len Mayerson, a Honeywell engineer. The first press was 13 ton Haller, followed by a 66 ton Bussman. By

2002, they had 5 presses and were making 3.5 million pieces per year.

Produce: regular P/M captive

Horizon Technologies, Inc.
St Marys, PA

Started in 2001 by Eric Wolfe and others from Windfall Products, which company had earlier been started by the Wolfe family and others.

Produce: regular P/M

Howmet Corp., HIP
Whitehall, MI

In 1981 they began to make abradable seals for jet engines. They are a HIP'd product that is a mixture of metal and oxides. They were attached to blade tips and wore a track in a mating outer casting. At temperature, the tips were very close to the outer casting's track, and sealed well, preventing gases from bypassing the rotor blades. The abradable tip was eventually replaced by a sprayed deposit. Currently make other HIP'd materials, for example sputtering targets.

Produce: HIP'd parts

Hydramatic Div., GMC
Ypsilanti, MI

Started in 1975-76 to make over running clutch races for automatic transmissions by powder forging. Technical guidance from GMC engineers Clyde Buxton and Al Stokes. Ceased operations in 1991

Produce: P/F

I

ICM/Krebsoge
Livonia, MI

Holding company that owned the Amplex VanWert OH plant, the St. Thomas plant and Salem, IN plant. Sold to Sinter Metals, Inc. in 1997

Ilikon Corp.
Needham, MA

In mid-1960's, Lazlo Bonis was President, and had technology to sell on specially alloyed P/M parts.

Imperial Sintered Metals, Inc.
St. Marys, PA

Started circa 1958-1961 by Arthur VanSlander who had worked at Keystone starting in 1936 , Kenny Renwick who had worked at Keystone Carbon Co. starting in 1942, and investors: Paul Gardner, George Young, Herb Schreiber, and Allie Wegemer.

Operations ended in 1990

Produce: regular P/M

Industrial Materials Tech
Andover, MA

Founded in Cambridge, MA, in 1968 by Dr. Robert Widmer, a Research Associate at MIT and a group of investors from MIT and Robert Wilder, Chairman of National Forge Co. By 1969, National Forge bought out all of the other investors and IMT became part of National Forge. The plant was then moved to Woburn, MA. The intent was to make HIP'd P/M parts but toll HIP'ing was a growth business and became the focus of the company. The plant was later moved to Andover, MA. In 1992, Bodycote International, plc bought IMT.

Produce: Toll HIP'ing and CIP and HIP parts.

Industrias Coronet, S.A. de. C. V.
Monterrey N. L. 66350
Mexico

Produce: regular P/M

Ingersoll Rand
Proto Tool Div.

Powder Forge Department

Wichita Falls, TX

Started in 1977 by Dr. Banerjee from the Ingersoll Rand Research Dept.

Difficulty competing with imported wrenches. Ceased operations in 1983.

Produce: P/F open end wrenches

Injectamax
Escondido, CA

Was AMAX-RISI

Founded in 1988 by Yancey Smith, an AMAX employee.

Produce: metal injection molding

Injection Molded Metal Products
Wilsonville, OR

See also its successor, Kinetics.

Produce: metal injection molding

Innovative Sintered Metals, Inc.
St. Marys, PA

Started in 1997 by Jay Draper who had managed American Powdered Metals and worked at Ridgway Powdered Metals.

Produce: regular P/M

Intech Metals, Inc.
Ridgway, PA

Started in 1985 by David Hanlin and Richard Phillips. In 1995, Mr. Hanlin acquired the whole company and changed the name to Intech P/M Stainless, Inc. Name change to Intech Stainless, Inc. in 2001

Produce: stainless steel in segregated, dedicated plant, also filters and magnetics

Intech P/M Stainless
Ridgway, PA

See also Intech Metals.

International Business Machines
Endicott, NY
Poughkeepsie, NY

During the early 1950's, a captive producer of P/M parts for IBM. Peter Schneider was hired from the Ford Motor Company Rouge River Plant and was in charge of the P/M lab. Harry Ambs was a young engineer in those days and has provided this information on the P/M operation: IBM made or used 250 - 300 different parts for typewriters and card carriers. Many of these parts were made at Endicott, NY. Iron parts were double pressed and double sintered electrolytic iron, HVA Star and later, Glidden A210. At that density, the case hardened parts had well defined cases. They made sintered bronze bearings, controlling sintered dimensions by varying the proportion of coarse and fine tin to suit each in-coming copper lot! Pete Schneider fostered the MPIF In-plant Powder Metallurgy

Association, trying to spread knowledge of P/M throughout industry. He encouraged the writing and publishing of detailed meaningful papers to help others to produce P/M parts. IBM had a separate P/M lab at Poughkeepsie, NY where Athan Stosuy played an early role. Harry Ambs saw the writing on the wall at IBM when part of their P/M plant was later used for a wave solderer for printed circuit boards!

Produce: regular P/M, captive

International Business Machine Corp.
Information Products, Div.
Lexington, KY

In 1981, they became a Parmatech licensee to make MIM parts for their Selectric Electric typewriter, e.g. the tilt ring, hammer and ribbon meter. Technical direction by George Wood and Ron Miracle. By 1988, the department was being phased out with the intention to purchase parts from outside vendors.

Produce: metal injection molded, captive

International Powder Metallurgy, IPM
Ridgway, PA

Started in 1948 by Gib Hoehn, who had worked for Keystone Carbon from 1933 on, and was joined a year later by Mario (Spike) Victor, a Salesman who had worked for Keystone Carbon Co. Sold in 1965-68 to Allegheny International. Sold in 1984 to the management team, James Delhunty, Rudy Ferragine, Howard Peterson, Dallas Kelley, (who had been at IPM since 1949) et al. Name changed to Clarion Sintered Metals in 1984. Part of it became the Columbus, OH, P/F facility for Keystone Powdered Metals

Produce: regular P/M and P/F

International Sintered Metals
Guelph, Ontario, Canada

Sold to ICM which became GKN Sinter

Metals. *See also* Bundy of Canada.

Produce: regular P/M and P/F

Isostatic Industries

cf. Bearing Sales

Isthmium Metals, Inc.
Waltham, MA

Active in the 1950's and 1960's with presses and furnaces, doing development work. One principal was Gregory Comstock. Held many P/M patents including ones covering infiltration of iron with copper, which delayed general acceptance of copper infiltration. Sold to Sintered Metals of Jamaica Plain, Boston, MA in early 1970's. Mike Jaffee has these patents.

Iverpac Corporation
Huntington Beach, CA

Started by John F. Iversen about 1974. He had P/M experience working at Kwikset and MDS.

Sold to Precision Powdered Metals, 1999

Produce: regular P/M

J

Jacobson Group

Kenilworth, NJ

Captive PM producer

Johnson Bronze

Newcastle, PA

Started in 1890 to make cast bronze for ship bearings. In P/M business in 1944. Purchased Ferraloy in 1960 , and successively absorbed by Gould, Pullman, ICM, Sinter Metals and GKN

K

Karsten Manufacturing

Phoenix, AZ

Started in 1981, based on technology obtained in purchasing a P/M house, in Sante Fe Springs, California, which had technology developed by Orville Bowen. Terry Driescher was active at Karsten. Used cold isostatic pressing techniques to make golf ball molds. Ceased operations about 1995-97. Now owned by golf club maker Ping.

Captive P/M for golf ball molds

Keystone Carbon Company

Started in Emporium PA in 1927 and moved to St. Marys, PA in 1931, by Ben Reuscher and Adolph Schmidthammer, who had worked at Speer Carbon and Jerome Lanzel.

Name Change in 1997 to Keystone Powdered Metal Company. Many P/M houses trace their roots to Keystone Carbon.

Produce: regular P/M, P/F

Kinetics Inc.

Wilsonville, OR

In 1982, Omark Industries, established a development program in metal injection molding with Fred Croydon, Darrel Teel, and Sil Malalan. They were using Brunswick Technetics technology and equipment at the start and then developed other technology. In 1985 Blount purchased Omark and changed the name of the MIM operation to IMMP (Injection Molded Metal Products) In 1988 they moved to a new facility and commenced making MIM parts for the merchant market. In 1995 Walter Weyler purchased the operation from Blount and renamed it Kinetics.

Produce: metal injection molding

Kodiak Metals

Grand Haven, MI

Started in 1979 by Earl Burlingame who had worked at Cleveland Powdered Metal, Southern Sintered Metals, and Bear Metals in Battle Creek, MI, which he founded in 1975. He left Kodiak Metals in 1983 and established Polar Tech in DuBois PA. About 1985, Kodiak was purchased by Herb Tews who was a principal in Midwest Sintered Products. Kodiak was renamed Partnership P/M which he operated with Bob Koch. Bob Koch had previously worked at Supermet. *See also* Pleuco.

Produce: regular P/M

Krupp Engineering
Dexter, MI

See also Consolidated Metallurgical.

Kulite Tungsten Corp.
Rutherford, NJ

Started in 1949 by Jacob Kurtz who had worked as Director of Research at Callite Tungsten in New Jersey. Callite processed tungsten from the ore, all the way to lamp filaments and ceased operations about 1949. Kulite was sold 1997 to CSM a spin-off from AMAX and Climax. Resold in 2000 to H. C. Starck

Produce: refractory metals

Kwikset Lock Co.
P/M Department
Anaheim, CA

Kwikset existed as a lock manufacturer making locks from wrought brass materials. In 1953, they set up a department to make P/M parts for locks, and also to sell P/M parts in the merchant market. In 1952, they had a contract to make military fuse parts from brass and wanted to make them via P/M. The P/M expertise came from Philip Tarr, who had worked at the Waterbury Button Co. Around 1958-59, they were making cone shaped oil well perforating devices by gravity sintering bronze powder around conical resin bonded sand cores that formed the ID's. The cores were made by Donald Paullin, who had worked at Kwikset but was then at the very start of his Pacific Sintered Metals P/M business. Illustrating the meaning of "What goes around comes around", Pacific Sintered Metals bought Kwikset in 1985!

Produce: brass P/M parts

L

Lane Engineering Services
Coldwater, MI

Started by Robert Pettibone who had worked at American Electro Metals and Eaton Yale and Towne.

Ceased operations in 1990's

Produced: specialized P/M parts

Laurel Manufacturing
DuBois, PA

Started in 1984 by Dennis Heindl who had worked at Atlas Pressed Metals, Compacted Powder Metals and St. Marys Carbon Co. Sold to GKN in 1998, renamed GKN Sinter Metals-DuBois

Produce regular P/M

Leco Carbides
Hendersonville, TN

Started by Dr. Patts. Active in late 1980's and 1990's. Gone by 2000.

Produce: injection molded carbides

Lionel Corp.
Irvington, NJ

In May of 1950 they were making P/M parts for their model trains. A collector roller (for current?) was made from Fe + 10% Cu + 3.5% graphite + .5% Ca stearate. Managed by Joe Bonnano and Robert Bouman

(Member in-plant P/M group in 1963)

Lovejoy Mfg.
Downers Grove, IL

Produce: regular P/M, captive

Lux Timer
Waterbury. CT

Started about 1955 with technical help from Robert Talmadge. They had 6 presses and a furnace. They produced gears, pinions and bearings for their timers. Henry Bender was active.

Purchased by Robertshaw Fulton and moved to the Southern US. No further information available.

Produce: regular P/M parts, captive

Lyon Powdered Metals
Lyon, MI

Started about 1985 by Robert Peterson who owned Major Gauge and Tool.

Ceased operations in 1999

M

Magnum Powdered Metals
Emporium, PA then Streetsboro, OH.

Started in 1994 by Richard Sebring, who was from the carbon industry.

Closed in 1999

Produce: regular P/M

Mahle Valvetrain
Holland, MI

See also Pleuco Technologies.

Mallory Metallurgical
Indianapolis, IN

See also CMW, its' successor.

Produce: Cu-W, and electrical contacts

Martin Sprocket and Gear
Paragould, AR

The parent company started to make P/M bushings and steel parts in 1971. Some technology came from a P/M operation, East West Industries in Piggot, AR, whose assets they purchased.

Produce regular P/M

Martin Technologies
Huntsville, AL

Started in 1976 by Tony Martin as a tool shop making P/M tooling. It was based on his prior company, Martin Bros, Toolmakers in Park Royal London, which he had started in 1954. About 70% of the P/M tools which he made in England were exported. In England, they also represented Höganäs for the sale of iron powder.

In 1983, they started Sinttech in Huntsville AL to make P/M parts. In 1998, the Sinttech name was discontinued because of a name and identification conflict with an unrelated company with a similar name. P/M parts were then made under the Martin Technologies name.

Produce: regular P/M

MascoTech Sintered Components
Ridgway, PA

See also Exotic Metals.

Massachusetts Powdered Metals, Inc.
Worcester, MA

Around 1946 Morris Boorky of Mass. Steel Treating and Carl Johnson of WPI began to make powder metallurgy parts. The company was later renamed PresMet Corp.

Master Powdered Metals
Southfield, MI (on 8 Mile Rd.)

Started about 1968 by Master Spline which was part of Michigan Spline. Arthur Wash active in 1970. In the early 70's they were making parts for Burroughs Corp. which had captive P/M operation at least up until 1963.

Produce: regular P/M

Materials Research Corp.
Orangeburg, NY

Started by Dr. Sheldon Weinig in 1957. Sold to Sony in 1995 and then to Praxair in 1998.

Produce: Specialty materials by hot pressing and vacuum hot pressing.

Maxtech Powder Metal
Quebec City, Canada

See also Precitech.

Maytag Company
Hampton IA, Newton, IA and later Jefferson City, MO.

Started in 1949 with early engineering help from Jim Gamble, George Otto and Arthur Irvine. At Jefferson City, they had a very highly automated plant Sold in 2002 to Porite.

Produce: regular P/M, captive

MDS
Anaheim, CA

Started prior to 1974 by Dr. Peter Schwarzkopf (son of the founder of American Electro Metals) and the Mark Steel Stamp Co. (Small world note: Peter Schwarzkopf and LFP were office mates in graduate school in 1960) The company was best known for making cold isostatically pressed chess pieces and whole sets, commissioned by Alcoa, to promote aluminum P/M.

Ceased operations at a time now unknown.

Megamet Industries
Earth City, MO

Started in 1989 in St. Louis, MO by Gregory M. Brasel and his wife, Susan Brasel. Initial efforts were to develop MIM technologies. After a serious flood in 1993, focus changed to making MIM parts. Later, Megamet LLC was spun off to manufacture MIM parts, predominantly for the firearm market. In 2001, Megamet Industries changed its name to USMIM, Inc., to continue to develop the MIM process and equipment.

Produce: metal injection molding

Merriman Brothers
Boston, MA and then Hingham, MA.

A producer of marine hardware. P/M division started between 1946 and 1949 with the inventive help of Andrew Wayson who had developed the graphite pocketed bridge plate bearings. The company bought a press and a furnace and then looked for business. Lionel Trains asked them to make a P/M helical gear. For his first job in 1950, he used his bridge plate technology to support the die and allow it to rotate down onto the helical lower punch. Sold to Union Twist Drill in 1963, who wanted the marine hardware line, believing that in the future, folks would work 30 hour weeks and have lots of time to own boats! Sold to Litton Industries in 1968, who promptly got rid of the marine hardware line. In 1984, leverage buy out by Jack Williams and the management, combining Merriman and five other Litton divisions, with the holding company called Quamco. Sold to James Horan and others in 1988. Sold to PCC Airfoils in 1995. Ceased operations in late 1998, with assets sold to Capstan Industries.

Produce: regular P/M and spinning rings

Metafusion
Kersey, PA

Started in 1991 by Pat Gaffey, Raj Ranadive and Richard Uhl Sr. to handle the overflow from their Sinterphase operation in Johnsonburg, PA

Sold 1999 to GKN, became GKN/Sinter Metals, Kersey.

Irishtown Rd,

Produce: Regular P/M

Metal Ceramics
Franklin Park, IL

Started in 1951 by John Skach, Sr. who had P/M experience at Chicago Powdered Metals. In 1948 he established a tool and die business, making tooling for pressing metal powders, which lead to an interest in making parts in 1951. He had been exposed to powder metallurgy in college at University of California in the 1940's and in the tool and die apprentice program of Vultee Aircraft, during World War II.

Produce: regular P/M

Metal Molding Technology, Inc.
Margate, FL

Started in 1992 by Eugen Dragan who had worked at New Industrial Techniques.

Produce: metal injection molding

Metal Powder Components
Coldwater, MI

See also RB&W.

Metal Powder Products
Dayton, OH, then moved to Logan, OH

Started in 1946 by John Hildebolt who had worked at Moraine Products Div. of GMC. Moved to Logan in early 1950's. Purchased by Spartec, a ceramic company. Spartec then bought by Quadrant in 1972. Arlan Clayton appointed to run the Logan operation in 1978.

Renamed Metal Powder Specialties as part of the Metal Powder Products holding company. Sold to Charter Oaks Partners in 1998.

Produce: regular P/M

Metal Powder Specialties, Inc.
Logan, OH

See also Metal Powder Products.

Metalkraft Industries Inc.
Wellsboro, PA

Started in 1984 by Wayne Singer who had worked at St, Marys Carbon Co.

Produce: regular P/M

Metallamics, Inc.
Traverse City, MI

Started in 1988 by Dr. Robert MacDonald who had experience in pseudo-isostatic compaction of borides and intermetallics at Dow Chemical. At first, it was intended that Dow would license this compaction technology to be used at Metallamics. Through a management change at Dow, no agreement could be reached. Metallamics went on to do HIP'ing of nickel aluminides which were sold to companies who were hot pressing NdFeB magnet materials at 750°C and needed tooling material with high temperature strength. That market dried up and Metallamics ceased fabricating metal powders in 1993.

Produce: metal injection mold, pseudo-HIP

Metals Consolidating
Long Beach, CA

Robert W. Haley, active in 1970

Metal Sintering Corp. of America
Nazareth, PA

Started about 1969-1970 by Arnold Poster who had worked at GAF. The first P/M house started with a public subscription of stock. Discontinued operations in 1972 and assets were sold to PSI Div. of Warner Electric.

Produce: regular P/M

Metaltech, Inc.
DuBois, PA

Started in 1989 by Anthony M. Zaffuto who had worked at Symmco and Robert Hanak, a lawyer colleague.

Produce: Regular P/M

Metco Industries Inc.
St Marys, PA

Started in 1982 by Walter Spencer who had started Elite and also worked at Powder Metal Products. Sold in 1990 to Tom Bauer and Ronald Brennen

Produce: regular P/M

Michigan Powdered Metals
Livonia, MI

Started by John Haller about 1946 and sold to Allied in 1951 They renamed it Michigan Powdered Metals. This Livonia facility was eventually sold to Bog Warner

Produce: regular P/M, P/F

Mi-Tech Metals, Inc.
Indianapolis, IN

Started in 1978 by Wayne Seagren and Tom Duggen who had worked at P.R. Mallory and Ted Leslie with tooling experience from work at Microcarbide.

Produce: Cu-W

Michigan Sintered Metals
Owosso, MI

Started in 1979 by Pat Gaffey and Raj Ranadive who had worked at Carbon City Products. Sold in 1999 to GKN

Produce: regular P/M

Micro Ferrocart Products
Stamford, CT

Active during World War II, 1940 to 1945 making ferrite radio cores. In 1945, P/M operations sold to Kenneth Gleszer and Raymond Engstrand, located in Bethel, CT, and re-named American Sintered Products. They began to make P/M parts. Mr. Gleszer had worked for Moraine Products Div. of General Motors, with Dr. Fritz Lenel and Roland Koehring. Sold to Yale Lock and Hardware in 1950. Yale Lock and Hardware's P/M operation was sold in 1954 Gerry Fellows and others who renamed it American Powdered Metals.

See also American Powdered Metals.

Micro Metals
Jamestown, TN

Started in 1976 by Carl Edwards who had worked 25 years at National Cash Register, Dayton, OH, captive P/M operation.

Produce: regular P/M

Micrometals, Inc.
Anaheim, CA

Started by Richard Barden in 1951. He had some P/M experience through selling Stokes Presses.

Produce: electronics

Midwest Sintered Products
Chicago, IL

Started in 1955 by Jennings Crawford who had a brass cold heading company. Technology from Philip Tarr who had worked at Kwikset and Waterbury Button Co. Philp Tarr eventually owned the company.

It was sold to Herbert Tews in 1978 and then to Sinter Metals in July 1994, thence to GKN/Sinter Metals, and moved to Richton Park, IL. about 1999

Produce: regular P/M, SS, filters

M.I.I. Technologies, LLC
West Lebanon, NH

Started in 1990's by Glenn Beane and Alan Beane whose family had an ownership in Arvid Engineering, Laconia, NH. Developed their own line of P/M presses.

Soft Magnetics, High Stress P/M

Millett Sights Inc.
Huntington Beach, CA

Started by Raymond Millett in 1975. He had been a partner in Parmatech.

Produce: metal injection molding

Missouri Pressed Metals, Inc.
Sedalia, MO

Started in 1976 by Norman Bankovich who had worked at Keystone Carbon Co. and had been a founder of both Elco and Alpha Sintered Metals.

Produce: regular P/M

MLX Specialty Friction Materials Group
Solon, OH

Produce: friction materials

Mold-Masters Limited
Georgetown, Ontario, Canada

Started to produce MIM parts around 1994. The company was originally started as a producer of hot runner systems for injection molding tools. The MIM operation came out of their R and D department.

Produce: metal injection molded

Mohawk Sintered Alloys
Cobleskill, NY

Started about 1969 by Robert Perry who worked in an unrelated P/M area at General Electric in Schnectady, NY.

Active in 1981

Operations ended in 1980's

Produce regular P/M

Monroe Auto Equipment Co.
Cozad, NE

Started in 1968 to make parts for shock absorbers. Successor was Tenneco

Produce: regular P/M, captive

Monroe Auto Equipment Co.
Paragould, AR

Started in 1974 to make shock absorber parts. Successor was Tenneco

Produce: regular P/M, captive

Monroe Auto Equipment Co.
Owens Sound, Ontario, Canada.

Started in 1978 to make shock absorber parts. Successor was Tenneco

Produce: regular P/M, captive

Moraine Products Div. GMC
Dayton, OH

Started in the 1920's as part of the Dayton Engineering Laboratories Company under "Boss" Kettering. Athan Stosuy worked at Moraine Products and believes that one of the early objectives of the Dayton Engineering Laboratory was to make starting motors for automobile engines (borrowing a page from the motors NCR had in cash registers.) Starting motors would require bearings, and possibly sintered bearings. Another idea was to make sintered bronze main bearings for automobile engines. That application did not work, but sintered bronze bearings had a multitude of other uses. During World War II, they produced lead infiltrated iron tracer bullets for machine guns. They made a hot pressed multi-layered, laminated, copper and iron and steel plate brake for the B-29 aircraft, using a 2,000 ton press. This solved a heat dissipation and cracking problem on the existing brakes. (The hot press consisted of a furnace with a bellows in the roof that allowed the force of the press to be transmitted to the hot parts inside the oven.) The Dayton Engineering Laboratory Company became Delco Moraine. (A very early example of an Acronym! The "Moraine" refers to glacial moraines or hills that were found near Dayton and marked the southern-most extent of glaciers in the last ice age.) In its heyday, Delco Moraine was one of the 5 largest P/M operations in North America. At the end, some equipment and business were sold to Stackpole Ltd in 1981. The Meridian, MS powder forging plant was sold to GKN in 1998. In 2002, Delco Moraine Div. of Delphi Chassis continues to make "semi-metallic" brake pads for auto industry disc brakes.

Produce: regular P/M, filters, P/F and friction

Motor Wheel Corp.
Lansing, MI

Alexander Brede active in 1970. Ceased P/M operations at an unknown date.

Produce: brake drums by P/M

Mott Metallurgical Corp.

Originally in Hartford CT and moved to Farmington, CT in 1968.

Started in 1959 by Lambert "Bud" Mott who had received a Masters Degree in Metallurgy at Stevens Institute of Technology. He was a Research Assistant and Assistant Director of the Industrial Powder Metallurgy Lab at Stevens from 1947-54 . In 1954, he joined Cuno Engineering Corp. and was instrumental in building up their Connecticut Metals Inc., that specialized in porous metals production. Ultimately, Cuno exited this business and Lambert Mott went on to start Mott Metallurgical Corp. in 1959. He sold part of the company to Harry Gray Associates in 1993.

Produce: filters, mainly stainless steel

Mueller Brass
Port Huron, MI

Their P/M plant was located in the rear of their aluminum foundry at Marysville, MI, where they made P/M parts and filters for applications like soda fountains and other plumbing applications. Sales were about $300,000/yr in 1967.

Purchased about 1967 by Stu Yntema and Dick Thompson as the basis for their new P/M operation, Thompson Products.

Produce: regular P/M and filters

Multi-Material Molding, Inc.
San Diego, CA

Started in 1981 by Raymond Wiech and Guillermo (Bill) Verduzco

Merged into Fine Particle Technologies in. Ceased operation in 1990's

Produce: metal injection molded

Multimetals Div.
Vermont American Corp.
Louisville, KY

In 1987 began to make MIM metal parts and carbides. In 1988 decided to concentrate on carbides, such as the tips for consumer item, spade drills, sold by a sister division. Technical help from Claus Hagelganz.

Produce: injection molding carbides

Muskegon Piston Ring Co.
Muskegon, MI

Produced cast piston rings for many years prior to starting a P/M department in 1956 in their Sparta Foundry Div. near Grand Rapids, MI . The P/M parts consisted of valve seat inserts for Briggs and Stratton small engines. They also made cast valve seat inserts for larger engines. Goetze from Germany bought the company in 1979, and the name became Muskegon Piston Ring, Goetze or just MPR Goetze. In the 1980's with the advent of aluminum automobile engines, valve seat inserts were needed to protect the aluminum castings from the engine valves. In 1984 MPR Goetze sold the P/M department to Advanced Products Corp. The idea was that Advanced Products would make the valve seat blanks and MPR Goetze would machine them and sell them. In 1984-85, Advanced Products decided to establish separate plants for their regular press and sinter P/M parts, and to have an independent valve seat blank operation at Waupun, WI. That facility was called Comtech. In 1985, Marshall Wells of Wells Manufacturing bought Comtech. In 1990, MPR Goetze repurchased Comtech from Wells Mfg. In 1993 T and N from the UK bought Goetze world wide, including the Waupun facility. In 1998, Federal-Mogul bought T and N, and Comtech became part of Federal Mogul Sintered Products. (At the end of 2001, Federal Mogul entered Chapter 11 bankruptcy, for reasons independent of any of its P/M operations)

Produce: valve seat inserts

N

National Cash Register , NCR
Dayton, OH

Started around 1950. Captive regular P/M parts operation with high density parts for the mechanical components of cash registers. Ceased P/M operations about 1970 with the advent of electronic calculators and registers.

National Molded Products
St. Marys, PA

Started in 1942 by Hans Kerner, who had built P/M tooling at Keystone Carbon Co., William Strawbridge from Keystone Carbon Co., Walter Rigard, Ed Meier, (both bankers), Ray Harrigan, who owned a forge shop, and Ray's father-in-law. Sold in 1972 to McCrosky Tool. In 1982, the facility was moved to Franklin, TN. and combined with Southern Sintered Metals

Produce regular P/M

National Sintered Alloys, Inc.
Clinton, CT

Started in 1964 by William West who had previously worked at Norwalk Powdered Metals. Now owned by Steve and Greg West.

Produce: regular P/M

Neosid (Canada) Limited
Toronto, Ontario, Canada

Started in 1936 in England by Ed Michaelis. Started the Canadian facility in 1952. The original company became Magnetic Material Group, through a public stock offering. In 1992 that was sold to TT Group PLC and became TT Electronics. The company operates a facility in Patterson, NJ to make cores. It is called MMG/NA.

Produce: magnetic iron and ferrite cores

Net Shape Technologies
Solon, OH

Started in 1999 by Mike Klingersmith, Jay Schable and Norris Sales Associates. Moved to Solon, OH in January 2000. Controlling interest purchased by Hawk in Nov, 2000. Name change to Hawk MIM by summer 2002.

Produce: Metal Injection Molded

New Britain Machine Works
New Britain, CT

Started somewhere between 1955 and 1965 as a captive operation to make P/M pawls for ratchet wrenches in the company's hand tool division. Started by employee Bill Randolph and consultant Robert Talmadge. After 1970, this P/M department was moved out, possibly to Orangeburg, SC.

Produce: regular P/M, captive

New Industrial Techniques
Coral Gables, FL

Started in 1968 by the Manning family who owned a machine shop and Gene Andreotti who had worked at Burgess Norton and Kwikset. Sold to Handy and Harman in 1974. Sold to Engineered Sinterings and Plastics in 1993, equipment moved to Watertown, CT

Produce: regular P/M and metal injection molded

Newmet
Connecticut

Started prior to 1965 by William Johnson. About 1990 sold to Sintermetallwerke Krebsoge. Merged with Thermet in 1992. Then became part of Sinter Metals and then GKN, who moved it to the GKN Richton Park, IL facility in 1998.

North American Clutch
Milwaukee, WI

Started P/M department in 1980 by hiring employees with some P/M knowledge. Discontinued that operation in 1983. Re-started a P/M department in 1999 to make high volume, small parts for their own product line.

Produce: regular P/M, captive

Northern Industrial Manufacturing
Mt. Clemens, MI

Started in 1998 by Harvey Hohlfeldt

Produce: regular P/M captive

Northern Powdered Metals
Ridgway, PA

Started in 1997 by David Chittester who had some P/M experience.

Produce: Regular P/M

Northern Precision Materials
Minneapolis/St Paul area of MN

Started in 1984 by Roger Brown, an owner of several precision machine shops in Minnesota and Arizona. P/M technical help from Al Dornisch. Sold in less than a year to Del Bogema who owned a machine shop servicing the computer industry in Minneapolis. He renamed the company Paradigm. During 1985, operations ceased and the equipment was sold to Western Sintering.

Norwalk Powdered Metals, Inc.
Norwalk, CT

See also Copress and Pressco, its predecessors.

O

Oklahoma Pressed Metals, Inc.
Miami, OK

Started in 1992 by Norman Bankovich, as a division of his Missouri Pressed Metals Co.

Produce: Regular P/M

Omark Industries
Wilsonville, OR

See also its successor, Kinetics.

Produce: metal injection molding

Optomec Design
Albuquerque, NM

Started in 1998

Technical help from David Keicher

Produce: laser formed P/M parts

ORMCO, Inc.
Glendora, CA

Started a captive MIM plant in 1983. By the early 1990's, that work had ceased and they bought MIM orthodontic appliances from other MIM house and some investment cast orthodontic appliances from Mexico (based on a low labor cost).

Produce: metal injection molded, orthodontics, captive

Osram Sylvania
Towanda, PA

The Towanda facility for Sylvania Electric Products opened in 1941 making phosphors for radar screens and shortly, tungsten wire. In the 1950's, purchased by GTE. and in 1993 by Osram.

Produce: refractory metals.

P

P/M Krupp Technologies, Inc.
Chelsea, MI

See also Consolidated Metallurgical.

Produce: tool steel

Pacific Sintered Metals
Los Angeles, CA

Started in 1956 by Donald Paullin who had worked at Kwikset.

Produce: regular P/M

Pall Trinity Micro Corp.
Cortland, NY

Produce: filters

Paradigm
Minneapolis/St. Paul area of Minnesota

Successor to Northern Precision Materials.

Paradigm Sintered Metals
Battle Creek, MI

See also Bear Metals.

Paramount Sintered Products, Inc.
Zion, IL

Started in 1982 by Rolf Anderson who owned a tool shop making tooling for compacting P/M parts.

Produce: regular P/M

Parker Nichols Portland
Portland, ME

Successor to Sinterbond Corp. of Gloucester, MA in 1973

Produce:regular P/M, gerotors

Parker White Metal Co.
Fairview, PA

Active in 1970 with Tom Parker.

Produce: ?

Parmatech Corporation
Petaluma, CA

Started in 1973 by Karl Zueger, Raymond Wiech Raymond Millett and Philip Roth . It was one of the earliest producers of metal injection molded parts. Raymond Wiech may have gained some familiarity with metal powders and binders in earlier work at Thiokol, where he was involved in making solid rocket propellant, which is a mixture of aluminum powder and binders. Also, prior to 1971, he and Raymond Millett operated an MIM business called Cer-Plas in Huntington Beach, CA. Parmatech was Purchased from Karl Zueger by Carpenter Technology in 1995.

Produce: metal injection molding

Particulate Materials Technology
Pomona, CA

Started in 1991 by David Connelly, a principal in Precision Powdered Metals, who had worked at Ceromet, and before that, Asco Sintering Co.

Produce: metal injection molding, MIM

Paxol Limited
St. Thomas Ontario, Canada

Started in 1952 as a joint venture between Clevite and Manganese Bronze in England to make bronze bushings with the English technology. In 1956 Clevite bought out Manganese Bronze. The plant ultimately became the St. Thomas facility of GKN Sinter Metals. *See also* Clevite.

Produce: regular P/M

Pennsylvania Powdered Metals
St. Marys, PA

Started in Dec. 2001 by Dan Geci and George Heigel who had worked at Abbott Furnace.

Produce:regular P/M

Pennsylvania Pressed Metals
Emporium, PA

Started in 1966 by Kenneth Gerg who had worked at International Powdered Metals and Cleveland Powdered Metals, Tom Shields who had worked at Powder Metal Products, Bob Capello who had worked at IPM and Hans Arnold. Sold to Gleason Works, then to Consolidated Metal Holdings in 1992, then ICM. Became part of Sinter Metals in 1994, sold to GKN in 2000. Probably the seed from which sprang GKN/A-Z.

Produce: regular P/M and P/F

Pennsylvania Sintered Metals, Inc.
Emporium, PA

Started in 1990 by Steve Brown, whose formal education had included some P/M. Later founded Brownco.

Produce: regular P/M

Permaflow

Part of Capstan/Industries

Gardena, CA

Produce: Filters

Perry Tool and Research Co.
Hayward, CA

Started in 1962 by Perry Fusselman and his partner. They had been operating a tool and die business and happened to see a P/M press at a local trade show. They bought a press and commenced to make P/M parts.

Produce: regular P/M

Perth Metals
Stratford, Ontario, Canada

Started by Amplex Div. of Chrysler Corp. in 1958 and sold in 1991 to Stackpole Limited.

Produce: regular P/M

Philips Orien Powder Metal Molding
Menomonie, WI

In 1991 they began to make MIM parts with technical help from their employee, Roger Gehring. They developed their own binder system, making parts for their own product line of laparoscopic equipment, (tips, needle holders, and jaws). In 1995 they began to sell parts on the open market.

Produce: metal injection molding, captive plus merchant

Picco Corp.
South Elmonte CA.

Started P/M parts production in 1955. It had existed as a corporation making investment cast parts like jewelry and aircraft parts. It was owned by Lynn Smith and Mary McKenna Smith (daughter of the founder of Kennametal). They joined with Robert Remington to start to make P/M parts based on the assets of a pre-existing California P/M business, which name has been lost. The predecessor company was owned by George Vermeil (sp) who was in the plumbing hardware business. In 1948-49 he had set up in the Los Angeles area to make sintered brass plumbing parts such as nuts and bib caps. In 1962, Picco was merged into ASCO Sintering Co.

Produce: regular P/M

Pitney Bowes, Inc.
Stamford, CT

P/M Dept. started in 1948

Technical help from consultant Robert Talmadge, and employees Dr. Y. T. Chen and Frank Leone. Produced some very intricate P/M parts for the mechanical parts of postage meters.

Out-sourced parts and equipment in 1998 and 1999 to Engineered Sinterings and Plastics, closed operation in 1999

Produce: regular P/M, captive

Pittenger Powdered Metals
Tulsa, OK

Started in 1968 by Richard Pittenger, an electrical engineer by training, but with no other P/M experience. Earliest products were the gears for Zebco fishing reels. Sold in 1988 to Powder Metal Products, a company that was unrelated to Powdered Metal Products of St. Marys, PA. Sold in 1996 to William E. Martinson, Jr. and renamed Precision Sintered Parts, L.L. C.

Produce: regular P/M

Pleuco Technologies, Inc.
Holland, MI

Started in 1984 as Partnership P/M Corp. by Herbert Tews, a principal in Midwest Sintered Products and Robert Koch, who had worked at Supermet. Sold in 1995 to Pleuco and then to (MWP) Mahle Weitman Pleuco and finally renamed Mahle Valvetrain about 1999.

Produce: regular P/M and valve seat inserts

Polar Tech Corp.
Falls Creek, PA

Started by Earl Burlingame about 1983. Originally was a sales agency, and then began to produce parts in 1983.

Produce: regular P/M

Polaris Powder Technologies
Colorado Springs, CO

In 1984, Mark and Robert Anderson were active.

Produce: Tool steel parts

Polese Company, Inc.
San Diego,

Produce: Cu-W

Polyalloys
Los Angeles, CA

Started by Pacific Sintered Metals in 2001 with technical help from Pat Bhave and John Saunders. Pat Bhave had worked in MIM at Omark Industries and Thermal Technology.

Produce: metal injection molded

Porous Metals Ltd.
Guelph, Ontario, Canada,

Started by Burt Johnson in 1966. He had recently moved to Canada from England.

Produce: filters

Poud Press
Pointe-aux-Trembles, Quebec, Canada

Started in 1991 by Lucien Parenteau, who had worked at Perth Metals and at Domfer Metal Powders.

Produce: Regular P/M

Powder Alloys Corp.
Clifton, NJ

Started in 1958 by Jerome Kuzmick and Glen Goodfellow. Jerry Kuzmick was a Stevens graduate under Gregory Comstock. He had worked for Ekstrand and Tholand who were the exclusive importers of Hoganas sponge iron powder from Sweden, prior to the building of the Hoeganaes plant at Riverton, NJ in 1953. He also worked at Lionel. Jerry Kuzmick received the MPIF Pioneer Award in 2000. Sold to Teledyne in early 1970's. Re-named Teledyne Powdered Alloys. In 1997, there was a merger of Teledyne and Allegheny Industries, becoming Alldyne. In 1999, the plants were split up and the vestiges of Powder Alloys became the Tungsten Products Div. of Alldyne, in Madison, AL.

Produce: heavymetals

Powdercraft Corp.

Spartanburg, SC

Started in 1952 by Thomas Robinson . Frank Kacsandy, G. B. Smith, and R. E. Watling, all of whom worked at, and managed, Wel-Met in Kent, Ohio.

Later sold to Samuel Lewis at a time when the employment totaled 35.

Sold to Koppy Corp. in 1998, became Powdercraft LLC. Assets sold in 2002.

Produce: regular P/M

Powder Metal Corp. of America

In business in 1944, later sold to American Powdered Metals

Powder Metallurgy, Inc.
Long Island, NY

Richard Seelig was there in 1940 and up to 1946. He was making fancy structural iron parts. Around 1946, the business was discontinued and Seelig joined the Schwarzkopf family at American Electro Metals in Yonkers, NY

Powder Metallurgy Company, Inc.
Lewisville, TX

Started by Larry Bobo in 1973 in Carrolton, TX, and then moved to Lewisville. He had P/M experience at Ceromet.

Produce: regular P/M

Powder Metal Products, Inc.
St Marys, PA

Started in 1947 by James M. Schloder, who had worked at Keystone Carbon Co. and Gerald Schloder

Assets sold to Metal Powder Products, 1999

Produce: regular P/M

Powder Metal Products Co. of America
Franklin Park, IL

Founded in the early 1940's by the Bagner family, who were well known in molded plastics. During World War II, their son-in-law, John E. Miller, Sr. developed a way to make molded porous nickel cups that were part of the arming mechanism for the proximity fuse on artillery shells. This allowed the shell to explode when it neared a metal object such as an airplane or tank. (cf. Keystone Carbon Co.) During World War II, the development of this arming device was the number one military priority until it was pushed into second place by the Manhattan Project. (cf Innovation and Technology Magazine) In the Battle of the Bulge, proximity fused shells would be shot just over advancing enemy tanks, explode, and kill the soldiers who were marching behind the tanks. In 1948, John E. Miller Sr. left the company to start Chicago Powdered Metals. The Powder Metal Products Co. of America ceased operations in the early 1960's

Produce: regular P/M and filters

Powder-Tech Associates, Inc.
Burlington, MA

Started in 1973 by Raymond L. Collette and Leander F. Pease, III, both having worked at Sinterbond Corp in Gloucester, MA and Wakefield Bearing Corp.

Produce: prototype P/M discs and blocks

Powder Tech Company
Clifton, NJ

Started by Jerome Kuzmick in the late 1960's as part of Powder Alloys. (This was prior to 1973 when Messrs. Pease and Collette named their fledgling business listed above, in a similar manner. This is reason why this fine list of P/M houses should be considered prior to naming any new P/M venture!) In 1984, James J. Kuzmick, a son, was President. We believe they ceased operations by 1990.

Produce: tungsten and refractory metal parts

POW-MET
Oklahoma City, OK

Victor Redgrave active in 1970

Powmet
"Southern", CA, perhaps San Diego

Owned by John MacPherson prior to his starting Form Physics in 1984.

Produce: metal injection molded

PowMet, Inc.
Dayton, OH

Started in the early 1950's by a group who had worked at the captive P/M operation at NCR (National Cash Register), in Dayton. In 1955, sold to Globe Motors and it became Supermet. John Powers was an early manager at Supermet

Powmetl Co.
Ripley, TN

Robert Dickinson active in 1981.

Produce: regular P/M

Precision Compacted Components
Wilcox, PA

Started in 1994 by Frank Beimel who had worked Pennsylvania Pressed Metals.

Produce: regular P/M

Precision Metal Products Co.
Elwood City, PA

Founded in 1945 by Bill Strawbridge who had been a founder of National Molded Products and who worked at Keystone Carbon in 1935.

Produce: regular P/M

Precision Powdered Metals Inc.
Pomona, CA

Started in 1978 by David Connelly who had worked at Ceromet, and before that, Asco Sintering Co.

Produce: regular P/M

Precision Pressed Powder Metal
Dayton, OH

Started by Ed Harwod about 1978, and originally called Ramparts. He had worked at NCR's captive P/M operation in Dayton. Sold in 1984 with name change to Precision Pressed Powder Metal.

Produce: regular P/M

Precision Sintered Parts, LLC.
Tulsa, OK

Was Powdered Metal Parts, not to be confused with Powder Metal Parts in St. Marys, PA. Earlier, was Pittenger Powdered Metal Parts

Produce: regular P/M

Precitech
Quebec, QC,

Canada G1P-4P4

Started in 1991 as Precitech Advanced Materials by Marc Veaugeois, Ghislain Theberge, and Louis Desrosiers. None had prior P/M experience but M. Desrosiers had a degree in technical ceramics. Sold in 1998 to Maxtech Manufacturing Inc, owned by Kacee Vasudeva and his wife. Name changed to Maxtech Powder Metal in 2001.

Produce: regular P/M, specialization in stainless steel

PresMet Corp.
Worcester, MA

Started in 1943 as a part of Mass Steel Treating, owned by the Boorky family. It began as a laboratory operation headed by Prof. Carl Johnson of Worcester Polytechnic Institute, a consultant to Mass. Steel Treating. Some lab work was done at WPI. By 1946, production of parts had started. Morris Boorky formed a corporation called Massachusetts Powdered Metals, Inc. with Carl Johnson as Vice President, to make P/M parts. The name was changed to PresMet about 1949. The company was first sold to GKN in 1973 and then Morris Boorky bought it back in 1980. See also Bound Brook Oilless Bearing Co. In 2001, GKN re-purchased PresMet from the Boorky estate. Renamed GKN SinterMetals-Worcester

Produce; regular P/M

Pressco, Inc.
Norwalk, CT

See Copress

Produce: regular P/M

Proform Division Alberox
New Bedford, MA

Started in 1991 by Dr. Ian Clark who had worked at Form Physics, Parker Nichols Portland Div. and Ametek. Alberox is part of Morgan Crucible.

Produce: metal injection molding

Proform Powdered Metals
Punxsatawney, PA

Started in 1998 by Gerry Duffell who had worked at Masten Corp. and Laurel Manufacturing.

Produce: regular P/M

Prometech
Division of National Forge.

St. Marys, PA

See also Carbon City Products.

PSI Division
Warner Electric Brake and Clutch
Pitman, NJ and Hurffville, NJ

Began operations in 1972 using equipment purchased from Metal Sintering Corp. of America in Nazareth, PA. Warner was purchased by Dana Corp. in 1984. Dana sold warner to Colfax. Colfax sold PSI to Danaher Corp. in 2001. In 2001, PSI joined the Deltran Div. of Danaher. Plant closed in 2002.

Produce regular P/M

Pure Industries
St. Marys, PA

Owned Brownsville Bearing Div., *See also* Bound Brook Oilless Bearing Co.

Purolator Products
Rahway, NJ

They made metallic filters but never managed to break the dominance of paper filters, for brewing coffee.

Member in-plant P/M group in 1963

Pyroferric, International, Inc.
Toledo, IL

Started in 1935

Produce: cores

Q

Quality Compacted Metals
Emporium, PA

Started in 2000 by Scott Horning, a Product/Process Engineer from GKN Sinter Metals, Emporium, Don Smith, Quality Management who had worked at the same company as well as MascoTech Sintered Components, and Don Reed from the private sector.

Produce: regular P/M, mid-tonnage

Quest Technology, Inc.
San Diego, CA

Started in 1987 by Stan Zalkind who had been President of Form Physics. Made injection molded ceramic parts for hard disc drives. Liquidated in early 1990's and a second iteration was later created.

Produce: metal injection molding, CIM

R

Ramparts Ltd.
Kettering, OH

Started in the 1970's by Edward M. Harwood who had worked at NCR

Produce: regular P/M

Raybestos Div.
Raybestos Manhattan, Inc.
Bridgeport, CT

Technical direction from Clyde Batchelor. One unusual product that they produced in the 1950's was carnival bullets. They were frangible, so that they would break apart on hitting the target or backstop, and not go aimlessly through the carnival! (In 2002, several other companies are making lead-free frangible bullets for use in law enforcement and military target ranges, to reduce lead contamination. {Take note of how delicately we have referred to reduction in lead content!})

Member in-plant P/M group in 1963

RB and W
Russell Birdsall and Ward
Port Chester , NY

Started around 1960 by RB and W with Ray Engstrand who had founded American Sintered Products in 1945 and John Polametz, to make sintered nuts. Moved to Coldwater, MI around 1967. Merged with Eaton Yale and Towne captive P/M facility at Coldwater, MI about 1967. The Eaton Yale and Towne operation had been managed by Robert Pettibone. The whole RB&W corporation was purchased by Camcar. The Coldwater P/M operation was later sold to Arlan Clayton of Metal Powder Products, in 1988 and later to Charter Oak Partners, maintaining the name MPP.

Produce: regular P/M and C/F

Reese Metal Products Corp.
Bird-in-Hand, PA

Started in 1955 by Eber Reese whose father owned the Reese Padlock Co. Prior to 1955, the padlock company was buying lock parts from Yale Lock Hardware's P/M operation in Connecticut. The padlock company was sold in 1954, and Eber Reese's new P/M company then made P/M parts for the new owner of the padlock business, Waterbury Lock and Specialty Co. of Milford, CT. Reese Metal Products was sold to Ames of Spain in 1997.

Reliance Electric Company
Weaverville, NC

Started a department to make P/M parts in 1983 with employees Dennis Davidson and Alan Turza, with technical help starting in 1984 from Byron Belden, who had worked at Cincinnati, Inc. Reliance Electric was purchased by Exxon and then bought itself back in 1980. It was then purchased by Rockwell Automation in 1995. It is referred to as the Dodge Mechanical Drives Div.

Produce: regular P/M, captive, e.g tapered bushings.

Remington Arms Company
Hi-Dense Parts Div.
Ilion, NY

Remington first purchased parts like the front gun site in 1945. There were problems with impact strength and porosity and so they started a research department with DuPont help in late 1940's. They had consulting help from Fred Rhines and then Fritz Lenel and employees Bud Miller, Phil Johnson and Dick Morris.. The research became a captive operation for firearms division until 1964, when parts were sold in the open market and Louis Baum joined the company. A MIM operation was established in the late 1970's. In 1979, a separate P/M plant was set up in Hazen, AR. In 1993, venture capitalists Cornell, Dubilier and Rice were preparing to buy Remington Arms from DuPont. CD&R did not want the separate P/M plant that Remington Arms had set up in Hazen, AR. That plant was sold to James McMillan and James Grant, who were Remington's sales representatives, and Capstan Industries. The facility was renamed Hazen-Tech LLC and has specialized in sintered stainless steel parts.

Produce: metal injection molding and regular P/M, both captive and merchant market

Remington-Rand Div.
Sperry Rand Corp.
Elmira, NY

As a student in 1957, Louis Baum, who later worked at Remington Arms, visited the plant. It was a small captive operation with two presses and a furnace making parts for typewriters and business machines. The operation was discontinued in the late 1960's with the advent of electronic calculators.

Member in-plant P/M group in 1963

Produce: regular P/M, captive

Ridgway Powdered Metals Inc.
Ridgway, PA

Started in 1978 by James Olay and Joe Schaefer, the latter having worked at IPM.

Produce: regular P/M

Robertshaw Fulton Controls
Lux Timer Div.
Waterbury, CT

Henry Bender Active in 1970

See Lux Timer

Produce: regular P/M, captive

Rocky Mountain Metals Division of IPM.
Colorado Springs, Colorado.

Started in 1953 Gib Hoehn and Spike Victor as a division of International Powder Metallurgy. All the founders had P/M experience. John Gerzina was hired in 1954 as General Manager. He had worked at Keystone Carbon, Moraine Products Div. GM, American Electro Metals, Ford Rouge River Plant, Pressco and Copress. Gib Hoehn worked at Keystone Carbon and IPM, Spike Victor worked at Keystone Carbon and IPM and Paul Phillips worked at National Molded Products and IPM. Closed by Allegheny Industries after they bought IPM in 1968.

Produce: regular P/M parts.

Rocky Mountain Orthodontics
Denver, CO

Started a MIM operation as a Brunswick Licensee in 1982 to make orthodontic appliances.

Produce: metal injection molded, captive

Rockwell Meter
Rockwell Manufacturing.
DuBois, PA

Started a captive P/M operation in 1952-3 and then made both custom and captive P/M parts. The department was managed by Jim Mitchell who had experience at Symmco and Powder Metal Products, and later by Bob Carney. Became part of Rockwell International. Sold in 1972 to Bob Carney, Samuel Payne, Peter Payne, and John Roberts, the last three of whom were not from the P/M industry. Re-named Dubois Powdered Metals, Inc. Sold to Anthony Zaffuto in 1994-5, who owned the P/M house Metaltech in DuBois, PA. Later merged into Metaltech in DuBois, PA.

Produce: regular P/M

Roper Outdoor Products
Bradley, IL

P/M parts department, captive, started about 1975. In 1982, the Bradley plant was closed and the P/M operation was moved to Orangeburg, SC. They made parts for the six speed transaxle for lawn and garden tractors. Around 1986-7, GE bought Roper, to obtain the range or stove business. GE sold the outdoor part of the business to Electrolux from Sweden. As late as 1989, P/M parts were still being made at Orangeburg. Technical help from Joseph Y. Dress who worked at Burgess-Norton and John Padwojski

Produce: regular P/M, captive

Rush Metals Inc.
Cloyes Gear and Products
Billings, OK

Started by William Rush in 1967. Bill Rush and his father had operated an automotive service station prior to starting Rush Metals, and had no prior P/M experience. Their partner was Henry Bellman, former Governor of Oklahoma. In 1974, Henry Bellman sold his interest to the Rush family and became the U.S. Senator from Oklahoma. First parts were face gears for Zebco Fishing Reels. Sold to Cloyes Gear and Products in 1979.

Produce: regular P/M

S

Sealed Power Corp.
Sintered Metals Div.
Zeeland, MI

Sealed Power started a division to make automotive parts in 1971 through the purchase of Accraform, a P/M house in Palatine, IL. This operation was moved to Zeeland, MI. The P/M plant was sold to Sinteris Canada about

1986, and renamed Sinteris. In 1993, Sinteris was sold to a management group: Tom Weaver, Richard Carmichael, Nancy Lundgren and Scott Paige (sp). The name was changed to Sinterform in 1993. In turn, they sold it to Sinter Metals of Cleveland, OH in 1996. In 1997, Sinter Metals was sold to GKN, resulting in the new name for this plant: GKN Sinter Metals/Zeeland, lending additional credence to the possibility of GKN/A-Z!

Produce: regular P/M

Sherbrooke

Started in 1993 by a person who had worked at Hogen Industries. Later sold to Mi-Tech Metals.

Produce: copper tungsten alloys

Sherritt Gordon
Fort Saskatchewan, Alberta, Canada

In Chapter 12 we recount how Sherritt Gordon began, in 1966, to convert their nickel powder into strip. The strip was then used in Canadian coinage. That process continues to this day, except that the branch of the company that makes the strip is now called Veridian.

Produce: powder rolled strip

Shieldalloy Metallurgical Corp.
Newfield, NJ

Produce: heavy metals

Simet Inc.
Twinsburgh, OH

Started in 1985 by Dr. Vladimir Moxson who had worked at PCC Air Foils and Stackpole (Pure Industries, Brownsville, TN)

Produce: regular P/M

Singer Sewing Machine Company

Denville, New Jersey, then Elizabeth, NJ. Captive P/M started about 1970. Used some P/M aluminum bearing technology from Clevite. Operated second P/M plant in Scotland. Fred Wiehl active in 1970 -1981. Ceased operations in 1981-82. Produced connecting rods for sewing machines as a captive operation.

Sinteral Corp.
Jamaica, NY

Started around 1965-1970 by Samuel Storchheim to make use of his patents on pressing aluminum powder. For a time, the patents restrained other companies from beginning to produce aluminum P/M parts. Ralph Black, active in 1970. Ceased operations at an unknown date.

Produce: P/M aluminum parts

Sinter Alloys Inc.
Pequabuck, CT

Started in 1967 by John Shenuski and William Randolph who had worked at New Britain Machine

Ceased operations about 1998

Produce: Regular P/M

Sinterbond Corp.
Gloucester, MA

Started in 1970 by Raymond L. Collette and Leander F. Pease, III both formerly of Wakefield Bearing, and William H. L. Hooper, formerly of Kennecott Copper.

Sold to W. H. Nichols Co. in 1973. A portion became Parker Nichols Portland Division in Portland, ME, and the remainder was sold by the Nichols Co. to Sintered Metals, Jamaica Plain, Boston, MA around 1977.

Produce: regular P/M

Sintered Alloys Ltd.
Markham, Ontario, Canada

Started in 1966 by Franz Emes and known as Canada's senior P/M house, being the oldest of the Canadian started and owned P/M businesses. Franz Emes was trained as a tool and die maker in Austria and then experienced P/M production at Brassinter in Sao Paulo, Brazil.

Produce: regular P/M

Sintered Metals, Inc.
Jamaica Plain, Boston, MA

Started in 1943 by Abe Stone, an Attorney, George Gibbs, an ordnance inspector in World War II, and a third person, now unknown. First job was the cam for the Zippo lighter. In 1943, the addition of powdered lubricants like Zinc Stearate was not well known and the first parts were made using die wall lubrication. After 1952, managed and later owned by Mike Jaffee. Sold in 1987 to Pacific Sintered Metals and renamed Atlantic Sintered Metals, and later, Capstan/Atlantic.

Produce: regular P/M, filters

Sintered Metal Products, Inc.
Coldwater, MI

Started in 1984 by Randall Heller who had worked at Canadian Stackpole.

Produce: regular P/M

Sintered Precision Components
Decatur, IL

In 1984-5 by Gene Stelzriede owned a machine shop, Decatur Machine and Tool Co. In 1989, name changed to Decatur Compacted Products. Later, may have been purchased by Supermet, Dayton, OH.

Produce: regular P/M

Sintered Products, Inc.
Cleveland, OH

Louis Esposito active in 1970

Sintering Technologies, Inc.
Greensburg, In

Started in 1989 as a joint venture between Cummins Engine and Hitachi. Now wholly owned by Hitachi.

Produce: regular P/M, large valve guides for diesel engines

Sinterings Ltd.
Guelph, Ontario, Canada

In 1970, employees were David Throne (President), Fred Archer and Brian Silver.

Produce: regular P/M

Sinteris
Zeeland, MI 49464

Leveraged buyout by employees in 1993, with name change to SinterForm Inc. (*See also* details at Sealed Power.)

Produce: regular P/M

Sinteris Canada, Ltd.
Blenheim, Ontario, Canada

Started about 1977 by Ing. Mantegazzini who owned Sinteris in Italy, a P/M parts producer.

Produce: regular P/M

Sinteris Magnetics Corp.
Grand Haven, MI

Sinteris in Grand Haven had developed magnetic applications for a Digital Equipment Corp. hard disc drive. They spun off this division in 1983 to do assemblies of the soft magnetic parts made at Zeeland and permanent magnets, purchased elsewhere.

Another "Unbegun Symphony", they never actually made P/M parts at this division.

Produce: magnetic assemblies of P/M parts

Sinterloy Corporation
Spring Grove, IL

Started in 1969 by Bill Randolph who had helped start the captive P/M operation of New Britain Machine, in CT. Sold in 1988-89 and resold in 1997 to Hawk Group.

Produce: regular P/M

Sinter-Met Corp.
North Brunswick, NJ

Donald Bennett active in 1970.

Produce: ?

Sintermet Inc.
Homestead, PA

Produce: carbide

Sinter Metals, Inc.
Cleveland, OH

Owned several and then purchased several more P/M parts producers including Pennsylvania Pressed Metals and ICM/Krebsoge plants.

Sold to GKN Plc in 1997

Produced regular P/M and P/F

Sinterphase, Inc.
Johnsonburg, PA

Started in 1983 by Raj Ranadive and Pat Gaffey from Michigan Sintered Metals and Richard Uhl, Sr. who had worked at St Marys Carbon, Keystone Carbon, IPM, and Carbon City Products. Sold to GKN/Sinter Metals in 1999 and the operations were soon merged into other GKN plants.

Produce: regular P/M

Sinttech, Inc.
Huntsville, AL

Started in 1983 by Tony Martin

Merged into Martin Technologies in 1998. See Martin Technologies

Produce: regular P/M

S. K. Wellman Ltd., Inc.
Bedford, OH

Founded by Samuel Wellman in 1925. Originally made copper based friction materials for transmission clutches for Henry Ford's automobiles. A spin-off by Fran Loewy resulted in Sintermet in Brook Park section, near Cleveland, OH , making metallic aircraft brakes. Sintermet was sold to American Brake Shoe. Fran Loewy then left and started Friction Products in Medina, OH in 1961. S. K. Wellman bought back Sintermet from American Brake Shoe, closed the Bedford, OH plant, and moved the operations in with Sintermet at Brook Park. Hawk Group then closed the loop by first buying Friction Products, and then, about 1995 buying S. K. Wellman.

Produce Friction Materials

SMC Powder Metallurgy
St. Marys, PA

Name change from St. Marys Carbon, in 1999. *See also* St. Marys Carbon.

Smith Metals
Lindstrom, MN

Started in 1995-96 as a division of a company that did molded plastics. Used the BASF technology. Purchased Thermal Technology in 2001.

Produce: metal injection molded

Southern Sintered Metals, Inc.

Started in Lafayette Franklin, TN by Jim Rose about 1968. Moved to Franklin. TN about 1977 and sold to Joe Swanson and Jim Smythe about 1978. Later sold to McCrosky Tool and National Molded Products 1982. Ceased operations about 1986

Produce: regular P/M

Spang and Co.
Butler, PA

By 1984, they were atomizing special alloys, e.g. for soft magnetic applications, and rolling them into strip.

Produce: powder rolled strip

Specialty Pressed Components, Inc.
Kersey, PA

Started in 1989 by Mack Schreiber who had worked at Keystone Carbon, Mascotech and American Powdered Metals.

Produce: regular P/M

Spectra-Mat, Inc.
Watsonville, CA

Produce: W-Cu, Mo-Cu, refractory metals

Speer Carbon Co.
St. Marys, PA

A manufacturer of carbon brushes for electric motor. Started in 1899 by John Speer, a chemist and Andrew Kaul, a financier. The founders of Keystone Carbon Co., Ben Reuscher and Adolph Schmidthammer, worked here, prior to leaving in 1927 to produce brushes with integrally molded copper wire shunts.

Produce: brushes for electric motors

Square D Company
Cedar Rapids, IA

Produce: electrical contacts, captive

SSI Technologies, Inc.
Janesville, WI

In the early 1950's. Parker Pen Co. started a P/M department where they made the tungsten carbide balls for their ball point pens and P/M platinum alloy writing nibs for fountain pens. In 1960 they expanded their interest in P/M and set up a plant called Gillman Engineering to make porous media such as sound attenuating acoustic discs used in telephone hand sets. In 1963, Parker Pen sold Gillman Engineering and it was renamed Panoramic Corp. It made structural P/M parts such as the mirror mount bracket. In the early 1980's, six people, including John Anderson, Don Dyke, Bill Mossner and David Baum bought the assets of the P/M operation, and April 1, 1982 named it Sintered Specialties, Inc. In 1984, the name was changed to SSI Technologies to reflect the increased manufacture of non-P/M parts. In 1987, David Baum purchased the shares of the other original owners and became the sole owner.

Produce: regular P/M, filters, metal injection molding

St. Marys Carbon, Co.
St Marys, PA 15857

Started in 1939 by Jerome Lanzel who had worked at Keystone Carbon and C. Eberl. Mr. Lanzel had a preference for making pressed and sintered (baked) carbon based materials. In 1947, he hired Gib Hoehn, who had worked at Keystone Carbon, to install a sintering furnace and to learn to make P/M parts. In 1965, they established a separate P/M parts operation at Galeton, PA.

Produce: regular P/M

St. Marys Pressed Metals Inc.
Kersey, PA then Ridgway, PA

Started in 1968 by Walter Schwabenbauer and Dave DeLullo.

Produce: regular P/M

Stackpole Carbon
St. Marys, PA

An early producer of carbon graphite and metal brushes for electrical motors. Probably was producing brushes before 1925.

Produce: carbon brushes for electric motors

Stackpole Ltd.
Toronto and Mississauga Ontario

Stared in 1965 with David Knapp as first President. Based on the technology that existed in Stackpole Corp, in St. Marys, PA, where copper-graphite brushes were being produced.

Produce: regular P/M

Standard Bearing Company
Norwalk, Connecticut

Started in January 1948, by John Caputo and John Gerzina.

John Gerzina had P/M experience as follows: 1940 to VJ Day in 1946 at Keystone Carbon Co., 4 months at Moraine Products Div. of GMC, a short stint with Richard Seelig at Powder Metallurgy Corp., Long Island, NY and from 1946 to 1948 worked with the Schwarzkopf family at Yonkers, NY at American Electrometal Products. Standard Bearing was merged into Copress in 1949.

Produce: regular P/M

Starmet
Was Nuclear Metals
Concord, MA

Fabricated nuclear fuel by extrusion, in the early 1950's. Some of these extrusions were based on mixtures of metal powders, e.g. Zircaloy and boron carbide. Textron and Whittaker owned it in 1970. Developed the rotating electrode process to make clean spherical metal powder. Some of this powder was canned and extruded into mill shapes.

Produce: extruded alloy mill shapes from metal powder.

Stellex
Palo Alto, CA

Started as a spin off from Watkins Johnson around 1998-99, to make microwave packages for military uses.

Produce: metal injection molded

Sterling Sintered Technologies Inc.
Winsted, CT

A tooling manufacturer and machine shop which started around 1942-44, diversified into P/M in 1960 with technical help from Charles DelRossi, who had worked at Wakefield Bearing Corp. Roland Royer was General Manager of the P/M operation and, along with Ed Werner, bought it from Sterling Engineering in 1985. They renamed it Sterling Sintered Technologies.

Produce: regular P/M

Sundstrand Aviation Div., Sundstrand
Rockford, IL

Started a department in 1974 to make parts for aircraft hydraulic systems. Ceased operations in 1981.

Produce: regular P/M

Supermet Div. Globe Industries
Dayton, OH

Purchased PowMet about 1955. (*See also* PowMet, Dayton, OH) Sold to TRW in 1967, then to Stanadyne in 1977, then to T and N in 1988 who changed the name to Brico, which was then bought by Federal Mogul in 1998 and renamed Federal Mogul Sintered Products. Federal Mogul sold the Marshall, IL plant to GKN in 2000, and then filed for bankruptcy in 2002 (but not related to its P/M businesses!).

Produce: regular P/M

Superwear Technology
Houston, TX

Started by Jay Ponchal who had worked for Alloy Technology in NY.

Sold to Pacific Sintered Metals in 2000.

Symmco, Inc.
Sykesville, PA 15865

Started in April 1952 by investors Cal Bean, Americo Torretti, Blair Weber and Henry Righi with no prior P/M experience. A coal mine had closed at that time, and a "town meeting" was held in the back of Weber Hardware Store to raise money and start a business to create employment. $50,000. was pledged that night. Founders with some P/M experience included John Reitz, George Gasbarre from IPM, Jim Mitchell from Powder Metal Products, and John Murgash from Powder Metal Products. In 1953 George Gasbarre sold his stock to John Reitz and in 1954, Jim Mitchell sold his stock to John Reitz. John Reitz became President in 1955 and served until 1969 when he was succeeded by Robert Vannat

Produce: regular P/M

Symphony P/M, Inc.
Providence, RI

Started in 1992-93 by John J. O'Meara and Richard Phillips of Engineered Pressed Components.

Operations discontinued in 2000.

Produce: Tool Steels

T

Taurus International Mfg, co.
Miami, FL

Started as a department in 1996.

Produce: metal injection molding, captive

TecSyn PMP, Inc.
Huntsville, AL

Started in 1991 as an outgrowth of a CIP'ing operation at Höganäs, Sweden, aimed at making sintered iron cylinder liners for diesel engines. Early technical direction in the US by Ola Pettersson.

Produce: cold iso-pressed parts

Teledyne Powder Alloys
Clifton, NJ

See also Powder Alloys.

Produce: heavy metals

Thermal Technology
Metal Injection Molding Div.
Santa Rosa, CA

Started about 1996 with technical help from Philip Booker and Pat Bhave. Discontinued operations of MIM plant in 2001, with sale to Smith Metals of Lindstrom, MN

Produce: metal injection molding

Thermat, Inc.
Corry, PA

Started in 1994 by Karl Hens, who had MIM experience as a Research Professional at Penn State. Sold to Medsource in 2000, became Medsource Corry.

Produce: metal injection molding

Thermet Inc.
Gloucester, MA

Started in 1963 by Jack Tarr who had worked at Sintered Metals in Jamaica Plain, Boston, MA. Sold to Krebsoge in 1989 then Sinter-Metals and GKN, Richton Park.

Produce: filters

3 M Company
St. Paul, MN

In 1969 and for 20-30 years thereafter, 3 M operated a merchant facility that made multiple copies of injection molding tools by infiltrating copper into a Stellite or tool steel matrix. The latter was formed by pressing a mixture of the alloy powder and a binder around a model of the part that needed a die. They debound the die and then infiltrated the pores with copper. Active in this operation was Dr. Richard Terchek. Ultimately, in the late 1990's, the business was sold to KelTools. It plays a significant role in rapid prototyping of MIM and other injection molded materials.

Produce: copper infiltrated high alloy molds.

Thompson Products, Inc.
St. Charles, MI

Started in 1967 by Richard Thompson and Stuart H. Yntema, both whom had worked at U.S. Graphite

Produce: regular P/M

TPI Powder Metallurgy
St. Charles, MI

Name Change from Thompson Products, Inc

Torrington Co.
Torrington, CT

Captive operation started with assistance of consultant Robert Talmadge. Technical help from employee Archie Deloge. Active during the 1960's

Trico Technologies, Corp.
Trico Products,
Buffalo, NY

Started after World War II and prior to 1949 when Joseph Drzewiecki went to work there. It was a department of the company, making P/M steel gears and stainless steel parts for windshield wiper motors. Early management of the department was under V. P. Raymond Diebel. Department was moved to Brownsville, TX in 1988.

Produce: regular P/M, captive

Tungsten Technology Inc.
Hendersonville, TN

Produce: metal injection mold carbides

See Leco, Hendersonville, TN

U

Ultraclad

Andover, MA

Started in 1992, by Joseph Runkle, as an outgrowth of Industrial Materials Technology in Andover, MA. HIP techniques are used to clad industrial objects like steel mill rolls with special wear resistant powder alloys. They have a patent on a special cladding for locomotive railway wheels that will have greater fatigue resistance and greater gripping power. Such wheels are currently under test on railroad trains.

Produce: composites, tool steel and WC

Underwood Typewriter Co.

Hartford, CT

Started a small captive P/M department in the 1950's, with assistance from consultant Robert Talmadge. They made parts for the old manual typewriters. Underwood never made the transition to electric typewriters, and the department passed into history, at a time now unknown.

Produce: regular P/M, captive

U.S. Graphite

Saginaw, MI

Started by the Wickes family in 1900.

The Wickes family distributed lumber processing equipment, built machine tools and also boilers (vertically integrated in lumber processing). They used to hunt in Sonora Mexico and found an amorphous graphite mine in 1888. That lead to making carbon brushes in 1910, then copper based bushings by 1930. In 1935, Ralph Zemanek joined the company and worked out the methods to make copper tin graphite bearings, called by the trade name Gramix. A holding company, Wickes Corp. was formed in 1947. Wickes changed U.S. Graphite name to Wickes Engineered Materials Corp. Wickes Corp. became a very large conglomerate which ultimately failed. All of this is chronicled in a book, "The Wild World of Wickes", 1975. The P/M company was closed in 1983.

Produce: regular P/M

V

Valform Corp.

Monument, CO

Started about 1979 by Robert Anderson and Valenite Corp.

Purchased by Fred Torretti of Symmco in 1990. Ceased operations in 1992.

Produce: CIP'd tool steel parts and Ti -6-4 alloys

Vanguard Componics

Colorado Springs, CO

Started about 1983 By Robert Anderson with assistance from Andrew Major of National Forge. In existence about one year.

Produce: Cold isostatically pressed and sintered Ti alloys.

Veeder-Root Inc.
Hartford, CT

Very large manufacturer of metering pumps for gasoline and fuel oil. Started a captive P/M operation in the 1950's with technical assistance from consultant Robert Talmadge. Presumably made small gears for their gasoline meters, before the advent of electronics. Member in-plant P/M group in 1963.

Produce: regular P/M, captive

Vertx
Denver, CO

Started by Mark Skrotzki with John and Lucy Gaspervich in 1984. Ceased operations about 1986.

Produce: tool steel parts, and Mark Skrotzki marketed software to estimate costs on P/M parts.

Victor Comptometer Corp.
Chicago, IL

Member in-plant P/M group in 1963

Volunteer Sintered Products Corp.
Lafayette, TN

Started in 1986 by Rich Ludick who had worked at Southern Sintered Metals.

Produce: regular P/M

W

Wakefeld Bearing Corp.
Wakefield, Massachusetts

Started as the Neveroil Bearing Co. in 1904 in Worcester, MA. The company is listed as a sponsor of the 1941 printing of a collection of papers from the 1940 powder metallurgy conference at M.I.T. In 1945, Nathaniel Clapp bought part of this business which had also been making wooden pulleys and sheaves. In 1952 he purchased the rest of the company, changed the name to Wakefield Bearing Corp., and began to make P/M bearings and parts. Name later changed to The Wakefield Corp.

Produce: regular P/M

Waterbury Button Co.
Waterbury, CT

Started in 1950 to make P/M parts for their own product line. The plant was set up by Phil Tarr, and we are not sure what his P/M experience was, prior to 1950. The P/M operation was purchased by Dixon Sintaloy in 1955 at the time when Dixon was managed by Kenneth M. Gleszer.

Produce: regular P/M, captive

Watkins Johnson
Palo Alto, CA

P/M operation started by Dr. Philip Booker and Roger Dirstine at their facility, Flatirons. Technology transferred to Watkins Johnson in 1993-94. P/M operation spun off as Stellex in 1998-99.

Produce: metal injection molded microwave circuit packages

Webster-Hoff Corp.
Originally in Odessa and then moved to Glendale Heights, IL

Started in 1970-71 by Jack Webster and Robert Hoff. Neither was from the P/M industry but

Jack Webster had worked in diecasting at Alcoa and was interested in producing sintered aluminum parts at their new company.

Produce: regular P/M, Al

Wellman Friction Products

Medina, OH

Was called Friction Products, name changed by Hawk Group in late 1990's

Produce: friction materials

Wel-Met Corp.

Kent, Ohio

Founded by Thomas L. Robinson and S. K. Wellman Co. in 1945. (Relates to S.K. Wellman Ltd. history, somehow.) Robinson had worked at the Amplex Division of Chrysler Corp as Chief Engineer and Assistant to the President. He left Wel-Met in 1952 to start Powdercraft Corp. in Spartanburg, South Carolina. Another employee in 1945-46 was Jess Helsel.

Ferro Corp. bought Wel-Met in 1954 and changed the name to Ferro Powdered Metals. In 1957 Jess Helsel and Andrew Rose bought the company from Ferro, changed the name to Ferraloy and moved it to Salem, IN. In 1960, Waltham Industries' Johnson Bronze P/M operation in Newcastle, PA bought Ferraloy and moved the Newcastle operation to Salem in 1961. In 1968, Ferraloy bought the P/M parts division of P.R. Mallory in Angola Indiana and moved it to Salem, IN. (We normally think of P. R. Mallory as making refractory materials like W-Cu or contacts. But, a Dr. Hentzel (sp) had helped them establish a conventional P/M parts business at Angola, IN in 1965)

In 1967 Johnson Bronze was acquired by Waltham Industries.

In 1970, as Clevite was preparing to buy Waltham Industries' Johnson Bronze Salem and St. Thomas plants, Gould bought Clevite. For the events after 1970, *see* Clevite.

Produce: friction materials, then regular P/M and P/F

West Coast Sintered Products

Costa Mesa, CA

Started in 1976 Robert Frazier who had been a metallurgist at Asco Sintering Co. and had worked at Remington arms. Ceased operations around 1981.

Produce: aluminum P/M

Western Sintering Company,Inc.

Richland, WA

Started by John Rector who had worked for duPont on the building of the graphite moderated reactor at Hanford, WA during World War II. He developed the first sintered Al-B control rods for a nuclear reactor. In 1955 he began building presses and furnaces, and by 1960 was producing P/M parts.

Produce: regular P/M

Weyburn Assembled Camshafts
Became Federal Mogul Assembled Camshafts
Grand Haven, MI

Produce: P/M Cams

Wheeling Pittsburgh

Ceased P/M operations about 1970. Had patented process to use sand as a pressurizing medium in pseudo-isostatic hot pressing, to consolidate preforms to full density without need for a can. Succeeded by Ceracon.

Produce: pseudo-HIP'd materials

Whitney Chain Co.
Hartford, CT

Started a captive department to make the rollers in roller chain in the 1950's. Technical help from Robert Talmadge. If you have read this far, you probably have lost track of the number of captive and custom P/M operations he assisted. No information on how long this department lasted.

Produce: regular P/M, captive

Windfall Products, Inc.
St Marys, PA 15857

Started in 1976 by Richard Wolfe, Arnold Wolfe who had worked at Powdered Metal Products, Jim Ryan and Joe Holjencin who had worked at Powder Metal Products.

Sold to Mascotech in 1999. Renamed Metaldyne Sintered Components, in 2001.

Produce: regular P/M

Y

Yale Lock Hardware
New Haven or North Haven, CT

P/M department started in 1950 based on purchase of American Sintered Alloys from Kenneth Gleszer and Raymond Engstrand. Sold in 1954 to Gerry Fellows and others, *see also* American Powdered Metals.

Z

Zenith Sintered Products, Inc.
Started in Menomonee Falls, Wisconsin, then Germantown, WI

Started in 1965 by the Tanner family. For many years it was successfully managed by Ed Gibson and then by Armour Swanson and Todd Tanner. Sold to GKN in 1999

Produce: regular P/M

Zentrix
Tucson, Arizona

Name change from Brush Wellman in 2001.

Produce Cu-W for thermal management

In 1944, Professor Wulff used the analogy of the Merry-Go-Round for the P/M Industry. 58 years later, we think of the long standing P/M houses as being like our favorite cartoon character: "I'm here at the finish, 'cuz I eats me spinach, I'm Popeye the Sailor Man! Toot! Toot!"

Possible answer to "Bassick" question: "castor oil"?

APPENDIX A

Alphabetical Index of Companies Listed in the P/M Genealogy

M

N

O

P

APPENDIX B

Primer References

The following references apply to the whole of this primer. We have not always tied our text's statements to particular reference books.

1. *Advances in Powder Metallurgy & Particulate Materials*, Metal Powder Industries Federation, Princeton, N.J.
2. *ASTM Volume 02.05*, ASTM, West Conshohocken, PA, latest edition has all the ASTM P/M standards and test methods in one volume.
3. Cai, Haimian, Zenger, David, *The Common Cracks in Green P/M Compacts*, MPIF, Princeton, NJ, 1997.
4. Chandler, Harry, *Metallurgy for the Non-Metallurgist,* ASM International, Materials Park, OH, 1998.
5. Cullity, B. D., *Introduction to Magnetic Materials*, Addison-Wesley, Reading, MA, 1973
6. Dowson, Gordon, *Powder Metallurgy, The Process and its Products*, Adam Hilger, Bristol, England, 1990. (The inspiration for this current text!)
7. *Fifty Years of Service to Powder Metallurgy, 1944-1994*, Metal Powder Industries Federation, Princeton, NJ., 1994.
8. German, R.M., *Powder Injection Molding*, Metal Powder Industries Federation, Princeton, NJ, 1990.
9. German, R.M., *Powder Metallurgy of Iron and Steel*, John Wiley and Sons, New York, 1998.
10. German, R.M., *Powder Metallurgy Science*, 2nd Edition, Metal Powder Industries Federation, Princeton, NJ, 1994.
11. German, R.M., *Sintering Theory and Practice*, John Wiley and Sons, New York, N.Y., 1996.
12. *Höganäs Handbook for Sintered Components,* North American Höganäs, Bethlehem, PA., 1997.
13. Lall, Chaman, *Soft Magnetism*, Metal Powder Industries Federation, Princeton, NJ, 1992.
14. Lawley, Alan, *International Journal of Powder Metallurgy*, APMI International, Princeton, NJ.
15. Lenel, F., *Powder Metallurgy, Principles and Applications*, Metal Powder Industries Federation, Princeton, NJ, 1980.
16. *Material Standards for P/M Structural Parts*, MPIF Standard 35, latest edition, also separate volumes of *Standard 35 for MIM Parts, Bearings, and Powder Forged Parts,* Metal Powder Industries Federation, Princeton, NJ.
17. *Metal Powder Report*, Elsevier, Kidlington, Oxford, UK.
18. MPIF website, at "www.mpif.org", with links to other P/M sites
19. *Powder Metallurgy Design*

Manual, 3rd edition, Metal Powder Industries Federation, Princeton, NJ, 1998.

20. *Standard Test Methods for Metal Powders and Powder Metallurgy Products*, latest edition, MPIF, Princeton, NJ.
21. *Volume 7, ASM Handbook, 10 edition, Powder Metal Technologies and Applications.*, ASM International, Materials Park, OH., 1998.

APPENDIX C

Unit Conversion, or How to Make Use of a Conversion Table

If we remember that 12 in.=1 ft. then we can convert quantities where we want to go from feet to inches or inches to feet:

12 in.=1 ft. so dividing both sides by 1 ft., we get 12 in./l ft.=1, or 1 ft./12 in. = 1.

Remember, we can multiply any quantity by 1 and change only its units, and not its magnitude.

For example, if you know that 3 ft.=1 yd. and you want the number of inches in a yard, multiply the ft. by the 12 in./ft. That gets rid of the ft. unit by unit cancellation:

3ft. x 12 in./l ft. cancels the ft. in numerator and denominator and leaves:

3 x 12 in. = 1 yd or 36 in. = 1 yd.

We now have a new conversion: 36 in./l yd =1, and 1 yd/36 in.= 1

To change m.ph.= miles/hour to ft./sec, if we know:

1 mile=5280 ft., we therefore know the conversions or unit multipliers,

1 mi/5280 ft. = 1 and 5280 ft./l mi.=1
and 1 hour 3600 sec. therefore: 1 hr./3600 sec.=1 and 3600 sec./l hr.=1

we have only to multiply the numerator in miles/hr. by ft./mi. to get rid of the miles unit and multiply the denominator by sec./hr., to get rid of the hour units:

or mi./hr. x 5280 ft./mi / 3600 sec/hr or mi./hr. x 5280/3600 = mi./hr. x 1.47 = ft./sec
e.g. 60 mi./hr. = 60 x 1.47 = 88 ft./sec.

The acceleration due to gravity is 32 ft./sec^2. A young physicist from England wondered what this would be in furlongs/ fortnight^2! .

We learn that 1 furlong = 1/8 mi, 1 mi.= 5280 ft., and 1 fortnight = 14 days and 1 day = 24 hours and 1 hour = 60 min. and 1 min.= 60 sec.

Therefore 32 x mi./ft. x furl/mi./ (min/sec^2 x hr/min^2 x day/hr^2 x ft. night /day^2)
5280 x 8^2 / 1/ 60^2 x 1/60^2 x 1/24^2 x 1/14^2 or ft./sec.^{2} x 6.18 x 10^{16} = furlongs/fortnight^2

and so for an acceleration of 32 ft./sec^2 we get 32 x 6.18 x 10^{16} furl/ft. night^2 or 1.98 x 10^{18} furl/ ft. night^2

What about in microns/ ft. night^2 ?.

1000 micron = l mm, 25.4 mm = 1 in., so:
furl x mi./furl x ft./mi. x in./ft. x mm/in. x micron/mm
1/8 x 5280 x 12 x 25.4 x 1000= 2.01 x 108 conversion to microns or g = 3.98 x 10^{26} microns/ft. night^2

It is important that some conversions be committed to memory but acceleration due to gravity in microns/fortnight^2 on Earth is not one of them. That said, in the "Table of Conversions" adopted from Prof. German, we see that the power conversion for watts and horsepower is given as: 1W = 1.34×10^{-3} hp. I can never remember that one, but I surely remember the reverse conversion learned in 1959 from my roommate in graduate school, Dr. Harry Keith now a retired Prof. from The Naval Academy at Annapolis, which went something like this:

In fourteen hundred ninety two (1492), Columbus sailed the ocean blue, divide the son of gun in two (2) and that's the Watts in a Horsepower, (746)!.

And you thought that you had heard the last of mnemonics!

SYSTEM INTERNATIONAL UNITS (SI) AND CONVERSION FACTORS

Source: R.M. German, Powder Metallurgy Science, Second Edition, Metal Powder Industries Federation, Princeton, NJ 1994

Length Conversions:
1m = 39.4in (inch)
1m = 3.28ft (foot)
1m = 1.09yd (yard)
1cm = 0.394in (inch)
1mm = 0.0394in (inch)
1μm = 39.4μin (microinch)
1nm = 10Å (angstrom)

Area and Volume Conversions:
$1cm^2 = 0.155in^2$ (square inch)
$1m^2 = 1550in^2$ (square inch)
$1cm^3 = 0.061in^3$ (cubic inch)
$1m^3 = 35ft^3$ (cubic foot)
$1L = 1000cm^3$ (cubic centimeter)
1L = 0.264gal (gallons)
1L = 1.06qt (quart)

Amount of Substance Conversion:
$1mol = 6.022 \bullet 10^{23}$ molecules

Density Conversions:
$1Mg/m^3 = 1g/cm^3$
$1g/cm^3 = 0.0361lb/in^3$ (pound per cubic inch)
$1kg/m^3 = 10^{-3}$ g/cm^3

Temperature Conversion:
to convert K to °F (fahrenheit), multiply by 1.8 then subtract 459.4°F
to convert °C to °F (fahrenheit), multiply by 1.8 then add 32°F

Heating and Cooling Rate Conversions:
1K/s = 1°C/s = 1.8°F/s
1 K/min = 1.8 °F/min

Mass Conversions:
1g = 0.035oz (ounce)
1kg = 2.2lb (pound)
1Mg = 1.1ton (ton = 2000 pounds)

Force Conversions:
$1N = 10^5$ dyne
1N = 0.225lb force (pound force)

Pressure, Stress and Strength Conversions:
1Pa = 0.0075 torr (millimeter of mercury)
$1Pa = 10$ $dyne/cm^2$ (dyne per centimeter square)
1kPa = 0.145psi (pounds per square inch)
1MPa = 9.87 bar (atmosphere)
1MPa = 145psi (pounds per square inch)
1MPa = 0.145kpsi (thousand pounds per square inch)
1Gpa = 145kpsi (thousand pounds per square inch)

Energy Conversions:
$1J = 9.48 \bullet 10^{-4}$ btu (British thermal unit)
1J = 0.737ft•lb (foot pound)
1J = 0.239cal (calorie)
$1J = 10^7$ erg
$1J = 2.8 \bullet 10^{-7}$kw•h (kilowatt hour)
$1J = 6.24 \bullet 10^{18}$ eV (electron volt)
1J = 4.83hp•h (horsepower•hour)
1J = 1 W•s (watt second)
1J = 1 V•C (volt coulomb)
1kJ = 0.239kcal (kilocalorie)

Power Conversions:
1W = 0.737ft•lb/s (foot pound per second)
$1W = 1.34 \bullet 10^{-3}$ hp (horsepower)

Thermal Conversions:
1J/(kg•K) = $2.39 \cdot 10^{-4}$ btu/(lb•°F) (British thermal unit per pound per degree fahrenheit)
1J/(kg•K) = $2.39 \cdot 10^{-4}$ cal/(g•°C) (calorie per gram per degree celsius)
1W/(m•K) = 0.578 btu/(ft•h•°F) (British thermal unit per foot per hour per degree fahrenheit)
1W/(m•K) = $2.39 \cdot 10^{-3}$ cal/(cm•s•°C) (calorie per centimeter per second per degree celsius)

Viscosity Conversions:
1Pa•s = 1kg/(m•s)
1Pa•s = 10P (poise)
1Pa•s = 10^3 cP (centipoise)

Stress Intensity Conversion:
1MPa•$m^{1/2}$ = 0.91kpsi•$in^{1/2}$ (kilopounds per square inch times square root inch)

Magnetic Conversions:
1T = 10^4 G (gauss)
1A/m = $1.257 \cdot 10^{-2}$ Oe (oersted)
1Wb = 10^8 Maxwell

INDEX